# THE WILDERNESS CONDITION

# Essays on Environment and Civilization

Edited by Max Oelschlaeger

ISLAND PRESS

*Washington, D.C.*  □  *Covelo, California*

## ABOUT ISLAND PRESS

Island Press, a nonprofit organization, publishes, markets, and distributes the most advanced thinking on the conservation of our natural resources—books about soil, land, water, forests, wildlife, and hazardous and toxic wastes. These books are practical tools used by public officials, business and industry leaders, natural resource managers, and concerned citizens working to solve both local and global resource problems.

Founded in 1978, Island Press reorganized in 1984 to meet the increasing demand for substantive books on all resource-related issues. Island Press publishes and distributes under its own imprint and offers these services to other nonprofit organizations.

Support for Island Press is provided by Apple Computer, Inc., Mary Reynolds Babcock Foundation, Geraldine R. Dodge Foundation, The Energy Foundation, The Charles Engelhard Foundation, The Ford Foundation, Glen Eagles Foundation, The George Gund Foundation, William and Flora Hewlett Foundation, The Joyce Foundation, The John D. and Catherine T. MacArthur Foundation, The Andrew W. Mellon Foundation, The Joyce Mertz-Gilmore Foundation, The New-Land Foundation, The J. N. Pew, Jr. Charitable Trust, Alida Rockefeller, The Rockefeller Brothers Fund, The Florence and John Schumann Foundation, The Tides Foundation, and individual donors.

COPYRIGHT © 1992 BY MAX OELSCHLAEGER

Originally published in hardcover by Sierra Club Books, San Francisco, in 1992.

LIBRARY OF CONGRESS CATALOGING IN PUBLICATION DATA

The wilderness condition : essays on environment and civilization / edited by Max Oelschlaeger.
    p. cm.
    Includes bibliographical references.
    ISBN 0-87156-642-7—ISBN 1-55963-190-2 (pbk.)
    1. Human ecology—Philosophy. 2. Wilderness areas. 3. Man—Influence on nature. 4. Environmental policy. I. Oelschlaeger, Max.
    GF21.W55 1992
    304.2—dc20                     92-52649
                                                          CIP

Production by Robin Rockey
Book design by Park Press

Printed in the United States of America on acid-free paper containing a minimum of 50% recovered waste paper, of which at least 10% of the fiber content is post-consumer waste.

10 9 8 7 6 5 4 3 2 1

*For A. E. Koenig, in memoriam*

# Table of Contents

Introduction: The Wilderness Condition Today                             1
  by Max Oelschlaeger

The Etiquette of Freedom                                                 21
  by Gary Snyder

A Post-Historic Primitivism                                              40
  by Paul Shepard

Ecocentrism, Wilderness, and Global Ecosystem Protection                 90
  by George Sessions

The Utility of Preservation and the Preservation of Utility:
Leopold's Fine Line                                                      131
  by Curt Meine

Perceiving the Good                                                      173
  by Erazim Kohák

A Brittle Thesis: A Ghost Dance: A Flower Opening                        188
  by Michael P. Cohen

The Disembodied Parasite and Other Tragedies; or:
Modern Western Philosophy and How to Get Out of It                       205
  by Pete A. Y. Gunter

Not Laws of Nature but Li (Pattern) of Nature                            220
  by Dolores LaChapelle

The Blessing of Otherness: Wilderness and the
Human Condition                                                          245
  by Michael Zimmerman

Wilderness, Civilization, and Language                                   271
  by Max Oelschlaeger

Appendices                                                               309
  by Paul Shepard

Notes                                                                    313

Notes on Contributors                                                    344

# The Wilderness Condition Today

## by Max Oelschlaeger

During 1989 the popular press rediscovered the environment. After almost a decade of neglect, during which time disasters such as "Bhopal" and "Chernobyl" or the misadventures of various Reagan appointments such as James Watt dominated environmental news coverage, less dramatic but substantive conservation issues again attracted attention. 1989 became the "Year of the Environment," with newspapers and magazines running races to stay on the breaking front of the news. Today the outlines of a global ecological crisis are well known (although scientists always hasten to warn that our knowledge remains sketchy and imprecise in some areas). With so much attention in the print media to "things environmental," especially to issues of pressing importance, the reader may wonder what function is served by this collection of articles on environment and civilization.

Reflective essays dealing with conservation issues have never reached a wide audience, and perhaps understandably so. The press has constitutional safeguards for its freedoms, and the for profit context in which publishers compete largely precludes any simultaneous combination of mass appeal and intellectuality. The media provide what the reading public will buy, and sober intellectual deliberations therefore have a relatively small market, since what sells newspapers and magazines tends to be crisis or scandal oriented. Clearly this collection of essays on "the wilderness condition" does not compete with publications that deal with issues of the moment, with the factual underpinnings of global ecological malaise, with considerations of public policy, and with detailed proposals for a "green future." And yet just as clearly it is in relation to the "re-greening of consciousness," engendered in large measure by the popular press, that *The Wilderness Condition* finds its justification.

The print media play an important role in shaping public awareness, we have suggested, by presenting facts, exploring issues, discussing policy proposals, and exploring alternative futures. Since policy change and consequent modification of human behavior depend upon societal consensus, and since American politicians show a tendency to follow majority opinion, it follows that the press serves a vital social function. However, those who have been involved in the conservation movement are immediately sobered by recent history. Two decades ago there was a greening of consciousness in America (similar to the one of 1989), a phenomenon that symbolically culminated in the first Earth Day (1970). Subsequent legislative action, such as passage of the Endangered Species Act (1973) and the Alaska National Interest Lands Conservation Act (1980), reflected in part the public opinion of the time. But as the 1970s gave way to the 1980s the cultural stage upon which our democratic theater was played out changed. The print media shifted attention away from conservation issues, in part because the world did not come to an end as some pessimistic environmentalists had predicted. To the popular press *the crisis* of the 1980s seemed to be economic rather than environmental. There was an almost concomitant shift in national politics toward a conservative, even *laissez faire* approach to ecological issues.

Earth Day 1990 now looms on the horizon, and will have been celebrated by the time this book is in print. We will likely see a spate of environmentally oriented legislation in the early parts of the 1990s. Such laws are sorely needed and consequently desirable, and in large measure they will be passed *only* because an overwhelming majority of Americans support such change at the ballot box. Yet we must remember that apparent law of all media: "What goes up must come down." The press will almost inevitably shift its focus elsewhere, and that buzzing, blooming sprawl which is industrial civilization will continue its relentless humanization, and thus destruction, of things wild and free. The most *reasonable* conjectures, based on the evidence of the last century, and baring any significant restructuring of ideological and institutional frameworks, are that population levels will continue to expand, public transportation will remain a dream, there will be no greening of the cities, scientific arguments for maintenance of biodiversity will not forestall massive extinctions, and so on. In short, public opinion will coalesce in the early 1990s around an ecological agenda for action, but as the decade goes on that agenda will remain largely unimplemented.

Future historians will be in a position to see how accurate these forecasts are. I hope that I am wrong, but the words of John Graves always come back to haunt me—"Men increase; country suffers." Therein lies the irony of our most recent greening of consciousness, since in many ways there is little new therein. John Muir, for example, died in 1914, and left a legacy of biocentric thought far greener than the writings found in our contemporary media. But despite the popularity of his conservation writings (and they were widely published in his time), and despite his enormous influence, he was unable to stem the encroachment of civilization on wilderness. Slowed? Yes. Stopped? No.

Much of the conservation problem now, as then, remains that of translating the popular will into meaningful policy initiatives and institutional change. Assayers of public opinion say that American values have been green throughout the last decade. Yet our basic ideology and institutions have not changed, and consequently the environmentally destructive ways in which we live have not been modified. What is truly remarkable about our contemporaneous eco-crisis is that it was forestalled for so long by so few. Today, as one pundit wrote, "The environment is back." True enough. The question is, "For how long?" When the attention of the media shifts elsewhere, and with it the public's eye, then the political will to change is diminished. Any other conjecture implies that we assume politicians are motivated more by reason and argument than by the ballot box. Alternatively stated, no matter how green American values, the political process largely responds to issues defined and shaped by the media.

Thus *The Wilderness Condition* finds its role primarily in the context of the re-greening of American consciousness. Here John Muir serves as a model, for he communicated with the public in a way which few serious thinkers do. While newspaper and magazine journalists tend to focus on the dramatic surface of environmental crisis, and not undertake substantive discussions of fundamental questions, more specialized academic writers delve more deeply but seldom communicate in a way that reaches large audiences. Such considerations are

essential to the reasoned understanding that necessarily underlies socio-
cultural transformation based on anything more than a groundswell of
public opinion. The genius of democratic process is not mere majority
rule, but rather the ever present potential that the citizens will carefully
evaluate the underlying issues. Insofar as fallible human beings are to
make decisions that literally affect the future of human life on earth, we
can only hope the public's attention remains focused on the interrelations between environment and civilization. We also hope that this
collection of essays might sensitize its readers to some of the many
issues raised within but seldom discussed outside scholarly circles.

The essays comprising this volume were presented at an interdisci-
plinary conference on Wilderness and Civilization held in August, 1989,
amid a mountain setting near Estes Park, Colorado. The papers were
unified in focusing on a common subject, defined broadly as the
wilderness condition, and by being written in a way designed to facilitate
communication across disciplinary boundaries.[1] While all the authors
have spent a considerable portion of their lives considering conservation
and wilderness related issues, their backgrounds are different. Some live
in the mountains amid wild areas of unsurpassed beauty, and others in
large cities where virtually all the land has been humanized. There are
poets, historians, philosophers, anthropologists, biographers, and even
a mountaineer. Some are academics, and others have never been in a
formal classroom except as a student. Some are accomplished outdoors
people: deep powder skiers, rock and mountain climbers, backpackers
and day-hikers, hunters and fishers, wilderness photographers and
painters. Most have been actively involved in efforts to protect things
wild and free, and are members of the major national wilderness
organizations as well as numerous regional organizations dedicated to
preservation interests. Some have been involved in consequential leadership roles in conservation.

The essays form a coherent if diverse whole focusing on the fixed
but ever changing relations between wild nature and civilization. Some
of the essays look to the seminal figures of wilderness philosophy for
new meaning and guidance in difficult times. Others look farther back
into Western culture, to the beginnings of modern science and the
industrial revolution, in an attempt to understand the grounds of today's

environmental crisis. Some look even farther into the past, turning their gaze onto the Paleolithic and the evolution of human nature itself. A few look outside our Western culture toward the East in an endeavor to better assess our own possibilities for tomorrow.

One way of viewing these essays collectively is to set them in an evolutionary context. Today the work of such nineteenth century luminaries as Charles Darwin and Rudolf Clausius is understood as having initiated a second scientific revolution. The first scientific revolution, and therefore modern science itself, began with the work of Galileo and Newton. René Descartes, from his mid-seventeenth century vantage point, epitomized the modern era's megalomania when he proclaimed that through the power of science humankind might render itself the "masters and possessors" of nature. But through the lens of modern science per se, unleavened by evolution, time was not real and humankind was outside the scientific picture of nature. Darwin and Clausius changed all that: for humankind itself was a product of nature, and the product of thinking human beings—the scientific world view— was itself subject to change through time. Part of the genius of America's great wilderness thinkers, Thoreau, Muir, and Leopold, lies in the fact that their thinking reflects the verity of evolution. And recognition of that reality gives cogency to most of the essays herein.

Another unifying theme running throughout the essays is the question of language. An ancient Chinese sage observed that the fish is the last to discover that it swims in water. So, also, we have only recently discovered that we swim in a linguistic milieu which for too long remained transparent to us. The woof and warp of thought itself is largely determined by the structure of language, and basic definitions of words strongly influence the related processes of conception and per- ception. Late twentieth century scholarship, especially of the so-called postmodern variety, dwells on language. Many of the papers herein attend closely to the meaning of words, and show how new possibilities for life in harmony with nature can be opened up by examining more closely how we speak of the world.

Finally, most of the essays allude in one way or another to the looming *possibility* of a so-called paradigm shift: that is, a sudden and dramatic transformation in human consciousness and institutions. The idea of such a shift is problematic, and scholars devote entire books to technical questions concerning the basic idea or category of revolution itself. Such issues are beyond the role and scope of this book: what our authors perceive is the ongoing reality of evolutionary process, and the possibility that a revolutionary change in human consciousness is in the offing, a development that would ultimately change the ways in which humankind lives and relates to the wild world from which it came.

Gary Snyder's "The Etiquette of Freedom" exemplifies natural history writing, and is perhaps the best prose that he has written. His essay seems a particularly fitting way to initiate our study of the wilderness condition, since it has been previously published in what might be considered a popular format.[2] "The Etiquette of Freedom" communicates some very difficult ideas about the relations between environment and civilization in a clear and concise fashion, thus demonstrating that one can do such a thing without producing that "byzantine artifact known as the professional paper." In the tradition of the natural history genre, the essay is leavened throughout with an unsurpassed richness of personal insight: into wild nature and things ecological, into archaic culture and mythology, and into the contemporary context of advanced industrial society and eco-crisis. The reader finds no abstruse theorizing and arcane speculation, but rather a virtual ecology of Gary Snyder's relations with the wild world, and his insightful perceptions as to how that wildness can, if we will but let it, renew our culture. While he is no Cassandra prophesying doom and destruction, he does touch the heart of darkness that has led our world to its perilous state, and he sees that the remedy is as simple as good manners: that is, minding nature's ways.

He accomplishes this in part through exploration of language that revolves around the meanings of words like "nature," and a pair of associated words, the "wild" and "free." The discussion reveals that our prevailing definitions of the "wilderness" preclude recognition of nature as a spontaneous and naturally organized system in which all parts are harmoniously interrelated; in consequence, humankind has believed itself compelled to impose order on nature. Snyder contends that the relation between the human species and the "other" is not and should not be simply one of the inexorable humanization and domination of nature.

Rather, wild nature and culture are organically related, and the destruction of those things wild and free will inevitably entail the collapse of our civilization. As he argues, "in a fixed universe there would be no freedom." If Snyder is correct, then the modern project, which has long promised the total humanization of the earth's surface, is paradoxically destined to fail through its own success. This point, I believe, is the central thrust of "The Etiquette of Freedom." To enjoy the benefits of freedom human beings are obligated to practice good manners, and that etiquette begins with preservation of the *natural grounds* that underlie the *cultural expression* of our political and economical individuality.

Readers familiar with Snyder's previous poetry and prose will recognize that he is traversing familiar ground, as in *Turtle Island* (for which he won the 1975 Pulitzer Prize in poetry), where he envisioned an "old/new way of life," or in *The Real Work*, where he discussed bioregionalism. What was a shamanistic vision of Turtle Island has now manifested itself in a new and vivid metaphor: the etiquette of freedom. And bioregionalism, a concept yet foreign to the American public, now finds a new voice in a seeming oxymoron, "wild culture." But there is no paradox in a culture of wildness, for a wild culture is a bioregional culture such that human beings live on a piece of land without spoiling it. By minding the etiquette of nature humankind might cease rending the fabric of life through economically short-sighted and rude exploitation.

Paul Shepard's essay, "A Post-Historic Primitivism," is dedicated to the recovery of our human roots that sink down into the fertile soils of the Paleolithic. His thesis that the seeds of our eco-crisis were planted during the agricultural revolution, argued in his many books such as *The Tender Carnivore and the Sacred Game* and *Nature and Madness*, has been and remains controversial. "A Post-Historic Primitivism" can be read as a forceful restatement of that thesis, buttressed by almost a decade of further research. In many ways Shepard remains a scholar ahead of his time, for few contemporary thinkers want to consider objectively the implications of evolution for human nature. And his work is seldom appreciated for what it is: of a piece with the monumental evolutionary thought of the nineteenth and twentieth centuries. But given the essay's sweeping implications for sociocultural change, it is a short statement of major importance for the conservation movement.

The essay begins with a careful analysis of the cultural mosaic. Discussions of cultural evolution usually rest on the hidden assumption that each succeeding stage or moment of culture represents an improvement over the past. "A Post-Historic Primitivism" suspends the presupposition that advanced industrial culture is the culminating achievement of human beings, and that it therefore provides a yardstick by which cultures in other times and places might be judged. On Shepard's view the basic cultural forms, from hunting-gathering to advanced industrial societies, do not form any linear sequence of evolution, but are rather mosaics built over time and place by countless generations. Once we understand that our perception of prehistory has been conditioned by the particular accidents of our history, then a post-historic primitivism becomes a possibility. Leavened by its understanding that culture is a mosaic of ever changing and yet recoverable parts that can be reintegrated into the present, it follows that humankind can recognize its many affinities with the Paleolithic past.

Having cleared away the undergrowth obscuring our interrelatedness with the archaic past, Shepard proceeds to discuss the still living and vital relations between wildness and human beingness, and draws a pointed distinction between our prevailing cultural notions of "wilderness" and wildness itself. Like Snyder, Shepard has seen through the disguise of language to the reality that "wildness" is the basis of a world system of species which sustains the biosphere. The "wilderness" is a cultural artifact, exemplified in the idea of a landscape, which Shepard argues is "intrinsically distancing, a science and art that reduces nature to pictures." Similarly, the ideal of domestication is a cultural product which repudiates the prehistoric and wild except as exemplification of the savage, primitive, barbaric, and therefore inferior. Once we grasp, however, our predicament—that is the use of the present as the measure of all forms of human existence—then we can at one and the same time recognize our affinities with the primitive while also acknowledging the differences. Paleolithic hunters and foragers, who lived in harmony with nature's economy, thus become an inspiration for the reform of our own culture. This is the condition which Shepard calls a post-historic primitivism.

George Sessions's paper, "Ecocentrism, Wilderness, and Global Ecosystem Protection," brings together a wide array of current thinking on environmental issues. The essay is framed by an initial discussion of two factions within the American conservation movement: one gravitating around the ideology defined by Gifford Pinchot and the other around the thought of John Muir. According to Sessions, the former group essentially views nature through an utilitarian and anthropocentric lens. From such a perspective wild nature—land, plants, animals, entire ecosystems—possesses no intrinsic value, and the only relevant consideration in its management is human values, be these economic, scientific, aesthetic, or recreational. The other perspective, called a deep ecology approach, sets human values within a larger natural context. From this vantage point inherent natural goods, such as environmental integrity or stability, might necessitate the modification or even elimination of human goods (such as ever increasing material standards of living). The essay then moves into an extended discussion of reasons for protecting wilderness and wild species, arrayed along a continuum that begins with purely instrumental and utilitarian values, and ends with arguments based on inherent or intrinsic values: that is, the idea that wild nature merits consideration on its own sake independent of and sometimes in conflict with instrumental values.

The final and longest section of the paper attempts to transcend the fissure between instrumental and inherent values by exploring theoretical questions involving the foundation of a postmodern society in an ecological world view. Sessions is also concerned with concrete steps that need to be taken, and particularly the questions of global ecosystem conservation. Thus the essay anticipates the possibility of a "paradigm shift" from "modernism" to "post-modernism," a far reaching change in consciousness (*theorie*) and culture (*praxis*). Sessions draws on a wide variety of scholars to buttress his argument, such as the eco-philosophy of Arne Naess and the new field of conservation biology associated with E. O. Wilson. Sessions believes that a Naess-like eco-philosophy is an adequate undergirding for a truly *deep ecology*, and that conservation biology provides solid scientific support for the premise that humankind must deliberately protect vast tracts of the earth as free nature. An implicit theme that runs throughout the essay is the question of balance between *the natural alien* and the *green world*. Sessions argues that if a postmodern cultural form is to be achieved, then civilization and the biosphere will necessarily exist in a harmonious and stable relation. Although he does not address the issue directly, the paper can be seen as a reasoned response to the charge that deep ecology promotes "green fascism": that is, the preservation of wild lands and life over and above all human interests, particularly those of the Third World.

Curt Meine's "The Utility of Preservation and the Preservation of Utility: Leopold's Fine Line" further extends the debate over the relative importance of inherent and instrumental values. The essay gives us reason to carefully reconsider Leopold's thinking in the context of our present re-greening of consciousness. Meine points out that more than twenty years before the land ethic, even while virtually all conservationists looked at the wilderness through utilitarian spectacles, Leopold was nonetheless aware of the need to convert popular opinion into meaningful political action that protected *environmental values in their own right*. In 1925, long before our own post-World War II lessons in environmental history, he wrote that unless "the wilderness idea" became a focal point for political action, then the "steam roller" of relentless economic advance would inevitably humanize all wild lands and life.

Yet, Meine insists, Leopold was at this time and always remained a realist, recognizing that considerations of utility—that is, the use value of wild nature—must be necessarily reflected in any policy proposal that hoped to be politically efficacious. Thus he avoided the extremes which might be characterized as "green fascism" on the one hand, and "industrial fascism" on the other, adopting instead an ecological perspective which recognized the mutual interdependence of the human economy on nature, and of wild lands and life on humankind's economic considerations. The problem involved more than anything else the question of balance between cultural and natural economies. Leopold always wanted to know how human beings could live on a piece of land without spoiling it, and he came to realize that some wilderness preservation was necessary in order to even begin formulating an answer. He wrote that "A science of land health needs, first of all, a base-datum of normality, a picture of how healthy land maintains itself as an organism...." While Leopold never used the term bioregionalism, one can see again his remarkable anticipation of things to come. The "dynamic of wilderness" was just as important as the "dynamic of civilization," and any culture which ignored the question of balance was destined to end sooner rather than later.

But the conservation of natural ecosystems had more than a scientific value. As Meine argues, Leopold grew increasingly sensitive to aesthetic considerations, and the land ethic was the mature statement of his integration of aesthetic with economic and ecologic values. Leopold even defined conservation as seeking to preserve both the utility and the beauty of the landscape. Society, by divorcing utility and beauty, had planted the seeds of its own ruin, for economic considerations dominated the collective consciousness. "We learn, in ecology at least, that all truths hold only without limits." This "ecological foresight" confirms the thoroughgoing nature of Leopold's evolutionary perspective: not only nature, but also culture itself and human consciousness were evolutionary phenomena. Seen from this perspective the history of the conservation movement reflects, as Meine puts it, "a process of continual evolution of the human race's ability to perceive and, hence, to anticipate and race to, new self-generated environmental considerations." Consequently, the early utilitarian oriented dominance, so characteristic of the conservation movement in the days of Theodore Roosevelt and Gifford Pinchot, can now be understood as inevitably giving way to a balance between utility and preservation.

Erazim Kohák's "Perceiving the Good" is yet another variation on the theme of utility vs. preservation. The paper advances a provocative *philosophical* thesis that our perceptions of beauty and value in nature reveal an intrinsically meaningful order of being that is prior to any and all human value judgments, including either *the beautiful* or *the useful*. Using the techniques of phenomenological description, Kohák discloses a radical value in nature itself that is prelinguistic, prereflective, prepredicative: in essence an intrinsic value entirely independent of and unconditioned by human subjectivity. Yet we live in an age which has lost this ability of *perceiving the good*; indeed, the modern mind believes the residual glimpses we have of intrinsic worth are mere sentimentalities. The question then is how we in the modern world can regain a clear vision of the intrinsic worth of being.

Kohák identifies three strategies by which human beings have attempted to account for the perception of intrinsic good in nature, and whereby we might recover it: the speculative, the contemplative, and the practical. While each of these strategies has its merits and proponents, he finds each wanting. The *speculative attempts* of philosophers (such as Edmund Husserl) to ground intrinsic worth in transcendental subjectivity fail in denying the "the hardness of reality," the "prevenient horizon that is radically other." The *contemplative strategies*, unlike the rationalizing strategies of speculation, stand in mute awe before the wonder of being, and encourage an evocation of similar experiences in others. The problem is that such experiences remain ineffable, beyond description. While contemplation leads us to stand in awe of nature, a "sense of wonder, stripped of all categories of utility, sinks into a poetic impotence." Which is precisely the appeal of *practical approaches*, which point to the virtue of work. By doing what is useful, the pragmatist contends, we also thereby encounter the good. But doubts emerge here also, since the pragmatist always runs the risk that functional categories become ends in themselves and thus lose moral significance by falling into a crass materialism.

At this juncture the essay ties back to issues raised by Snyder and Shepard, especially the question of perception and its unconscious shaping by culture. Kohák's main conclusion is that our contemporary greening of consciousness manifests an evolution in human sensibility, a "shift from a basic posture of taking to giving, caring, from nature as being-for-us to nature as being-for-itself, or, in the jargon of the trade, from an anthropocentric to an ecocentric perspective." Humankind is now learning to see nature in a new way, to see the good rather than only its utility. "Not in taking, but in giving of ourselves to the other do we learn to *see the good*." Kohák also points out that we have always taken wild nature for granted, as eternally there. There would always be fresh

water and air, and more whales and elephants. Suddenly we are confronted by the reality of an other that may not always be there. And thus,
Kohák believes, we loom on the edge of that same transformation that
Sessions anticipates, the paradigm shift from a shallow to a deep
ecology.

Michael P. Cohen's contribution, "A Brittle Thesis: A Ghost
Dance: A Flower Opening," revolves around the life and thought of John
Muir, and further extends the inquiry into questions of perception raised
by Kohák. Cohen's paper can be read as (in part) a reconsideration of his
Muir biography, *The Pathless Way* (1984), and his recent *History of the
Sierra Club* (1989). He begins with a question that biographers and
historians almost invariably come to: What is the relation (generally) of
the past to the present, and its implications for the future? Even more
specifically, Cohen is probing for further meaning in or a new perspective on his own interpretations of Muir and the Sierra Club. Still, the
reader may encounter difficulty in identifying Cohen's "brittle thesis."
 Just what is it that has apparently made him uncomfortable with his
reading of the past, especially as this involves conservation history
(Muir, the Sierra Club) and its implications for tomorrow? Is Cohen
repudiating any interpretation of environmental history as a linear
sequence of events that is inevitably progressive and will therefore lead
humankind toward a sustainable society, some postmodern return to the
Garden? Early in the paper Cohen seems to imply that John Muir
himself, despite his love for and defense of the land, was a prototypical
Euro-man, unaware that the California in which he revelled was in fact
a radically altered landscape. Are even the best of us, like John Muir,
truly oblivious to the fact that in environmental context we are nothing
more than weeds (a powerful metaphor Cohen uses, first used by Edgar
Anderson in *Plants, Man, and Life*)? Yet, on the other hand, Cohen
appears to repudiate this deconstructive reading of *The Pathless Way*
and John Muir, or the Sierra Club and the conservation movement, and
implies that sociocultural transformation may yet occur. From this
perspective Cohen sees humankind as an unfinished project that may
finally grow up, evolving into something more than invading opportunists (weeds) that thoughtlessly destroy nature's economy.

Such evolution is possible, Cohen tells us, because we still resonate to the same harmonics that Muir heard in his wilderness odyssey. Nature is so interesting! Cohen brings that point home with a poetic flair—a marvelous first person account of mountain climbing in the Tetons, written one suspects with a kind of Thoreauvian collapsing of time and events into a seamless narrative. The lesson for us is that, just as Muir met his social obligations at the cost of curtailing his wilderness sojourns, so all those who respond to the poetry of nature today are similarly obligated to confront pragmatic questions as well. Like Kohák, Cohen reminds us then that a "deep ecology" (or whatever else we might care to call it) which remains only a poetry is one inadequate to the tasks of living in a world where wilderness experiences are increasingly an opportunity for a privileged few.

Cohen's analysis, however subtle, is nonetheless forceful, and goes beyond the kinds of questions and issues that writers of environmental history alone must face. For he forces upon us questions of grave import: environmental history, and thus the history of our culture, is *in media res*. We are caught in midstream. As Cohen suggests, "If there is a paradigm shift coming, we will all find ourselves in an enlightened world, all deep ecologists, but...we are not there yet." Is history itself sound and fury, signifying nothing? Just what is it that those who seek to preserve wilderness actually preserve? Who and what are we, and where are we going? Cohen wants us to ask ourselves these questions as we contemplate the meaning of our contemporaneous greening of consciousness.

Pete Gunter's essay, "The Disembodied Parasite and Other Tragedies; or: Modern Western Philosophy and How to Get Out of It," turns our attention from the present and immediate future to the historical past. In particular the paper looks to Cartesian rationalism and its lingering effects on conservation. As Gunter points out, Descartes (along with coevals like Galileo and Francis Bacon) provided the underpinnings of our modern outlook on wild nature. For Descartes nature was a merely material, lifeless "other" subject to human domination, plants and animals were no more than unfeeling mechanisms or robots, and human beings were distinguished from all that was natural by their soul.

The consequences of Cartesianism, Gunter argues, still reverberate in our contemporary lives. Human beings think of themselves as removed from or above the natural world; nature itself is conceptualized as a geometrized and material entity devoid of feeling and value; and the modern mind believes that the only legitimate knowledge of nature is mathematical physics. It is this world view, Gunter argues, "that conservationists find themselves having to combat every time they appear at a Congressional hearing or on the evening news." Conservationists often are lead into defensive postures because Cartesianism has conditioned us to be skeptical of such qualities as environmental beauty or ecological integrity. Since such things are often incommensurable, environmentalists have been tempted to play cost-benefit games, calculating the economic benefits that flow from wilderness preservation. Gunter implies that this game reflects (in some measure) the underlying Cartesian grounds of environmental ruin, since cost benefit analysis assumes that dollars are a yardstick by which all things can be measured and so reduced to common, quantitative terms.

In the concluding section of his paper Gunter presents a succinct overview of nature from an evolutionary standpoint. Nature is not just matter-in-motion that has value only insofar as human beings use it. Rather nature is a living, self-organizing phenomenon of infinitely more complexity than Descartes ever realized. And human beings are not pure thinking things, but beings whose thoughts and feelings are embodied, centered in an organic human nature fashioned in the web of life over the longueurs of space and time. We are embodied in the whole world, Gunter explains, and "if there is anything that the wilderness experience conveys, indelibly, it is this universality of experience." Furthermore, legitimate modes of knowledge can be modelled on something other than mathematical physics without fear that we thereby become "green mystics." Gunter takes some pains to point out the perils of "green mysticism," but also points out that we must never let a Cartesian inspired concept of reason ground an "anti-green intolerance" that brands ecologism as primitivism. An ecological and evolutionary perspective on nature necessarily includes the human animal, and in consequence, Gunter believes, goes a long way toward resolving the unholy Cartesian trinity.

Dolores LaChapelle's paper, "Not Laws of Nature but *Li* (Pattern) of Nature," complements Gunter's critique of Descartes. She argues that the neo-Confucian concept of *li* might help us find our way back to a view of humanity as a part of nature. In part this is because the neo-Confucians were monists, and did not recognize any fundamental distinction (as did Descartes and Newton) between mind and matter. The neo-Confucians viewed the world as a single living organism in which all the various natural kinds functioned in accordance with natural patterns (*li*), and thereby maintained the harmony of the whole. Thus the Chinese, c. 1200 C.E., had already achieved what we today recognize as an ecological point of view.

There will be a tendency among some readers to think of LaChapelle's paper as irrationalist, outside the bounds of legitimate discourse. Gunter's arguments might help those readers to a more sympathetic perspective, and the work of scholars who have critiqued the ethnocentric tendencies of modern Western culture, such as Paul Feyerabend and Claude Lévi-Strauss, also lend credence. We prefer our concept of law to the idea of *li* since it is our own. But no *prima facie* case can be made that our concept of natural law is superior to the Chinese concept of *li*. Furthermore, and paradoxically, the apparent rejoinder to LaChapelle's thesis that China today is the most intensively humanized landscape in the world actually helps to confirm her argument. For modern China, in becoming a Marxist state, abandoned the neo-Confucian underpinnings of Chinese culture. Just as with capitalism, so with Marxism, for both presuppose the Western concept of natural law.

The West and modern China think of natural law as the cognitive vehicle by which the mastery and possession of nature can be achieved (through prediction and subsequent causal control). So viewed, nature is nothing more than raw materials for economic society. In contrast, the neo-Confucian East understood *li* as the vehicle through which human beings might achieve their full potential through integration with the pattern of nature itself. Furthermore, LaChapelle points out, unlike our Cartesian-Newtonian inspired conceptions of natural law as invariant universals, the *li* of things varies from place to place. From this perspective there would not be any one cultural model for all people in all places at all times; rather, culture would necessarily vary from place to place, reflecting the enormous variety of nature itself. What works, for example, on the prairies of the American Midwest would not be thought of as an appropriate agricultural model for the Amazon rain forest.

Perhaps even more importantly, the idea of *li* opens up a new perspective on our own human nature as having a pattern which, when allowed its free expression within the surrounding order of things, allows the individual to flourish. She suggests that by living in place, especially in proximity to unhumanized landscapes, we can experience *li*. Clearly the contingencies of modern culture—urbanization, burgeoning populations, the worship of Mammon—imply that a conscious effort is needed to put ourselves in a place where this can happen. But she argues that the effort is worthwhile by drawing parallels between this concept and Jung's notion of the collective unconscious, the wilderness deep inside us beyond the ego's ability to control.

Michael Zimmerman's "The Blessing of Otherness: Wilderness and the Human Condition" complements LaChapelle's inquiry into the relations between nature and human personality. However, he does not look to the East but to three twentieth-century thinkers who see the wilderness as constituting an otherness which plays an essential role in human identity. Zimmerman's essay also reflects the traditions of natural history writing, for his account begins with an intense, almost Thoreauvian, exploration of the fashioning of his own character amid the Ohio wilderness of his youth. He suggests that individuals in natural settings often become aware of rhythms and movements that are far vaster and older than the ego or sense of self, and thus personal encounters with wild nature play an important role in determining human personality structure. Wilderness is "the other," an other which "reminds humanity of its own dependency on the powers at work not only in wilderness, but also in humanity itself." So framed, "The Blessing Of Otherness" proceeds to explore the work of Martin Heidegger, Susan Griffin, and Hans Duerr.

Zimmerman, one of the leading American interpreters of Martin

Heidegger, pulls back from an earlier interpretation, where he was relatively sanguine about the prospects for a Heideggerian inspired deep ecology. True enough, he points out, in the Heideggerian deconstruction of our prevailing anthropocentric and dualistic outlooks there is much that attracts conservationists. But on his new reading Zimmerman finds Heidegger to be "fiercely anti-natural," and thus difficult to reconcile with any ecologically informed environmental ethic and wilderness philosophy.

Zimmerman also argues, in what seems to be a tangent but is actually a most significant point, that this anti-naturalistic attitude led Heidegger into the clutches of the Nazis. National Socialism represented itself in the early 1930s as championing the German *Volk*, as a party that would lead the *Volk* out of the clutches of capitalistic entrepreneurs to a preindustrial, agricultural world more closely in tune with nature. While we know the outcome of this sorry chapter in European politics, the point is of contemporary relevance. Can environmentalists defend themselves against the charge that they too are green fascists who promise restoration of a natural way of life, but in truth are little more than political opportunists with a popular cause? Zimmerman's essay does not resolve but sensitizes us to this complicated issue. Two things are clear: the human species cannot be seen as a natural kind through Heideggerian spectacles, and National Socialism remained enframed within the conventional political, economical, and historical categories of the modern West. The message for the conservation movement is that power alone can do little in helping resolve global eco-crisis: the crucial question is the one of underlying ideology, and particularly the grounds of a truly deep ecology.

Zimmerman also explores the work of Susan Griffin and Hans Duerr. While there is no definitive ecofeminist position, most would agree that wild nature has been dominated by patriarchal values and the subsequent androcentric organization of society. By introducing Griffin, a considerable thinker in her own right, Zimmerman invites us to consider more of the ecofeminist literature, as for example that of Rosemary Ruether or Mary Daly. Duerr serves to ground Zimmerman's thesis that the essential human function of the wilderness is to reveal to us what it means to be *civilized* human beings. Part of Duerr's point is that only through the recognition or perception of what it is that we are not (the negative) can we have any true understanding of what we are, in fact (the positive). Unlike Heidegger, Duerr affirms humankind as internally related to wild nature. Experience of the wilderness as an "other" is necessary to any grounded understanding of our distinctively human beingness. Civilization is, in effect, defined by its difference from wilderness.

My own contribution, "Wilderness, Language, and Civilization," concludes *The Wilderness Condition*, and perhaps brings the collection of essays full circle by returning to the questions of language and definition broached by Gary Snyder. The paper suggests that what often prevents us from thinking of nature as anything more than a source of goods or as a hostile force to be conquered—an environmental stage upon which the human drama is enacted—is the lack of a vocabulary for such expression. Any one familiar with at least some of the writings of Thoreau, Muir, Leopold, or Snyder should have little difficulty understanding the argument. I attempt to show that the writings of America's three great wilderness philosophers—Henry David Thoreau, John Muir, and Aldo Leopold—are not merely items of historical interest or antiquarian curiosity, but can and should be seen as contributing to our presently changing conceptions of the relation between wilderness and civilization. The study concludes with a similar treatment of America's prodigious poet of the wilderness, Gary Snyder.

The basic premise upon which "Wilderness, Language, and Civilization" turns is that who and what we are as a species distinct from all others largely involves language. Language is world-making. It brings forth a human space in which we dwell. Our present fissure from wild nature came about in part because of Hebraic and Greek traditions that drew radical distinctions between humankind and the rest of the natural world, and viewed nature not as something with which human beings were intrinsically related but as a source of material sustenance only. Through their writings the great wilderness thinkers and poets penetrate the vocabulary and rhetoric of the modern age, and reanimate an ancient, premodern way of being in the world. So viewed, they manage to pass through the second world, our created cultural world which conceals from us the rooting of the human project in a natural soil, to the first world, that green world from which we have come.

To the modern mind, of course, such analysis seems mere romantic nonsense, readily dismissed as a kind of puerile longing for the halcyon days of our Paleolithic childhood, when in fact what is called for is increased management of an unruly natural world that is turning against us. To the postmodern mind, such analysis confirms the validity of that thesis articulated in one form or another by all the essential wilderness writers: in wildness is the preservation of the world. For they look both behind and ahead, to a world that might be, where humankind has rediscovered its fundamental co-relatedness with nature, and advances toward a future where harmony, integrity, and stability characterize the relations between culture and nature. Such would be, as Snyder so compellingly suggests, an old-new way of being.

# The Etiquette of Freedom

## by Gary Snyder

### I

One June afternoon in the early seventies I walked through crackly gold grasses to a neat but unpainted cabin at the back end of a ranch near the drainage of the South Fork of the Yuba in northern California. It had no glass in the windows, no door. It was shaded by a huge black oak. The house looked abandoned and my friend, a student of Native California literature and languages, walked right in. Off to the side, at a bare wooden table, with a mug of coffee, sat a solid old gray-haired Indian man. He acknowledged us, greeted my friend, and gravely offered us instant coffee and canned milk. He was fine, he said, but he would never go back to a V.A. hospital again; from now on if he got sick he would stay where he was, he liked being home. We spoke for some time of people and places along the western slope of the northern Sierra Nevada, the territories of Concow and Nisenan people. Finally my friend broke his good news: "Louie, I have found another person who speaks Nisenan." There were perhaps no more than three people alive speaking Nisenan at that time, and Louie was one of them. "Who?" Louie asked. He told her name. "She lives back of Oroville. I can bring her here, and you two can speak." "I know her from way back" Louie said. "She wouldn't want to come over here. I don't think I should see her. Besides, her family and mine never did get along." That took my breath away. Here was a man who would not let the mere threat of cultural extinction stand in the way of his (and her) cultural values. To well-meaning sympathetic white people this response is almost incomprehensible. In the world of his people, never over-populated, rich in acorn, deer, salmon, and flicker-feathers, to cleave to such purity, to be perfectionists about matters of family or clan, were affordable luxuries. Louie and his

21

fellow Nisenan had more important business with each other than conversations. I think he saw it as a matter of keeping their dignity, their pride, and their own ways—regardless of what straits they had fallen upon—until the end.

Coyote and ground squirrel do not break the compact they have with each other that one play predator and the other play game. In the wild a baby black-tailed hare gets maybe one free chance to run across a meadow without looking up. There won't be a second. The sharper the knife, the cleaner the line of the carving. We can appreciate the elegance of the forces that shape life and the world, that have shaped every line of our bodies—teeth and nails, nipples and eyebrows. We also see that we must try to live without causing unnecessary harm, not just to fellow humans but to all beings. We must try not to be stingy, or to exploit others. There will be enough pain in the world as it is.

Such are the lessons of the wild. The school where these lessons can be learned, the realms of caribou and elk, elephant and rhinoceros, orca and walrus, are shrinking day by day. Creatures who have travelled with us through the ages are now apparently doomed, as their habitat—and the old old habitat of humans—falls before the slow-motion explosion of expanding world economies. If the lad or lass who knows where the secret heart of this monster is hidden among us, let them please tell us where to shoot the arrow that will slow it down. And if the secret heart stays secret and our work is made no easier, I for one will keep working for wilderness day by day, on every level, with every tool available, whether there is hope or not, and count it a great climb.

"Wild and free." An American dream-phrase, loosing images: a long-maned stallion racing across the grasslands, a V of Canada geese high and honking, a squirrel chattering and leaping limb to limb overhead in an oak. It also sounds like an ad for a Harley-Davidson. Both words, profoundly political and sensitive as they are, have become consumer baubles. I hope to investigate the meaning of "wild" and how that connects with "free" and what one would want to do with those meanings. To be truly free one must take on the basic conditions as they are: painful, impermanent, open, and imperfect; and then be grateful for impermanence and the freedom it grants us, for in a fixed universe there would be no freedom. With that freedom we improve the campsite, teach children, oust tyrants. The world is nature, and in the long run inevitably wild, because the wild, as the process and essence of nature, is also an ordering of impermanence.

Although "nature" is a term that is not of itself threatening, the idea of the "wild" in civilized societies—both European and Asian—is often associated with unruliness, disorder, and violence. The Chinese word for nature, *zi-ran* (Japanese *shizen*), means "self-thus." It is a bland and general word. The word for wild in Chinese, *ye* (Japanese *ya*), which basically means "open country," has a wide set of meanings: in various combinations the term becomes illicit connection, desert country, an illegitimate child (open-country child), prostitute (open-country flower), and such. In an interesting case, *ye-man zi-yu* "open-country southern-tribal-person-freedom" means "wild license." In another context "open-country story" becomes "fiction and fictitious romance." Other associations are usually with the rustic and uncouth. In a way *ye* is taken to mean "nature at its worst." Although the Chinese and Japanese have long given lip-service to nature, only the early Daoists might have thought that wisdom could come of wildness.

Thoreau says "give me a wildness no civilization can endure." That's clearly not difficult to find. It is harder to imagine a civilization that wildness can endure, yet this is just what we must try to do. Wildness is not just the "preservation of the world," it is the world. Civilizations East and West have long been on a collision course with wild nature, and now the developed nations in particular have the witless power to destroy individual creatures, whole species, whole processes, of the earth. We need a civilization that can live fully and creatively together with wildness. We must start growing it right here, in the New World.

When we think of wilderness in America today, we think of remote and perhaps designated regions that are commonly alpine, desert, or swamp. Just a few centuries ago, when virtually *all* was wild in North America, wilderness was not something exceptionally severe. Pronghorn and bison trailed through the grasslands, creeks ran full of salmon, there were acres of clams, and grizzlies, cougar, and bighorn sheep were common in the lowlands. There were human beings, too: North America was *all populated*. One might say yes, but thinly—which raises the question of according to who. The fact is, people were everywhere.

When the Spanish foot-soldier Alvar Nunez (Cabeza de Vaca) and his two companions (one of whom was African) were wrecked on the beach of what is now Galveston, and walked to the Rio Grande valley and then south back into Mexico between 1528 to 1536, there were few times in the whole eight years that they were not staying at a native settlement or camp. They were always on trails.

It has been a part of basic human experience to live in a culture of wilderness. There has been no wilderness without some kind of human presence for several hundred thousand years. Nature is not a place to visit, it is *home*: and within that home territory there are more familiar and less familiar places. Often there are areas that are difficult and remote, but all are *known*, and even named. One August I was at a pass in the Brooks Range of northern Alaska at the headwaters of the Koyukuk River, a green 3000 foot tundra pass between the broad ranges, open and gentle, dividing the waters that flow to the Arctic Sea from the Yukon. It is as remote a place as you could be in North America, no roads, and the trails are those made by migrating caribou. Yet this pass has been steadily used by Inupiaq people of the north slope, and Athapaskan people of the Yukon, as a regular north-south trade route for at least seven thousand years. All of the hills and lakes of Alaska have been named in one or another of the dozen or so languages spoken by the native people, as the researches of Jim Kari and others have shown. Euro-American mapmakers name these places after transient exploiters, or their own girl friends, or hometowns in the lower 48. The point is, it's all in the native story, and yet only the tiniest trace of human presence through all that time shows. The place-based stories the people tell, and the naming they've done, is their archeology, architecture, and *title* to the land. Talk about living lightly.

Cultures of wilderness live by the life and death lessons of subsistence economies. But what can we now mean by the words "wild" and for that matter "nature"? Languages meander like great rivers leaving oxbow traces over forgotten beds, to be seen only from the air or by scholars. Language is like some kind of infinitely inter-fertile family of species spreading or mysteriously declining over time, shamelessly and endlessly hybridizing, changing its own rules as it goes. Words are used as signs, as stand-ins, arbitrary and temporary, even as language reflects (and informs) the shifting values of the peoples whose minds it inhabits and glides through. We have faith in "meaning" like we might believe in wolverines—putting trust in the occasional reports of others, or on the authority of once seeing a pelt. But it is sometimes worth tracking this trickster back.

## II

**The words *nature* and *wild*.** Take "nature" first. The word gets two slightly different meanings. One is "the outdoors"—the physical world, including all living things. Nature by this definition is norm of the world that is apart from the features or products of civilization and human will. The machine, the artifact, the devised, or the extra-ordinary (like a two-headed calf) is spoken of as "unnatural." The other meaning, which is broader, is "the material world, or its collective objects and phenomena" including the products of human action and intention. As an agency nature is defined as "the creative and regulative physical power which is conceived of as operating in the material world and as the immediate cause of all its phenomena." Science and some sorts of mysticism rightly propose that "everything is natural." By these lights there is nothing unnatural about New York City, or toxic wastes, or atomic energy, and nothing—by definition—that we do or experience in life is "unnatural." (The "supernatural"? One way to deal with it is to say that "the supernatural" is a name for phenomena which are reported by so few people as to leave their reality in doubt. Nonetheless these events— ghosts, gods, magical transformations and such—are described often enough to make them continue to be intriguing and for some, credible.)

The word *nature* is from Latin *natura*, "birth, constitution, character, course of things"—ultimately from *nasci*, to be born. So we have nation, natal, native, pregnant. The probable Indo-European root (via Greek *gna*—hence cognate, agnate)—is *gen*—(Sanskrit *jan*) which provides "generate" and "genus," as well as "kin" and "kind."

The physical universe and all its properties—I would prefer to use the word in this sense.

**The word *wild* is like a gray fox** trotting off through the forest, ducking behind bushes, going in and out of sight. Up close, first glance it is "wild"—then farther into the woods next glance it's "wyld" and it recedes via Old Norse *villr* and Old Teutonic *wilthijaz* into a faint pre-Teutonic *ghweltijos* which means, still, wild, and maybe wooded (*wald*) and lurks back there with possible connections to "will," to Latin *silva* (forest, sauvage), and to the IE root *ghwer*, base of Latin *ferus*, (feral, fierce) which swings us around to Thoreau's "awful ferity" shared by virtuous people and lovers. The Oxford English Dictionary has it:

Of animals—Not tame, undomesticated, unruly. Of plants—not cultivated.

Of land—Uninhabited, uncultivated.

Of foodcrops—Produced or yielded without cultivation.

Of societies—Uncivilized, rude, resisting constituted government.

Of individuals—Unrestrained. Insubordinate, licentious, disso-lute, loose, "Wild and wanton widowes"—1614.

Of behaviour—Violent, destructive, cruel, unruly.

Of behaviour—Artless, free, spontaneous, "Warble his native wood-notes wild"—John Milton.

*Wild* is largely defined in our dictionaries by what—from a human standpoint—it is not. It cannot be seen by this approach for what it is. Turn it the other way:

Of animals—Free agents, each with their own endowments, living within natural systems.

Of plants—Self-propagating, self-maintaining, flourishing in ac-cordance with innate qualities.

Of land—A place where the original and potential vegetation and fauna are intact and in full interaction, and the landforms are entirely the result of non-human forces. Pristine.

Of societies—Societies whose order has grown from within and is maintained by the force of consensus and custom rather than explicit legislation. Primary cultures, which consider themselves the original and eternal inhabitants of their territory. Societies which resist eco-nomic and political domination by civilization. Societies whose eco-nomic system is in a close and sustainable relation to the local ecosystem.

Of individuals—Following local custom, style, and etiquette, with-out concern for the standards of the metropolis or nearest trading post. Unintimidated, self-reliant, independent. "Proud and free."

Of behaviour—Fiercely resisting any oppression, confinement, or exploitation. Far-out, outrageous, "bad," admirable.

Of behaviour—Artless, free, spontaneous, unconditioned. Expressive, physical, openly sexual, ecstatic.

Most of the senses in this second set of definitions come very close to being how the Chinese define the term "Dao," Way. The way of Great Nature: eluding analysis, beyond categories, self-organizing, self-informing, playful, surprising, impermanent, insubstantial, independent, complete, orderly, unmediated, freely manifesting, self-authenticating, self-willed, complex, quite simple. Both empty and "real" at the same time. In some cases we might call it "sacred." It is not far from the Buddhist term "Dharma" with its original senses of forming and firming.

**The word** *wilderness.* *Wyldernesse,* Old English *wildeornes,* possibly from "wild-deer-ness," (*deor,* deer and other forest animals), but more likely "wildern-ness," with the meanings:

A large area of wild land, with original vegetation and wildlife, ranging from dense jungle or rain forest to arctic or alpine "white wilderness."

A wasteland: as an area unused or useless for agriculture or pasture.

A space of sea or air, as in Shakespeare, "I stand as one upon a Rock, environ'd with a Wilderness of Sea." *Titus Andronicus.* The oceans.

A place of danger and difficulty: where you take your own chances, depend on your own skills, and do not count on rescue.

This world as contrasted with heaven. "As I walk'd through the wilderness of this world." *The Pilgrim's Progress.*

A place of abundance: John Milton, "A wildernesse of sweets." This usage of wilderness caught the very real condition of energy and richness that is so often found in wild systems. "A wildernesse of sweets" is like the billions of herring or mackerel babies in the ocean, the cubic miles of krill, wild prairie grass seed (leading to the bread of this day, made from the germs of grasses)—all the incredible fecundity of small animals and plants, feeding the web. But from another side, wilderness has implied chaos, the unknown, realms of taboo and the habitat of the demonic. In both senses it is a place of archetypal power, teaching, and challenge.

## III

So we can say that New York City or Tokyo are "natural" but not "wild." They do not deviate from the laws of nature, but they are habitat so exclusive in the matter of who and what they give shelter to, and so intolerant of other creatures, as to be truly odd. Wilderness is a place where the wild potential is fully expressed, a diversity of living and non-living beings flourishing according to their own sorts of order. In ecology we speak of "wild systems." When an ecosystem is fully functioning, all the members are present at the assembly. To speak of wilderness is to speak of wholeness. Human beings came out of that wholeness, and to consider the possibility of reactivating membership in the Assembly of All Beings is in no way regressive.

By the sixteenth century the lands of the Occident, the countries of Asia, and all the civilizations and cities from the Indian subcontinent to the coast of North Africa were becoming ecologically impoverished. The people were rapidly becoming nature-illiterate. Much of the original vegetation had been destroyed by the expansion of grazing or agriculture, and the remaining land was of no great human economic use, "waste," mountain regions and deserts. The lingering larger animals—big cats, desert sheep, serows and such, managed to survive by retreating to the harsher habitats. The leaders of these civilizations grew up with less and less personal knowledge of animal behavior and were no longer taught the intimate wide-ranging plant knowledge that had once been universal. By way of tradeoff they learned "human management"—administration—rhetorical skills. Only the most marginal of the *paysan*, people of the land, kept up practical plant and animal lore, and memories of the old ways. People who grew up in towns or cities, or on large estates, had less chance to learn how wild systems work. Then major blocks of citified mythology (Medieval Christianity and then the "Rise of Science") denied first soul, then consciousness, and finally even sentiency to the natural world. Huge numbers of Europeans were losing the opportunity for direct experience of nature, in the climate of a nature-denying mechanistic ideology.

A new sort of nature-traveller came into existence: They were men who went out as resource scouts, financed by companies or aristocratic families, penetrating the lightly populated lands of people who lived in and with the wilderness. Conquistadores and priests. Europe had killed off the wolves and bears, deforested vast areas and over-grazed the hills. The search for slaves, fish, sugar, and precious metals ran over the edge of the horizon and into Asia, Africa, and the New World. These over-refined and warlike states once more came up against wild nature and natural societies: people who lived without Church or State. In return for gold or raw sugar, the white men had to give up something of themselves: they had to look into their own sense of what it meant to be a human being, wonder about the nature of hierarchy, ask if life was worth the honor of a King, or worth gold. (A lost and starving man stands and examines the nicked edge of his sword and his frayed Spanish cape in a Florida swamp.) Some, like Nuño de Guzmán, became crazed and sadistic. "When he began to govern this province, it contained 25,000 Indians, subjugated and peaceful. Of these he has sold 10,000 as slaves, and the others, fearing the same fate, have abandoned their villages."[1] Cortes, the conqueror of Mexico, ended up a beaten, depressed beggar-to-the-throne. Alvar Nunez, who for eight years walked naked across Texas and New Mexico, came out transformed into a person of the New World. He had rejoined the old ways and was never the same again.

He gained a compassionate heart, a taste for self-sufficiency and simplicity, and a knack for healing. The types of both Guzmán and Nunez are still among us. Another person has also walked onto the Noh stage of Turtle Island history to hold hands with Alvar Nunez at the far end of the process—Ishi the Yahi, who walked into civilization with as much desperation as Nunez walked out of it. Nunez was the first European to encounter North America and its native myth-mind, and Ishi was the last Native American to fully know that mind, and he had to leave it behind. What lies between those two brackets is not dead and gone, for it is perennially within us. But it lies dormant as a hard-shelled seed, awaiting the fire or flood that awakes it again. In those intervening centuries, tens of millions of North and South American Indians died early and violent deaths (as did countless Europeans), the world's largest mammal herd was extinguished (the bison), and fifteen million pronghorn disappeared. The grasslands and their soils are largely gone, and only remnants survive from the original old growth eastern hardwood and western conifer forests. We all know more items for this list.

It is often said that the frontier gave a special turn to American history. A frontier is a burning edge, a frazzle, a strange market zone between two utterly different worlds. It is a strip where there are pelts and tongues and tits for the taking. There is an almost visible line that a person of the invading culture could walk across: out of history and into a perpetual present, a way of life attuned to the slower and steadier processes of nature. The possibility of passage into that myth-time world had all but been forgotten in Europe. Its rediscovery—the unsettling vision of a natural self—has haunted the Euro-American peoples as they cleared and roaded the many corners of the North American continent.

*Wilderness* is now—for much of North America—places that are formally set aside on Public Lands—Forest Service or Bureau of Land Management holdings or State and Federal Parks. Some tiny but critical tracts are held by private non-profit groups like the Nature Conservancy, or the Trust for Public Land. These are the shrines saved from all the land that was once known and lived on by the original people, the little bits left as they were, the last little places where intrinsic nature totally wails, blooms, nests, glints away. They make up only two percent of the land of the United States.

## IV

**But Wildness** is not limited to the two percent formal wilderness areas. Shifting scales, it is everywhere: ineradicable populations of fungi, moss, mould, yeasts and such—that surround and inhabit us. Deer mice on the back porch, deer bounding across the freeway, pigeons in the park. Spiders in the corners. There were crickets in the paint-locker of the *Sappa Creek* oil tanker, as I worked as a wiper in the engine room out in mid-Pacific, cleaning brushes. Exquisite complex beings in their energy webs inhabiting the fertile corners of the urban world in accord with the rules of wild systems, the visible hardy stalks and stems of vacant lots and railroads, the persistent raccoon squads. Bacteria in the loam and in our yogurt. The term culture, in its meaning of "a deliberately maintained aesthetic and intellectual life" and in its other meaning of "the totality of socially transmitted behaviour patterns," is never far from a biological root-meaning as in "yogurt culture"—a nourishing habitat. Civilization is permeable, and could be as inhabited as the wild is.

Wilderness may temporarily dwindle, but wildness won't go away. A ghost wilderness hovers around the entire planet, the millions of tiny seeds of the original vegetation are hiding in the mud on the foot of an arctic tern, in the dry desert sands, or in the wind. These seeds are each uniquely adapted to a specific soil or circumstance, each with its own little form and fluff, ready to float, freeze, or be swallowed, always preserving the germ. Wilderness will inevitably return, but it will not be as fine a world as the one that was glistening in the early morning of the Holocene. Much life will be lost in the wake of human agency on earth, that of the 20th and 21st centuries. Much is already lost: the soils and waters unravel—

"What's that dark thing in the water?
Is it not an oil-soaked otter?"—

Where do we start to resolve the dichotomy of the civilized and the wild?

## V

**The wild in human beings.** Do you really believe you are an animal? We are now taught this in school. It is a wonderful piece of information: I have been enjoying it all my life and I come back to it over and over again, as something to investigate and test. I grew up on a small farm with cows and chickens, and with a second-growth forest right at the back fence, so I had the good fortune of seeing the human and animal as in the same realm. But many people who have been hearing this since childhood have not absorbed the implications of it, perhaps feel remote from the non-human world, are not *sure* that they are animals. They would like to feel that they might be something better than animals. That's understandable, other animals might feel that they are something different than "just animals" too. But we must contemplate the shared ground of our common biological being before emphasizing the differences.

Our bodies are wild. The involuntary quick turn of the head at a shout, the vertigo at looking off a precipice, the heart-in-the-throat in a moment of danger, the catch of the breath, the quiet moments relaxing, staring, reflecting—are universal responses of this mammal body. They can be seen throughout the class. The body does not require the intercession of some conscious intellect to make it breathe, to keep the heart beating. It is to a great extent self-regulating; it is a life of its own. Sensation and perception do not exactly come from outside, and the unremitting thought and image-flow are not exactly outside. The world is our consciousness, and it surrounds us. There are more things in mind, in the imagination, than "you" can keep track of—thoughts, memories, images, angers, delights, rise unbidden. The depths of mind, the unconscious, are our inner wilderness areas, and that is where a bobcat is, right now. I do not mean personal bobcats in personal minds—the bobcat roams from dream to dream. The conscious agenda-planning ego occupies a very tiny territory, a little cubicle somewhere near the gate, keeping track of some of what goes in and out (and sometimes making expansionistic plots), and the rest takes care of itself. The body is, so to speak, in the mind. They are both wild.

**Language.** Some will say, so far so good. "We are mammal primates. But we have language, and the animals don't." By some definitions perhaps they don't. In any case they communicate extensively, and by call systems we are just beginning to grasp. But it would be a mistake to think that human beings got "smarter" at some point and invented first language and then society. Language and culture emerge from our biological-social natural existence, animals that we were/are. Language is a mind-body system that co-evolved with our needs and nerves. Like imagination and the body, language rises unbidden. It is of a complexity that eludes our rational intellectual capacities. All attempts at scientific description of natural languages have fallen short of completeness, as the descriptive linguists readily confess, yet the child learns the mother-tongue early and has virtually mastered it by six. Language is learned in the house and in the fields, not at school. Without having ever been taught formal grammar we utter syntactically correct sentences, one after another, for all the waking hours of the years of our life. Without conscious device we constantly reach into the vast word-hoards in the depths of the wild unconscious. We cannot as individuals or even as a species take credit for this power. It came from someplace else, from the way clouds divide and mingle (and the arms of energy that coil first back and then forward), from the way the many flowerlets of a composite blossom divide and re-divide, from the gleaming calligraphy of the ancient riverbeds under present riverbeds of the Yukon River out the Yukon flats, from the wind in the pine needles, from the chuckles of grouse in the ceanothus bushes. Language-teaching in schools is a matter of corralling off a little of the language-behaviour territory and cultivating a few favorite features, culturally defined elite forms that will help you apply for jobs, or give you social credibility at a party. One might even learn how to produce the byzantine artifact known as the professional paper. There are many excellent reasons to master these things, but the power, the *virtu*, remains on the side of the wild.

Social order is found throughout nature—long prior to the age of books and legal codes. It is inherently part of what we are, and its patterns follow the same foldings, checks and balances, as flesh or stone. What we call social organization and order in government is a set of forms that have been appropriated by the calculating mind from the operating principles in nature.

## VI

**The world is as sharp as the edge of a knife.** A Northwest coast saying. Now how does it look from the standpoint of peoples for whom there is no great dichotomy between their culture and nature, those who live in societies whose economies draw on uncultivated systems? The pathless world of wild nature is a surpassing school and those who have lived through her can be tough and funny teachers. Out here one is in constant engagement with countless plants and animals. To be well-educated is to have learned the songs, proverbs, stories, sayings, myths (and technologies) that come with this experiencing of the non-human members of the local ecological community. Practice in the field, "open country," is foremost. Walking is the great adventure, the first medita-tion, a practice of heartiness and soul primary to humankind. Walking is the exact balance of spirit and humility. Out walking, one notices where there is food. And there are first-hand true stories of "Your ass is somebody else's meal"—a blunt way of saying interdependence, inter-connection, "ecology," on the level where it counts, also a teaching of mindfulness and preparedness. There is an extraordinary teaching of specific plants and animals and their uses, empirical and impeccable, that never reduces them to objects and commodities.

It seems that a short ways back in the history of Occidental ideas there was a fork in the trail. The line of thought that is signified by the names of Descartes, Newton, and Hobbes (saying that life in a primary society was "nasty, brutish, and short"—all of them city-dwellers) was a profound rejection of the organic world. For a reproductive universe they substituted a model of sterile mechanism and an economy of "production." These thinkers were as hysterical about "chaos" as their predecessors, the witch-hunt prosecutors of only a century before, were about "witches." They not only didn't enjoy the possibility that the world is as sharp as the edge of a knife, they wanted to take that edge away from nature. Instead of making the world safer for humankind, the foolish tinkering with the powers of life-and-death by the Occidental scientist-engineer-ruler puts the whole planet on the brink of degrada-tion. Most of humanity—foragers, peasants, or artisans—has always taken the other fork. That is to say, has understood the play of the real world, with all its suffering, not in simple terms of "nature red in tooth and claw" but through the celebration of the gift-exchange quality of our give-and-take. "What a big potlatch we are all members of!" To acknowledge that each of us at the table will eventually be part of the meal is not just being "realistic." It is allowing the sacred to enter, accepting the sacramental aspect of our shaky temporary personal being.

The world is watching: One cannot walk through a meadow or forest without a ripple of report spreading out from one's passage. The thrush darts back, the jay squalls, a beetle scuttles under the grasses and the signal is passed along. Every creature knows when a hawk is cruising or a human strolling. The information passed through the system is intelligence.

In Hindu and Buddhist iconography an animal trace is registered on the images of the Deities or Buddhas and Bodhisattvas. Manjusri the

Bodhisattva of Discriminating Wisdom rides a lion, Samantabhadra the Bodhisattva of Kindness rides an elephant, Sarasvati the Goddess of Music and Learning a peacock, Shiva relaxes in the company of a snake and a bull. Some wear tiny animals in their crowns or hair. In this ecumenical spiritual ecology it is suggested that the other animals occupy spiritual as well as "thermodynamic" niches. Whether or not their consciousness is identical with that of the humans is a moot point. Why should the peculiarities of human consciousness be the narrow standard by which other creatures are judged? "Whoever told people that 'Mind' means thoughts, opinions, ideas, and concepts? Mind means trees, fence posts, tiles, and grasses," says Dōgen (the philosopher and founder of the Soto school of Japanese Zen) in his funny cryptic way.

We are all capable of extraordinary transformations. In myth and story these changes are animal-to-human, human-to-animal, animal-to-animal, or even farther leaps. The essential nature remains clear and steady through these changes. So the animal-icons of the Inupiaq people ("eskimos") of the Bering Sea (here's the reverse!) have a tiny human face sewn into the fur, or under the feathers, or carved on the back or breast, or even inside the eye, peeping out. This is the *inua*, which is often called "spirit" but it could just as well be termed the "essential nature" of that creature. It remains the same face regardless of the playful temporary changes. Just as Buddhism has chosen to represent our condition by presenting an image of a steady, solid, gentle, meditating human figure seated in the midst of the world of phenomena, the Inupiaq would present a panoply of different creatures, each with a little hidden human face. This is not the same as anthropocentrism or human arrogance—it is a way of saying that each creature is a spirit, with an intelligence as brilliant as our own. The Buddhist iconographers hide a little animal-face in the hair of the human, to remind us that we see with archetypal wilderness eyes as well.

The world is not only watching, it is listening too. A rude and thoughtless comment about a ground squirrel or a flicker or a porcupine will not go unnoticed. Other beings (the instructors from the old ways tell us) do not mind being killed and eaten as food, but they expect us to say please, and thank you, and they hate to see themselves wasted. The precept against needlessly taking life is inevitably the first and most

difficult of commandments. In their practice of killing and eating with
gentleness and thanks the primary peoples are our teachers: the attitude
toward animals, and their treatment, in twentieth century American
industrial meat-production, is literally sickening, unethical, and a source
of boundless bad luck for this society.

An ethical life is one that is mindful, mannerly, and has style. Of all
moral failings and flaws of character, the worst is stinginess of thought,
which includes meanness in all its forms. Rudeness in thought or deed
toward others, toward nature, reduces the chances of conviviality and
interspecies communication, which are essential to physical and spiritual survival. Richard Nelson has said that an Athapaskan mother might
tell her little girl "Don't point at the mountain! It's rude!" One must not
waste, or be careless, with the bodies or the parts of any creature one has
hunted or gathered. One must not boast, or show much pride in
accomplishment, and one must not take one's skill for granted. Waste-
fulness and carelessness are caused by stinginess of spirit, an ungracious
unwillingness to complete the gift exchange transaction. (These rules
are also particularly true for healers, artists, and gamblers.)

**Perhaps one should not talk** (or write) too much about the wild
world; it may be that it embarrasses other animals to have attention
called to them. A sensibility of this sort might help explain why there is
so little "landscape poetry" from the cultures of the old ways. Nature
description is a kind of writing that comes with civilization and its habits
of collection and classification. Chinese landscape poetry begins around
the fifth century A.D. with the work of Hsieh Ling-yun. There were 1500
years of Chinese song and poetry prior to him (allowing as the *Shi-jing*—
China's first collection of poems and songs, "The Book of Songs"—
might register some five centuries of folks-song prior to the writing
down) and there is much nature, but no broad landscapes: it is about
mulberry trees, wild edible greens, threshing, the forager's and farmer's
world up close. By Hsieh's time the Chinese had become removed
enough from their own mountains and rivers to aestheticize them. This
doesn't mean that old ways people don't appreciate the view, but they
have a different point of view.

The same kind of cautions apply to the stories or songs one might
tell about oneself. Malcolm Margolin points out that the original people

of California did not easily recount an "autobiography." The details of their individual lives, they said, were unexceptional: the only events that bore recounting were descriptions of a few of their outstanding dreams, and their moments of encounter with the spirit-world and its transformations. The telling of their life stories were, then, very brief. They told of dream, insight, and healing.

## VII

The etiquette of the wild world requires not only generosity, but a good-humoured toughness that cheerfully tolerates discomfort, an appreciation of everyone's fragility, and a certain modesty. Good quick blueberry picking, the knack of tracking, getting to where the fishing's good ("an angry man cannot catch a fish")—reading the surface of the sea or sky, are achievements not to be gained by mere effort. Mountaineering has the same quality. These moves take practice, which calls for a certain amount of self-abnegation, and intuition, which takes emptying of yourself. Some great insights have come to some people only after they reached the point where they had nothing left. Alvar Nunez Cabeza de Vaca became unaccountably deepened after losing his way and spending several winter nights sleeping naked in a pit in the Texas desert under a north wind. He truly had reached the point where he had nothing. "To have nothing, you must *have nothing!*" Lord Buckley says of this moment. After that he found himself able to heal sick native people he met on his way westward. His fame spread ahead of him. Once he had made his way back to Mexico and was again a civilized Spaniard he found he had lost his power of healing—not just the ability to heal, but the will to heal, which is the will to be whole: for as he said, there were "real doctors" in the city, and he began to doubt his powers. To resolve the dichotomy of the civilized and the wild, we must first resolve to be whole.

One may reach such a place as Alvar Nunez by literally losing everything. Painful and dangerous experiences have often transformed the people who survived them. Human beings are audacious, they set out to have adventures and try to do more than perhaps they should. So by practicing yogic austerities or monastic disciplines, some people make a structured attempt at having nothing. Some of us have learned much from travelling day after day on foot over snowfields, rockslides, passes, torrents, and valley floor forests, by "putting ourselves out there." Another—and most sophisticated—way is that of Vimalakirti, the legendary Buddhist layman, who taught that by directly intuiting our condition in the actually existing world we realize that we have had nothing from the beginning. A Tibetan saying has it "The experience of emptiness engenders compassion."

For those who would seek directly, by entering the primary temple, the wilderness can be a ferocious teacher, rapidly stripping down the inexperienced or the careless. It is easy to make the mistakes that will bring one to an extremity. Practically speaking, a life that is vowed to simplicity, appropriate boldness, good humour, gratitude, unstinting work and play, and lots of walking, brings us close to the actually existing world and its wholeness.

People of wilderness cultures rarely seek out adventures. If they deliberately risk themselves, it is for spiritual rather than economic reasons. Ultimately all such journeys are done for the sake of the whole, not as some private quest. The quiet dignity that characterizes so many so-called primitives is a reflection of that. Florence Edenshaw, a contemporary Haida elder who has lived a long life of work and family, was asked by the young woman anthropologist who interviewed her and was impressed by her coherence, presence, and dignity, "What can I do for self-respect?" Mrs. Edenshaw said, "Dress up and stay home." The "home" of course is as large as you make it.

The lessons we learn from the wild become the etiquette of freedom. We can enjoy our humanity with its flashy brains and sexual buzz, its social cravings and stubborn tantrums, and take ourselves as no more and no less than another being in the Big Watershed. We can accept each other all as barefoot equals sleeping on the same ground. We can give up hoping to be eternal, and quit fighting dirt. We can chase off mosquitoes and fence out varmints without hating them. No expectations, alert and sufficient, grateful and careful, generous and direct. A calm and clarity attends us in the moment that we are wiping the grease off our hands between tasks and glancing up at the passing clouds. Another joy is finally sitting down to have coffee with a friend. The wild requires that we learn the terrain, nod to all the plants and animals and birds, ford the streams and cross the ridges, and tell a good story when we get back home.

And when the children are safe in bed, at one of the great holidays like the Fourth of July, New Years, or Halloween, we can bring out some spirits and turn on the music, and the men and the women who are still among the living can get loose and really wild. So that's the final meaning of "wild"—the esoteric meaning, the deepest and most scary. Those who are ready for it will come to it. Please do not repeat this to the uninitiated.

# A Post-Historic Primitivism

## by Paul Shepard

### 1. The Problem of the Relevance of the Past

#### History as a Different Consciousness

H. J. Muller's classic *The Uses of the Past: Profiles of Former Societies* presented us with a paradox: "Our age is notorious for its want of piety or sense of the past…. Our age is nevertheless more historically minded than any previous age."[1]

Two decades later, with the publication of Herbert Schneidau's *Sacred Discontent*, the paradox vanished in a radical new insight.[2] For Schneidau History was not simply a chronicle, nor even an "interpretation," but a new way of perceiving reality, one that set out to oppose and destroy the vision which preceded it. It does not refer to readers' understanding but to a cognitive style.

History, he said, is the view of the world from the outside. It was "invented" by early Hebrews who took their own alienation as the touchstone of humankind. Especially did they conceive themselves as outside the earth-centered belief systems of the great valley civilizations of their time. Central to those beliefs was cyclic return and its paradigmatic and exemplary stories linking past, present, and future with eternal structure. Schneidau calls this the "mythic" way of life. Alternatively, the view created by the Hebrews and later polished by the Greeks and Christians was that time may produce analogies but not a true embeddedness. All important events resulted from the thoughts and actions of a living, distant, unknowable God. There could never be a return. The only thing of which we could be sure is that God would punish those deluded enough to believe in the powers of the mythic earth or who fell away from the worship of himself.

A perspective on Schneidau's concept of pre-history can be gained from recent studies of a style of consciousness among living, non-historical peoples. Dorothy Lee, describing the Trobriand Islanders, refers to the "non-linear codification of reality"; space which is not defined by lines connecting points: a world without tenses or causality in language, where change is not a becoming but a new are-ness; a journey, not a passage through but a revised at-ness. Walter Ong calls it "an event world, signified by sound," a world composed of interiors rather than surfaces, where events are embedded instead of reading like the lines of a book. Of Eskimos, Bogert O'Brien says, "The Inuit does not depend on objects for orientation. One's position in space is fundamentally relational and based upon activity. The clues are not objects of analysis.... The relational manner of orienting is a profoundly different way of interpreting space. First, all of the environment is perceived subjectively as dynamic, experiencing processes.... Secondly, the hunter moves as a participant amidst other participants oriented by the action."[3]

For the Hebrews who invented History, the record of the linear sequence of ever-new events would be the Old Testament. By the time we get to Herbert Muller that record has the density of civilized millennia, and could be projected back upon the whole 5,000 years of written words and such records as archaeology offers.

Muller's paradox, of our obsession with and obliviousness toward history, vanishes because we can begin to understand that the passion is an anxiety with our circumstances and our identity, which only grow thicker, like layers of limestone, as we burrow into that vast accumulation. The hidden truth of history is that the more we know the stranger it all becomes. It is human to want to know ourselves from the past, but History's perspective narrows that identity to portraits, ideology, and abstractions to which nation states committed human purpose. True ancestors are absent. Our search simply sharpens desire.

The meaning for our lives, of nature, of purposeful animals, of simple societies, of everything in this "past," is in doubt. We do not feel our ancestors looking over our shoulders or their lives pressing on our own. The past is the temporal form of a distant place. Our view is that you cannot be in two places or two times at once. I speak of this as a "view" in the sense of Ong's observation that the modern West is hypervisual, and my own conviction that what it considers a "view" is a perceptual habit. From this viewpoint we can see mere "oral tradition" as a nadir from which it was impossible to know that water in time's river runs its course but once and that you can no more recover the primordial sense of earth-linked at-homeness than a waterfall can run backward. And further, once we have shaken off that mythic immersion, and put on the garment of dry History, we are unable to shed the detachment and skepticism that define the Western personality, embodied in the written "dialogues" which Robert Hutchins defined as the central feature of the Western civilization.[4]

History not only envisioned, it created sense of the moment. Its content is sometimes delectable, sometimes horrible, but always irretrievable except as beads on the string from which we now dangle. It deals with an arc of time and of measured location; its creative principle being external rather than intrinsic to the world; deity as distant, unknowable and arbitrary. Central to History is a subjectivity which also distances us from our ancestors.

The legacy of History with respect to primitive peoples is threefold: (1) primitive life is devoid of admirable qualities, (2) our circumstances render them inappropriate even if admirable, and (3) the matter is moot, as "You cannot go back."

"You can't go back" shelters a number of corollaries. Most of these are physical rationalizations—too many people in the world, too much commitment to technology or its social and economic systems, ethical and moral ideas that make up civilized sensibilities, and the unwillingness of people to surrender to a less interesting, cruder, or more toilsome life, from which time and progress delivered us. This progress is the work of technology. When technology's "side effects" are bad, progress becomes simply "change," which is, by the same rote, "inevitable." Progress is a visible extension of the precognitive habit of History that influences concept and explanation by modulating understanding. It was not only the mathematicians, astronomers, and philosophers of the modern era who gave us the theoretical basis of progress.

All of these objections—and they seem insurmountable—seem to me to imply a deeper mindset which does not have to do with the content of history. It is more a reflex than a concept. We care little for its theories or inventions since the time of Francis Bacon or for the moods in Christendom which reversed the older view that things only get worse.

Its true genesis lies in the work of Hebrew and Greek demythologizers. They created a reality focused outside the self, one that could be manipulated the way god-the-potter fingered the world. In rooting out the inner-directed, cyclic cosmos of gentiles and naive barbarians, they destroyed the spiraled form of myth with its rituals of eternal return, its mimetic means of transmitting values and ideas, its role in providing exemplary models, its central metaphor of nature and culture, and most of all as a way of comprehending the past. It began the deconstruction of the empirical wisdom of earlier peoples, and culminated in the monumental Western view of reality whose central theme was the outwardness of nature.

Along with pictorial space and Euclidean time goes the phonetic alphabet as inadvertent "causes" of estrangement.[5] But these are not simply inventions of the post-medieval West. They are markers in the way the world is experienced. Their antecedents occur in the Bronze Age Mediterranean where much of what we call "Western" has its roots.

Elsewhere I have tried to describe this history as a crazy idea, fostered not as a concept so much as the socially sanctioned mutilation of childhood, the training ground of perception, by the blocking of what Erik Erikson called "epigenesis."[6] But, whatever its dynamic, History alters not our interest in the past (witness Muller's observation that we moderns seem more interested than ever), but the work of attention itself, the deep current of precomprehension that runs silently beneath our spoken thoughts.

### History and Ambiguity

If we attempt to recover the difficult and "distant" art of tool-flaking we may do so over the objections of modern rationality that denies that the pterodactyl can fly since no one has seen it do so. That is, you cannot know the ancient technique. Not only does History define it as beyond access, but incomprehensible. History thinks its own process is an evolution separating us by our very nature from our past—medieval, Neanderthal or primate.

Central to History is the notion of a fixed essence, an inner state that persists in spite of the contradictions of appearance, that our visible form not only fails to inform but can be made to deceive. Shifting appearance is dangerous, larval forms signify evil. The question of our primate or Neanderthal past cannot be addressed except as alternatives to our present identity. We are predisposed by the immense cultural momentum of History to dismiss such ambiguous assertions as one of a larger class of moot points in which categorical contradiction, the simultaneous reality of two opposing truths about ourselves, is denied.

Equally paradoxical is the matter of being in two times at once, even though our senses tell us that we are not today what we were yesterday. This movement from one state or one thing to another is not so much a problem for human consciousness as for meaning. The liminal or boundary area of categories heightens cognitive intensity. In the historical world, such transformations have been handled by accepting reality as made up of fixed identities, oppositions, and beyond them, transcendent meaning, declaring one of the appearances to be illusory, or by seeing them as good and evil. In all cases except the last the surface or apparent contradiction is cast into doubt in favor of some deeper, hidden, more real reality. Mostly this problem has been met in the West by denying appearance—especially when it shifts or is a larval state—as the true identity and instead postulating essences and spirits within or seeking principles and abstractions as the enduring, unchanging reality, despite outward shape.

In non-Western, non-industrial, and largely non-literate (hence non-historical) societies, external form is dealt with quite differently. Edmund Carpenter cites our difficulty with the visual duck/rabbit pun as our loss of the "multiplicity of thought," a collapse of metaphor in a mind-set related to phonetic writing.[7] A. David Napier has traced the matter in elegant detail in connection with the ritual use of masks as the perceptual means of assenting to a universal principle of shape-shifting. Coupled with dance, this is humankind's central means of reconciliation with a world of changes.[8] The many shapes in such masked dances testify also to a world in which abstractions are given lively form. Ahistorical peoples usually live in worlds where power is plural, as in egalitarian small societies in which leadership is not monopolized but changing and dispersed. The concrete or given model for this discontinuity of emphatic and exemplary qualities is the range of natural species. To varying degrees the animals and plants are regarded as centers, metaphors, and mentors of the different traits, skills, and roles of people. In polytheistic worlds there is no omniscience and no single hierarchy, although there may be said to loom a single creative principle behind it all. Insofar as they model diversity and the polytheistic cosmos, the animals provide metaphors of forms and movements that can be brought ceremonially into human presence, as interlocutors of change. Their heads as masks, the animals in such rites become combinational figures created to give palpable expression to transitional states. The animal mask on the body of a person joins in thought that which is otherwise separate, not only representing human change but conceptualizing shared qualities, so that unity in difference and difference in unity can be conceived as an intrinsic truth. And some animals, by their form or habit, are boundary creatures who signify the passages of human life. Finally, in dance these bodies move to deep rhythms that bind the world and bring the humans into mimetic participation with other beings.

The sophisticated Greeks after the time of Pericles ridiculed these predications, and the Jews and Christians rejected them. The thinness of music and dance in temples, churches, and mosques indicates the minimalizing of what was and is basic to hundreds of different, indigenous religions marked by "mythic" imagination.

The nature of the primitive world is at the center of our dilemma about essence, appearance, and change. Since we are not now what we once were—we are not bacteria or quadruped mammals, or apish hominids, or primitive people living without domesticated plants and animals—the dichotomy is clear enough. We each know as adults that we are no longer a child, yet we are not so sure that our being doesn't still embrace that other self who we were. We are attached to that primitive way of understanding, of double being, in spite of our modern perspective. Depth psychology has led us to understand that this going back is going into ourselves, into what, from the civilized historical view, is a "heart of darkness." Clearly a threat of the loss of self-identity is implied, swallowed by a second nature which is hidden and unpredictable.

As born anti-historians, our secret desire is to explicate the inexplicable, to recover that which is said to be denied. It is a yearning, a nostalgia in the bone, an intuition of the self as other selves, perhaps other animals, a shadow of something significant that haunts us, a need for exemplary events as they occur in myth rather than History. If not a necessity, it is a hunger that can be suppressed and distanced. The experience of that past is in terms of something still lived with, like fire, that still draws us. We cannot explain it, but it is there, made fragile in our psyche and hearts, drowned perhaps in our logic, but unquenchable.

It has been said that those who do not learn from history are doomed to repeat it, and yet by definition it cannot be repeated. Presumably such repetition means analogy. One does not really "go back," but merely discovers similar patterns. To ask the question in the perspective of pre-history: what are we to learn from history? The answer: history rejects the ambiguities of overlapping identity, space and time, and creates its own dilemmas of discontent and alienation from Others, from non-human life, primitive ancestors, and tribal peoples. Failing to enact pre-history, we can live only in history, caught between captivity and escape, afflicted with Henry Thoreau's "life of quiet desperation," now called neurosis. Since history began, most people most of the time have lived under tyrants and demagogues (Mr. Progress, Mr. Collectivity, Mr. Centralized Power, Mr. Growthmania, and Mr. Technophilia). No empire lasts, and when states collapse their subjects are enslaved by other states.

The crucial question of the modern world is "How are we to become native to this land?" It is a question that history cannot answer, for history is the de-nativizing process. In history "going native" is a madman's costume ball, a child's romp in the attic, a misanthrope's escape.

Unlike History, prehistory does not participate in the dichotomy that divides experience into inherited and acquired. Nor does it imply that our behavior is instinctive rather than learned. It refers us to mythos, the exemplifications of the past-in-the-present. Ancestors are the dreamtime ones, and their world is the ground of our being. They are with us still.

The real lesson of history is that it is no guide. By its own definition, History is a declaration of independence from the deep past and its peoples, living and dead, the natural state of being which is outside its own domain. Indeed, History corrupts the imperatives of pre-history.

What are the imperatives? What are we to learn from pre-history? Perhaps as Edith Cobb said of childhood, "The purpose is to discover a world the way the world was made."[9]

## 2. Savagery—Once More

After 2,500 years of yearning for lost garden paradises in Western mythology perhaps one of the most outrageous ideas of the 20th century was the advocacy of a hunting/gathering model of human life. Much of the world is still caught up in making a transition from an agrarian civilization. A writer for *Horizon* proclaims that "An epoch that started ten thousand years ago is ending. We are involved in a revolution of society that is as complete and as profound as the one that changed man from hunter and food gatherer to settled farmer."[10] He alerts us to the colossal struggle to go forward from the tottering institutions of agricultural life, and I am suggesting that we do move ahead to—of all things—hunting/gathering!

Among the problems that plague the "uses of the past," as H. J. Muller called it, the search for a lost paradise seems to resist the "facts" of history. One wonders whether it is even possible to write about the deep past without nostalgia, or without creating a world that never existed. Its images are a mix of dreams and visions, infantile mnemonics, ethnographic misinformation, and attempts to locate mythological events in geographical space and recorded history. History, indeed, is not exactly anti-myth, dealing as it does with "origins" and recitations of the significant events of the past. But its "past" is radically different from the one shaping human evolution.

It was great fun working on a book on hunter-gatherer people in the early 1970s because almost everything that the layman generally thought to be true of them was wrong. In writing *The Tender Carnivore*[11] I tried to avoid the snare of idealism by disarming my critics in advance. I avoided the beatifying language of Noble Savagery and I engaged Fons von Woerkom to draw chapter headings, as his art was anything but romantic. Even so, the incredulity with which it was greeted was puzzling. Looking back, I now see that the objection was not only that primitive life was inferior and irrelevant, but in the lens of historical memory, inaccessible.

For two centuries the ideology of inevitable change had set its values in contrast to fictional images of the lost innocence of deprived and depraved savage. Forty years ago George Boas traced the history of that idea of the primitive over the last 2,000 years, from early attempts to associate tribal peoples with Biblical paradise, various views of perfection, and the saga of evolving mankind.[12] For the Greeks anyone who lacked civil life in a polis and spoke incoherently (babbled) was a barbarian. Hostility to the idea that we have anything to learn from savages has as long a tradition as the dream itself. Skepticism about the full humanity of the Hyperboreans and Scythians among some Classical authors was opposed by the idealizing of the Celts, the Getae, and the Druids by Herodotus and Strabo.

The Christians got their ideas on prehistory from Plato's *Laws* via the Romans, which portrayed the pagans as childlike. Spanish endeavors to associate American Indians with European *sylvestres homines*, the wild man, and the legacy of the Greek *barbori* have been reviewed by Anthony Pagdon. He makes some distinctions between Franciscan and Jesuit perception of the Indians, the Franciscans determined to destroy Indian culture in order to Christianize and the Jesuits ignoring the "secular" side of the culture as irrelevant—an ironic twist on holism.[13] Oddly enough, it was the "unnaturalness" of the native peoples rather than their "naturalness" that justified decimation. Natural men, for example, did not eat each other.

In neo-Classical times Dr. Johnson observed that the hope of knowing anything about the people of the past was "idle conjecture." Horace Walpole derided antiquarians' fantasies. Locke and Hume gave us images of slavering brutes as an alternative to Rousseau's fictions of innocence and integrity. Admiral Cook's Polynesia would not look benign after the untamed sons of Adam did him in on the beach at Oahu. The images were part of the heritage of the Roman idea of barbarians, the Christian notion of pagans, and 18th century political philosophy of the benighted savage. Von Herder, Hegel, Compte, and Adelung all strove to disassociate mankind from the "laws of nature," to identify culture with History, to see conscious intellect identified with urban life, property, law, government, and "great art," as the final flowers in the human odyssey. The tradition continues. As M. Navarro said as late as 1924 of the South American Campa, "Degraded and ignorant beings, they lead a life exotic, purely animal, savage, in which are eclipsed the faint glimmerings of their reason, in which are drowned the weak pangs of their conscience, and all the instincts and lusts of animal existence alone float and are reflected...."[14] Or, closer to home, is the testimony of Will Durant, the historian: "Through 97 per cent of history, man lived by hunting and nomadic pasturage. During those 975,000 years his basic character was formed—to greedy acquisitiveness, violent pugnacity and lawless sexuality."[15] Quite apart from anthropology this conglomerate idea of the primitive remains the central dogma of civilization held by modern humanists.

By the end of the 19th century there emerged in the United States a substantial body of admiration for Indian ways. I remember as a boy in the 1930s meeting Ernest Thompson Seton in Santa Fe. He ran a summer camp in which boys came to his ranch to be tutored by local Navajos, bunked in tepees, and lived out the handcraft and nature study ventures of *Two Little Savages*.[16] The image of the American Indians in this dialectic has been reviewed by Calvin Martin, who observes that by the late 1960s the image of the "ecological Indian" was being articulated by Indians themselves, notably Scott Momaday and Vine Deloria. Arrayed against them in postures of "iconoclastic scorn" are experts who pursued an old line in anthropological guise—debunkers of the image of the Noble Savage, which they said merely masked a knave who was not nature's friend but who typically over-killed the game at every opportunity.[17]

Oddly enough, science did not rapidly resolve what seemed to be a question of facts. Geology after Lyell, evolution after Darwin, and archaeological time after Libby's atomic dating complicated but did not settle much. With a slight twist evolution could be the handmaiden of Progress. "It began to look," says Glyn Daniel, "as if prehistoric archaeology was confirming the philosophical and sociological speculations of the mid-nineteenth century scholars."[18] Anthropology idealized value-free science and cultural relativism, thwarting European chauvinism but throwing out the baby with the bath water.

I was, of course, not the first to try to formulate the meaning of hunting-gathering for our own time. But not all efforts to clarify the description of hunters were applied to ourselves. Knowledgeable writers tiptoed among the ferocious critics, pretending that hunting signified only a remote past, as in Robert Ardrey's *Hunting Hypothesis*[19] or John Pfeiffer's *The Emergence of Man.*[20] Nigel Calder's *Eden Was No Garden*[21] and Gordon Rattray Taylor's *Rethink*[22] stirred the pot, but could hardly be said to have influenced, say, the civilized dogma of the modern university. Scholarly silence greeted the English translation of Ortega y Gasset's *Meditations on Hunting*[23] as though an imposter had inserted an aberration in his works.

The message is clear: Advocacy of a way of life that is both repulsive and no longer within reach seems futile. Time is an unreturning arrow. The hunting idea is a barbaric atavism, unwelcome at a time when aggression and violence seem epidemic. The idea is obviously economically impractical for billions of people and incongruent with the growing concern for the rights of animals. Animal protectionists and many feminists seem generally to feel that hunting is simply a final grab at symbolic virility by insensitive, city-bred male chauvinists, or one more convulsion of a tattered and misplaced nostalgia. Less and less, however, is hunting condemned as the brutal expression of tribal sub-humans, for that would conflict with modern ethnic liberation.

The idea of inherent "nobility" of the individual savage was laughed out of school a century ago, properly so. Hunter-gatherers are not always pacific (though they do not keep standing armies or make organized war), nor innocent of ordinary human vices and violence. There is small-scale cruelty, infanticide, inability or unwillingness to end intratribal scuffling or intertribal vengeance. From the time of Vasco da Gama Westerners have been fascinated by indigenous punishment for crimes and by cannibalism (although cannibalism is primarily a trait of agri-cultures). Hunter-gatherers may not always live in perfect harmony with nature or each other, being subject to human shortcomings. Nor are they always happy, content, well-fed, free of disease, or profoundly philosophical. Like people everywhere they are, in some sense, incompetent. In "Little Big Man" the Indian actor Dan George did an unforgettable satire on the wise old chief who, delivering his rhetoric of joining the Great Spirit, lies down on the mountain to die and gets only rain in the face for his trouble. Given a century of this kind of scientific dis-illusioning, what is left?

It has been uphill and downhill for the anthropologists all along. The 19th century "humanist anthropologists" like Edward Tylor and Malinowski dismissed native religious rites as logical error, although they allowed that ritual may work symbolically. As to the veracity of their religion, an "embarrassed silence" has marked anthropology ever since, say Bourdillon and Fortes.[24]

Against these relativists there has also been an eccentric group of anthropologists who were not neutral about the tribal cultures. A. O. Hallowell, W. E. H. Stanner, Carleton Coon, and Julian Steward walked a narrow line between science and advocacy. Claude Lévi-Strauss rescued the savage mind. Coon's courage was exemplary. He scorned the "academic debunkers and soft peddlers," including those who spoke of "the brotherhood of man" as contradicting the reality of race.[25] Stanner was perhaps the most eloquent, describing Aboriginal thought as a "metaphysical gift," its idea of the world as an object of contemplation, its lack of omniscient, omnipotent, adjudicating gods—a world without inverted pride, quarrel with life, moral dualism, rewards of heaven and hell, prophets, saints, grace or redemption. All this among Blackfellows whose "great achievement in social structure" he said was equal in complexity to parliamentary government, a wonderful metaphysic of assent and abidingness, "hopelessly out of place in a world in which the Renaissance has triumphed only to be perverted and in which the products of secular humanism, rationalism and science challenge their own hopes."[26] If any modern intellectuals read him they must have thought he had "gone native" and left his critical intelligence in the outback.

After twenty centuries of ideological controversy it may be impossible to enter the dialogue without trailing some of its biases and illusions. But there is perspective from different quarters—from the study of higher primates, hominid paleontology, paleolithic archaeology, ethology, ecology, field studies of living hunter-gatherers and direct testimony from living hunter-gatherers.

THE WILDERNESS CONDITION

A turning point was a Wenner-Gren symposium in Chicago and its publication as *Man the Hunter* in 1968.[27] The essays therein reported scientific evidence that the cave man as well as the noble savage was so much urban moonshine. It was a meeting of field workers who had studied living tribal peoples in many parts of the world, coming together and finding common threads that linked diverse hunter-gatherer cultures to one another and to paleolithic archaeology. This shift toward species-specific thinking benefitted from "the new systematics," an evolutionary perspective based on genetics and natural selection articulated by G. G. Simpson, Ernst Mayr, Julian Huxley, and others. *The Social Life of Early Man*[28] was indicative of the new level of continuity among primitive societies, afterwards given cross-cultural generalizations in George Murdock's ethnographic atlas.[29]

Although a few bold voices had been heard among them, such as Marshall Sahlins' *Stone Age Economics*,[30] their own evidence did not make anthropologists into advocates of a new primitivism. Their restraint was no doubt the result of a hard-won professional posture, the 20th century effort to overcome two centuries of ethnocentrism. But it was also the outwash of three generations of cultural relativism by mainstream social science, pioneered by Boas and Kroeber,[31] recently voiced with imperious assurance by Clifford Geertz that "there are no generalizations that can be made about man as man, save that he is a most various animal."[32] Catch them saying that any culture is better than another!

In any case, such a judgment would be irrelevant, since even present-day hunter-gatherers are, by its historical logic, part of an irrecoverable past. Melvin Konner, a Harvard-bred anthropologist, spent years studying the !Kung San of the Kalahari desert of Africa, wrote a fascinating account of his study showing the marvelous superiority of their lives to their counterparts in Cleveland or Los Angeles, and then pulled the covers over his head by saying, "But here is the bad news. You can't go back."[33] One can only be grateful for Loren Eiseley[34] and Laurens van der Post[35] in their admiration of the same Kalahari Bushmen. Perhaps they anticipated what Roger Keesing calls the "new ethnography," which seeks "universal cultural design" based on psychological approaches. "If a cognitive anthropology is to be productive, we will need to seek underlying processes and rules," he says, observing that the old ethnoscience has been undermined by transformational linguistics and its sense of "universal grammatical design." He concludes that "the assumption of radical diversity in cultures can no longer be sustained by linguistics."[36]

So to return to the question—just what is it that is so much better in hunter-gatherer life? How does one encapsulate what can be sifted from an enormous body of scientific literature? It is not only, or even mainly a matter of how nature is perceived, but of the whole of personal existence, from birth through death, among what history arrogantly calls "pre-agricultural" peoples. In the bosom of family and society, the life cycle is punctuated by formal, social recognition with its metaphors in the terrain and the plant and animal life. Group size is ideal for human relationships, including vernacular roles for men and women without sexual exploitation.[37] The esteem gained in sharing and giving outweighs the advantages of hoarding. Health is good in terms of diet as well as social relationships.[38] Interpenetration with the non-human world is an extraordinary achievement of tools, intellectual sophistication, philosophy, and tradition. There is a quality of mind, a sort of venatic phenomenology. "In a world where diversity exceeds our mental capacity nothing is impossible in our capacity to become human."[39] Custom firmly and in mutual council modulates human frailty and crime. Organized war and the hounding of nature do not exist. Ecological affinities are stable and non-polluting. Humankind is in the humble position of being small in number, sensitive to the seasons, comfortable as one species in many, with an admirable humility toward the universe. No hunter on record has bragged that he was captain of his soul. Hunting, both in an evolutionary sense and individually, is "the source of those saving instincts that tell us that we have a responsibility towards the living world."[40]

To make such statements is to set out the game board for the dialectics of our intellectual life. Graduate students, religious fundamentalists, economists, corporate executives, and numerous others, including a gleeful band of book reviewers, will leap to prove differently. I have a wonderful set of newspaper book reviews of *The Tender Carnivore* with headings like "Professor Says Back to the Cave" and "Aw, Shoot!" And there is always an anthropologist somewhere to point to a tribe which is an exception to one or another of the "typical" characteristics of hunter-gatherers, hence there can be no "universals," and so on.

The most erudite essay on hunting, ancient or modern, is José Ortega y Gasset's *Meditations on Hunting*. He conceives the hunt in terms of "authenticity," especially in its direct dealing with the inescapable and formidable necessity of killing, a reality faced in the "generic" way of being human. He also refers to the hunter's ability to "be inside" the countryside, by which he means the natural system—"wind, light, temperature, round-relief, minerals, vegetation, all play a part; they are not simply there, as they are for the tourist or the botanist, but rather they function, they act." Ultimately, this function is the reciprocity of life and death. The enigma of death and that of the animal are the same, and therefore "we must seek his company" in the "subtle rite of the hunt." In all other kinds of landscape, he says—the field, grove, city, battleground—we see "man travelling within himself," outside the larger reality.

The humanized and domesticated places may have their own domestic reality, but Ortega refers to generic being. Ortega's is a larger understanding; he attends to human "species-specific" traits, and escapes the cultural relativism and social reduction that have dominated anthropology. A biologist turned philosopher/historian, Ortega links "primitive" hunter-gatherers to ourselves. This is because there are characteristics of humankind, as Eibes-Eibesfeldt tells us,[41] as well as shared characteristics of hunter-gatherers, present and past.

What has been learned about the nature of our own problems in the past twenty years?

*Item:* Health disorders are increasingly traced to polluting poisons and to a diet of domesticated (i.e., chemically altered or chemically treated) plants and animals. More people every year eat the meat of wild animals, seek "organic" vegetables, and seek alternatives to chemicalized nature.

*Item:* Evidence indicates that the small, face-to-face, social group works better in the quality of social experience and decision-making for its members and in its efficacy as a functional institution.[42]

*Item:* Percussive music and great intervals of silence are evidently conducive to our well-being. A meditative stillness, suggests Gary Snyder, was invented by waiting hunters.[43] Perhaps this reflected the poised and ruminating hush of mothers of sleeping infants. High levels of sound have been directly linked to degenerative disease in urban life.

*Item:* Regular exercise, especially jogging, rare in 1965, was common by 1980. The sorts of exercise for men and women (aerobics, jogging, stretching) correlate with certain routines of life in cynegetic societies. The benefits are not only physical but mental.[44]

*Item:* One of the hardest stereotypes about the savage to die is gluttony. In arguing that Pleistocene peoples were responsible for the extinctions of large mammals, Paul Martin projected urban greed on the ancient hunters.[45] This preposterous theory ignores fundamental ecology, comparative ethnography, and the anthropological distinctions between people who maximize their take and those who optimize it.[46] Given the whole range of Pleistocene extinctions it is a poor fit in the paleontological and archaeological record.

*Item:* Childhood among hunter-gatherers better fits the human genome[47] in terms of the experience and satisfaction of both parents and children. I refer to the "epigenetic" calendar, which is based on the complex biological specialization of neoteny, to which human culture is in part mediator and mitigator.

*Item:* That advanced intelligence not only arrived with hunting and being hunted, but continues to be the central characteristic of the hunt is still hard to accept for those who think of predation as something like a dogfight. Knowledge is of overwhelming importance in accommodating the whole of society to a "watchful world" and structuring the mentality of the hunter. There are three evolutionary correlates of large cerebral hemispheres: large size, predator-prey interaction, and intense sociality.[48]

*Item:* The cosmography of tribal peoples is as intricate as any, and marked by a humility which is lacking in civilized society. For example, two of the "principles of Koyukon world view" are "each animal knows way more than you do," and "the physical environment is spiritual, conscious, and subject to rules of respectful behavior."[49] The essays in Gary Urton's *Animal Myths and Metaphors in South America*[50] describe myths of the sort depicted in Huichol yarn paintings of Mexico—visual evocations of stories that integrate the human and non-human in dazzling, sophisticated metaphor.

## The Paradox of the Civilized Hunter

There is no room here to review current ideas about hunting by modern, urban people, except to observe that the argument for hunting links primitive and civilized people, past and present. One can split this distinction and say with Barry Lopez that hunting is OK for ethnic groups but not for modern people. I think that view is based mistakenly on the notion that there are vicarious alternatives and reflects a kind of despair over the practical question of how the sheer numbers of people now living could gain the benefits of hunting-gathering.

Anti-hunters are outraged by "sport killing" as opposed to ethnic tradition, pointing for example to the diminished presence of wildlife and to old photographs of white African hunters with numerous dead animals. Who would consider defending such "slaughter"? What is sometimes regarded as vanity needs to be understood in the context of the traditional laying out of the dead animals. One of the most thoughtful modern hunters, C. H. D. Clarke, writes, "The Mexican Indian shamanic deer hunt is as much pure sport as mine, and the parallels between its rituals, where the dead game is laid out in state, and those of European hunts, where the horns sound the 'Sorbiati,' or 'tears of the stag,' over the dead quarry, are beyond coincidence."[51]

Fanatic opposition to hunting suggests that some other fear is at work. Neither the animal protectionists, the animal rights philosophers, nor the feminists hostile to vernacular gender have ecosystems (including the wildness of humans) at heart. When anti-hunters heard that "a Royal Commission on blood sports in Britain reported that deer had to be controlled and that hunting was just as humane as any alternative, these people wanted deer exterminated once and for all, as the only way to deliver the land from the infamy of hunting." In America we have similar ecological blindness regarding the killing of goats on the coastal islands of California and wild horses in national parks. I once heard a nationally known radio commentator, Paul Harvey, complain that the trouble with the idea of national parks protecting both predators and prey animals was that "mercy" was missing. Clarke concludes that the "rejection of hunting is just one in a long list of rejections of things natural," and that hunting will linger as one of the human connections to the natural environment "until the human race has completed its flight from nature, and set the scene for its own destruction."[52]

## 3. Romancing the Potato

Seventeen years after the publication of *The Tender Carnivore* there is still only speculation among scholars about the "cause" of the first agriculture. It is clear now as it was then, however, that recent hunting-gathering peoples did not joyfully leap into farming. The hunter-gatherers' progressive collapse by invasion from the outside is typified in Woodburn's description of the Haida.[53] For ten millennia there has been organized aggression against hunters, who themselves had no tradition of war or organized armies. The psychology of such assault probably grew out of the territoriality inherent in agriculture and farmers' exclusionary attitude toward outsiders, land hunger growing from the decline of field fertility and the increase in human density, and, with the rise of "archaic high civilizations," social pathologies related to group stresses and insecurity in an economy of monocultures (i.e., grains, goats), and the loss of autonomy in the pyramiding of power. Hunting-gathering peoples have been the victims of these pressures that beset farmers and ranchers, bureaucratically amplified upward in the levels of government.

The old idea that farming favored more security, longer life, and greater productivity is not always correct. For example, Marek Zvelebil, in the *Scientific American* in 1986, says, "Hunting-and-gathering is often thought of as little more than the prelude to agriculture. A reevaluation suggests it was a parallel development that was as productive as early farming in some areas."[54] As for modern agriculture, C. Dean Freudenberger says, "Agriculture, closely related to global deforestation by making room for expanding cropping systems, is the most environmentally abusive activity perpetuated by the human species."[55]

At least six millennia of mixed tending and foraging followed the first domesticated wheat and preceded the first wheel, writing, sewers,

and armies. In varying degrees local, regenerative, subsistence economies blended the cultivated and gathered, the kept animal and the
hunted. Before cities, the world remained rich, fresh, and partly wild
beyond the little gardens and goat pens. Extended family, small-scale
life with profound incorporation into the rhythms of the world made this
"hamlet society" the best life humans ever lived in the eyes of many. It
is this village society of horticulture, relatively free of monetary com-
merce and outside control, that most idealizers of the farm look to as a
model.

Perhaps that image motivated Liberty Hyde Bailey in his turn-of-
the-century book, *The Holy Earth*. Yet, his feeling for the land seems
betrayed by a drive to dominate. Bailey says, "Man now begins to
measure himself against nature also, and he begins to see that herein
shall lie his greatest conquests beyond himself; in fact, by this means
shall he conquer himself,—by great feats of engineering, by complete
utilization of the possibilities of the planet, by vast discoveries in the
unknown, and by the final enlargement of the soul; and in these fields
shall he be the heroes. The most virile and upstanding qualities can find
expression in the conquest of the earth. In the contest with the planet
every man may feel himself grow."[56] Tethering the neolithic reciprocity with a nourishing earth, he suddenly jerks us into the heroic Iron Age.
In the same book, however, he says, "I hope that some reaches of the sea
may never be sailed, that some swamps may never be drained, that some
mountain peaks may never be scaled, that some forests may never be
harvested."[57] Inconsistent? No, it is an expression of the enclave
mentality, the same one that gave us national parks and Indian reserva-
tions, the same that gives us wilderness areas.

The ideal of hamlet-centered life is represented by *Mother Earth
News*, a search for equilibrium between autonomy and compromise. It
is difficult not to be sympathetic. So too do Wes Jackson and the
"permaculture" people seem to seek the hamlet life.[58] Their objective of
replacing the annual plants with perennials seems laudable enough. Yet
they are busily domesticating through selective breeding more wild
perennials as fast as possible. They are making what geneticist Helen
Spurway called genetic "goofies," the tragic deprivation of wildness
from wild things.[59]

Who among us is not touched by the idyll of the family farm, the Jeffersonian yeoman, the placeness and playground of a rural existence? Above all, this way of life seems to have what hunting-gathering does not—retrievability. The yearning for it is not from academic studies of exotic tribal peoples, but is only a generation or two away—indeed, only a few miles away in bits of the countryside in Europe and America. After all, it incorporates part-time hunting and gathering, as though creating the best of all possible worlds. Like many others, I admire Jefferson as the complete man and share the search for peace of mind and good life of its modern spokesmen like Wendell Berry.

Of course, most agriculture of the past five millennia has not been like that. The theocratic agricultural states, from the early centralized forms in ancient Sumer onward, have been enslaving rather than liberating. Even where the small scale seems to prevail, such conviviality is not typical in medieval or modern peasant life with its drudgery, meanness, and suffering at the hands of exploitive classes above it.[60]

The primary feature of the farmer's concept of reality is the notion of "limited good." There is seldom enough of anything. By contrast, the hunter's world is more often rich in signs that guide toward a gifting destiny in a realm of alternatives and generous subsistence. Since they know nature well enough to appreciate how little they know of its enormous complexity, hunter-gatherers are engaged in a vast play of adventitious risk, hypostatized in gambling, a major leisure-time activity. Their myths are rich in the strangeness of life, its unexpected boons and encounters, its unanticipated penalties and mysterious rewards, not as arbitrary features but as enduring, infinitely complex structure. Gathering and hunting are a great, complex cosmology in which a numinous reality is mediated by wild animals. It is a zero-sum game, a matter of leaning toward harmony in a system which they disturb so little that its inter-species parities seem more influenced by intuition and rites than physical actions. Autonomous, subsistence farming or gardening shares much of this natural reverence for the biotic community and the satisfactions of light work schedules, hands-on routines, and sensitivity to seasonal cycles.

But agriculture, ancient and modern, is increasingly faced with a matter of winners and losers, dependence on single crops. Harmony with the world is sustained by enlarging the scope of human physical control or by rites of negotiation with sacred powers, such as sacrifice. The domesticated world reduces the immediate life forms of interest to a few score species which are dependent on human cultivation and care—just as the farmers see themselves, dependent on a master with human-like, often perverse actions. Theirs is a cosmos controlled by powers more or less like themselves, from local bureaucrats up through greedy princes to jealous gods. No wonder they prefer games of strategy and folktales in which the "animals," burlesques of their various persecutors, are outwitted by clever foxes like themselves. The world does not so much have parts as it has sides substructured as class. From simple to complex agriculture these increase in importance as kin connections diminish.

The transition from a relatively free, diverse, gentle subsistence to suppressed peasantry yoked to the metropole is a matter of record. The subsistence people clearly long for genuine contact with the non-human world, independence from the market and the basic satisfaction of a livelihood gained by their own hands. But this distinction among agricultures has its limits and was not apparently in mind when Chief Washakie of the Shoshones said, "God damn a potato." Sooner or later you get just what the Irish got after they thought they had rediscovered Eden in a spud skin.

We may ask whether there are not hidden imperatives in the books of Wendell Berry obscured by the portrayal of the moral quality, stewardship syndrome, and natural satisfactions of farm life. He seems to make the garden and barnyard equivalent to morality and esthetics and to relate it to monotheism and sexual monogamy, as though conjugal loyalty, husbandry, and a metaphysical principle were all one. And he is right. This identity of the woman with the land is the agricultural monument, where the environment is genderized and she becomes the means of productivity, reciprocity, and access to otherness, compressed in the central symbol of the goddess. When the subsistence base erodes this morality changes. Fanaticism about virginity, women as pawns in games of power, and their control by men as the touchstone of honor and vengeance has been clearly shown to be the destiny of sub-equatorial and Mediterranean agriculture.[61] Aldous Huxley's scorn of Momism is

not popular today, but there are reasons to wonder whether the metaphors that mirror agri-culture are not infantile.[62] (For hunter-gatherers
the living metaphor is other species, for farmers it is mother, for
pastoralists the father, for urban peoples it has become the machine.)[63]

In time, events and people seem to come back in new guise. I keep
thinking that Wendell Berry is the second half-century's Louis Bromfield.
Bromfield was a celebrated author and gentleman farmer, known for his
conservation practices and the good life on his Ohio farm. He could
prove the economic benefit of modern farming by his detailed ledgers.
But it was his novels that made him wealthy, and the dirt farmers who
were invited along with the celebrities to see his showplace could well
ask, "Does Bromfield keep books or do the books keep Bromfield?"

Berry writes with great feeling about fresh air and water, good soil,
the sky, the rhythms of the earth, and human sense in these things. But
those were not invented by farmers. They are the heritage of the non-
domesticated world. Much that is "good" in his descriptions does not
derive from its husbandry but from the residual "wild" nature. He
accepts Biblical admonishments about being God's steward, responsible for the care of the earth. None of the six definitions of "steward"
in my dictionary mentions responsibility toward that which is managed. It refers to one who administers another's property, especially
one in charge of the provisions; another way of saying that the world
biomes need to be ruled, that nature's order must be imposed from the
outside.

Alternatively, one could pick any number of Christian blue-noses,
from popes to puritans and apostles to saints, who wanted nothing to do
with nature and who were disgusted to think they were part of it. The best
that can be said about Christianity from an ecological viewpoint is that
the Roman church, in its evangelical lust for souls, is a leaky ship.
Locally it can allow reconciliation of its own dogma with "pagan" cults,
as when the Yucatan Indians were Christianized by permitting the
continued worship of limestone sinks, or *cenotes*, making the Church
truly catholic.[64] Similar blending may be seen in eccentrics like St.
Francis or Wendell Berry, who voice a "tradition" that never existed.

The worst is difficult to choose, although its shadow may be
discerned behind the figure of Berry himself in *The Unsettling of*

*America*, humming his bucolic paeans to the land and clouds and birds as he sits astride a horse, his feet off the ground, on that domestic animal which more than any other symbolized and energized the worldwide pastoral debacle of the skinning of the earth, and the pastoralists' ideology of human dissociation from the earthbound realm. No wonder the horse is the end-of-the-world mount of Vishnu and Christ. As famine, death, and pestilence, it was the apocalyptic beast who carried Middle East sky-worship and the sword to thousands of hapless tribal peoples and farmers from India to Mexico.

### Dealing with Death

Joseph Campbell, who clearly understood the hunter-gatherer life, tried to have it both ways. The hunters' rituals, he said (capitulating to the 19th century anthropological opinion that primitive religion is simply bad logic) tried to deny death by the pretense that a soul lived on. "But in the planting societies a new insight or solution was opened by the lesson of the plant world itself, which is linked somehow to the moon, which also dies and is resurrected and moreover influences, in some mysterious way still unknown, the lunar cycle of the womb."[65] The planters did indeed lock themselves to the fecundity and fate of annual grains (and their women to an annual pregnancy). But according to Alexander Marshack the moon's periodicity had long since been observed by hunters. In any case it was not seen by the early planters of the Near East as a plant but as a bull eaten by the lion sun.

Campbell regards sacrifice as the central rite of agriculture's big idea that the grain crop is the soul's metaphor. Sacrifice—the offering of fruit or grain, or the ritual slaughter of an animal or person—is a means of participating in the great round. But in agriculture participation turns into manipulation. The game changes from one of chance to one of strategy, from reading one's state of grace in terms of the hunt to bartering for it, from finding to making, from a sacrament received to a negotiator with anthropomorphic deities. This transition can be seen in a series of North Asian forms of the ceremony of the slain bear, from an egalitarian, *ad hoc* though traditional celebration of the wild kill as a symbolic acceptance of the given to the shaman-centered spectacle of the sacrifice of a captive bear in order to deflect evil from the village.[66]

The transition from bear hunt to bull slaughter has been traced by Tim Ingold.[67] Sacrifice does not seem to me to accommodate the "problem of death" but to domesticate it. It reverses the gift flow idea from receiving according to one's state of grace to bartering, from the animal example of "giving away" to the animal's blood as currency.

The changes that take place as people are forced from hunting-gathering to agriculture are not conjectural, but observed in recent times among the !Kung.[68] Their small-group egalitarian life vanishes beneath chiefdoms, children become excessively attached and more aggressive, there are more contagious diseases, poorer nourishment, more high blood pressure, earlier menarche, three times as many childbirths per woman, and a loss of freedom in every aspect of their lives.[69] The farmer remains lean if he is hungry, but otherwise his body loses its suppleness. One might well wonder who benefits from all this, and of course the answer is the landholders, middlemen, bureaucrats, white-collar workers, and corporations. It is their spokesmen who echo C. H. Brown's blithe view that "a major benefit of agriculture is that it supports population densities many times greater than those that can be maintained by a foraging way of life." He adds, "Of course, this benefit becomes a liability if broad crop failure occurs."[70] He does not say who benefits from the bigger population density, and he is wrong about the "if" of crop failure—it is only a matter of "when."

Today most of us live in cities but the left-over ideology of farming is the basis, ever since the Greek pastoral poets, Roman bucolics, and later the European rustic artists, of the nature fantasies of urban dwellers. Its images of a happy yeomanry and happy countryside are therapeutic to the abrasions of city life. This potato romance is not only one of celebrating humanity surrounded by genetic slaves and freaks, but of perceiving the vegetable world as a better metaphor. The heritable deformity of cows and dogs is inescapable while carrots and cereal grains seem fresh from the pristine hand of nature. This post-Neolithic dream lends itself, for example, to the recovery of the paradisiacal ecological relations of a no-meat diet.

## The Vegetarians
The ethical-nutritional vegetarians, the zucchini-killers and drink-

ers of the dark blood of innocent soy beans, argue for quantity instead of quality. The Animal Aid Society's "Campaign to Promote the Vegetarian Diet" calculates that ten acres will feed two people keeping cattle, ten eating maize, twenty-four munching wheat, and sixty-one gulping soya.[71] The same space would probably support one or fewer hunter-gatherers. There is nothing wrong with their humane effort "toward fighting hunger in the Third World" of course, but what is life to be like for the sixty-one people and what do we do when there are 122 or 488? And what becomes of the Fourth World of tribal peoples or the Fifth World of non-human life?

The quantitative-mindedness links them philosophically with the nationalistic maximizers who assume that military advantage belongs to the most populous countries, with the politics of growth-economists and with the local greed for sales. Nutritionally, energy increase is no substitute for protein quality, nor adipose fats for the structural fats necessary for growth and repair, nor calories over immune system needs, or over the proportions of vitamins and essential minerals found in animal tissues.

Apart from their demographic and ecological short-sightedness, the vegetarians rightfully reject the fat-assed arrogance of piggish beefsteak-eaters, but they become slaves to protein hunger, by striving to get eight of the twenty amino acids that their own bodies cannot make and that meat contains in optimum amounts. The search leads to cereals and legumes, the first are low in lysine, the second in methionine. Humans with little or no meat must get combinations of legumes and grain (lentils and rice, rice and beans, corn and beans), and they must locate a substitute source for vitamin B-12, which comes from meat.

Just this side of the vegetarians are various degrees of meat eating, and the same chains of reasoning carry us from red to white meats and from meat to eggs and milk. Neither domestic cereals nor milk from hoofed animals are "natural" foods in an evolutionary sense; witness the high levels of immune reaction, cholesterol susceptibility, and the dietary complications from too much or too little milling of grains.

Except for a tiny minority, people everywhere, including farmers, prefer to eat meat, even when its quality has been reduced by domestication. Marvin Harris has summed up the evidence from ethnology and physiology: "Despite recent findings which link the over consumption of animal fats and cholesterol to degenerative diseases in affluent societies, animal foods are more critical for sound nutrition than plant foods."[72]

Nutritionally, little detailed comparison has been made between domestic and wild meats. Long-chain fatty-acids, found only in meat, are necessary for brain development. These come from structural rather than adipose fat. You can get them in meat from the butcher, but domestic cattle often lack access to an adequate variety of seeds and leaves to make an optimum proportion of structural fats.[73] The latter are richest in wild meats.

Theories that attempt to center human evolution around something like the role of female chimpanzees or to link gathering with a gender-facilitated evolution by reference to the "vegetarian" diets of primates, neglect the protein-hunger of primates and their uptake of meat in insect and other animal materials. The argument that humans are physiologically "closer" to herbivory than to carnivory, somehow placing women closer to the center of human being, is a red herring based on a mistaken dichotomy. It simply ignores human omnivory, signified not only in food preferences but physiologically in the passage time of food in the gut (longer in herbivores because of the slow digestion of cellulose-rich and fibrous foods, shorter in carnivores). In humans it is half-length between gorillas and lions.

Among most tribal peoples most of the time meat comprises less than fifty percent of the total diet, the bulk being made up of a wide variety of fruits and vegetables. But meat is always the "relish" that makes the meal worthwhile, and close attention is always paid to the way meat is butchered and shared. Vegetarianism, like creationism, simply re-invents human biology to suit an ideology. There is no phylogenetic felicity in it.

As for the alternatives in turning from the cholesterol of domestic meats, not everything comes up yogurt. Many European restaurants now offer a separate menu of game animals (reared but not domesticated). S. Boyd Eaton and Marjorie Shostak, an M.D. and an anthropologist, comment, "The difference between our diet and that of our hunter-gatherer forebears may hold keys to many of our current health problems.... If there is a diet natural to our human makeup, one to which our genes are still best suited, this is it."[74]

## 4. Cultural Evolution

The casual misuse of "evolution" in describing social change produced enough confusion to mislead generations of students. Every society was said to be evolving somewhere in a great chain of progress. Beginning in a Heart of Darkness in the individual and at the center of remote forests humankind advanced to ethics, democracy, morality, art, and the other benefits of civilization. This ladder probably still represents the concept of the past for most modern, educated people. It is a direct heritage of the Enlightenment and its industrial science, its spectatorship (as in the art museum or at the play), elitism, and the cult of the *polis*.

Recently there have appeared new versions of lifeways that refute a universal yearning toward civilization, from savagery through nomadic pastoralism and various agricultures to a pinnacle of urban existence.[75] The revised version also denies a hierarchy of inherent physical or mental differences among the peoples of different economies.

One modified view presents us with shifts in which societies are compelled to change not so much as an advance as a result of circumstances beyond their control—increased population density and the struggle for power and space. It offers a "circumscription theory." Societies at the denser demographic end show a hierarchical, imperial domain and the loss of local autonomy in which symbols of participation in the larger system replace real participation for the individual. Such societies subjugate or are conquered by others.

In a recent book Allen Johnson and Timothy Earle cite specific examples from first to last.[76] They begin with a description of hunting-gathering at the family-level of economy, characterizing them as low in population density, making personal tools, engaged in annual rhythms of social aggregation and dispersion, informally organized with *ad hoc* leadership, collectively hunting large game, lightly assuming tasks of gathering, without territoriality or war, and with numerous alternatives in "managing risk."

Such easy-going societies continue with minor introduction of domestic plants and animals, at the same time consciously resisting life in denser structures. In villages, however, men begin to fight over "the means of reproduction" and depart from the "modesty and conviviality" found in family-level societies. As "geographical circumscription" closes around them, leaving nowhere to go, there is more bullying, impulsive aggression, revenge, and territoriality. "Scarcity of key resources" and war become "a threat to the daily lives" of these horticulturalists and pig-raisers. As the economy "evolves" the "domestication of people into interdependent social groups and the growth of political economy are thus closely tied to competition, warfare and the necessity of group defense."

As villages get bigger, Johnson and Earle continue, "Big Man" power appears, ceremonial life shifts from cosmos-focused family activity to public affirmation of political rank. Dams and weirs and slaves and food surplus and shortage management occupy the leaders. But "the primary cause of organization elaboration appears to be defensive needs." Among typical yam-growers of the South Pacific "half a mile beyond a person's home lies an alien world fraught with sudden death."

Meanwhile, the pastoralists also "evolve." Their lives are increasingly centralized under patriarchal systems based on "friends" who "help spread the risk" of resource depletion and defense needs. As cattle become currency, raiding and banditry increase in a "highly unpredictable environment." Chiefdoms are subordinated by greater chiefs, who allocate pasture and travel lanes, manage "disagreement resolution" locally, and negotiate alliances and conflict externally. Life is lived in camp, i.e., "a small nucleus of human warmth surrounded by evil." Their equivalents in sedentary towns are concerned with crop monocultures and massive tasks of "governing redistribution," regulating the bureaucracy and management of field use and irrigation works.

When we get to the first true or archaic states, vassalage, standing armies, and taxes make their appearance. "Social circumscription" is added to geographical circumscription. Religion and staple food storage is centralized. As the state matures the peasants emerge with "no end of disagreement and even disparagement" among themselves. They often "live so close to the margin of survival that they visibly lose weight in the months before harvest." As we approach the modern state the authors say, "peasant economics provide a less satisfactory subsistence than the others we have examined," with poor diet, undernourishment, extreme competition, and a meager security experienced as vulnerability to markets controlled from the outside or the arbitrary will of patrons.

Johnson and Earle conclude, at the end of this long road to a "regional polity," that the record is one of endless rounds of population increase and "intensification," producing societies symbolized by their dependence on "starchy staples." All hail the potato.

The authors are careful to remain mere observers. If a book can have a straight face while taking off civilization's pants, here is a wonderful irony, although probably a competent synthesis of the record. Yet euphemisms and semi-technical phrases abound. For "diminished resources" one should read "collapse of life support" or "failed ecosystems." For "local slave management" read "tyranny," for "risk management" simply "debacle." The increasing need for "defense" is frequently mentioned, but who is doing all the offense? How casually and with value-free candor we move from many options in "risk management" to few, from personal tools to work schedules, from ad-hoc leadership to hierarchies of chiefs. Little is said about children, women, the source of slaves, the loss of forests and soil, the scale of tensions between farmers and pastoralists. One has to interpolate the relevant changes in the role and status of women, the lives of children, or the condition of the non-human fellow-beings. The book seems to achieve its objective of combining "economic anthropology and cultural ecology," making disaster humdrum and so inevitable. The recitation of the "evolution of culture" in such expressionless fashion is in fact enormously effective, for the authors seem oblivious to the horrors they describe. I am reminded of academics who reply to descriptions of the biotic costs of civilization with murmurings about how difficult life would be for them without Beethoven, cathedrals, and jurisprudence. But then, it was a tiny elite who benefitted from this "evolution" all along, and I suppose that they can easily imagine that others, in their benighted state, cannot possibly appreciate the gains.

For twenty years my students and colleagues have responded to this scenario by asking why people changed if the old way was better, and then refuse to believe that the majority were compelled by centralized force in which power and privilege motivated the few. Zvelebil says, "The stubborn persistence of foraging long after it 'should' have disappeared is one of the qualities that is contributing to a fundamental reassessment of post-glacial hunting and gathering."

The idea of cultural change as a paradoxical "development" can also be seen in a comparison of American Indian tribes. John Berry and Robert Annis studied differences in six northern Indian tribes using George Murdock's classifications of culture types, "a broad ecological dimension running from agricultural and pastoral interactions with the environment through to hunting and gathering interactions." They describe a corresponding psychological differentiation, defined along this axis.

Agriculture tends to be associated with high food accumulation, population density, social stratification and compliance. At the other end of the series are the low food accumulators—hunter-gatherers—with a high sense of personal identity, social independence, emphasis on assertion and self reliance, high self control, and low social stratification. Berry and Annis see these differences in terms of "cognitive style," "affective style," and "perceptual style."[77] These studies are consistent with the work of Robert Edgerton, who found distinct personality differences between farmers and pastoralists.[78]

What we come to is an uneasy sense of economic determinism. There is a profound similarity of hunter-gatherers everywhere. This convergence demonstrates the niche-like effect of a way of life. The possibilities for human cultural mixtures can be seen in the variety of peoples in the modern world. There seems to be no end to the anthropological exploration of their differences. Still, the surprising thing is not their dissimilarity but the extent of common style. Something enormously powerful binds living hunter-gatherers to those of the past and to modern sportsmen.

They are all engaged in a game of chance amid heterogeneous, exemplary powers rather than in collective strategies of accumulation and control. Their metaphysics conceives a living, sentient, and dispersed comity whose main features are given in narrations that are outside History. Their mood is assent. Their lives are committed to the understanding of a vast semiosis, presented to them on every hand, in which they are not only readers but participants. The hunt becomes a kind of search gestalt. The lifelong test and theme is "learning to give away" what was a gift received in the first place.

There are also convergent likenesses among subsistence farmers, pastoralists, and urban peoples. The economic constraints seem to transcend religions and ethnic differences, to surpass the unique effects of history, to overstep ideology and technology. The philosophies as well as the material cultures of otherwise distant peoples who have similar ecologies seem to converge.

## 5. Wilderness and Wildness

### Wilderness

How are we to translate the question of the hunt into the present? One road leads to the idea of wilderness, the sanctuaries or sacrosanct processes of nature preserved.

The idea of wilderness—both as a realm of purification outside civilization and as a place of beneficial qualities—has strong antecedents in the Western world. In spite of the recent national policies of designating wilderness areas, the idea of solace, naturalness, nearness to fundamental metaphysical forces, escape from cities, access to ruminative solitude, and locus of test, trial, and special visions—all these extend Biblical traditions. As for wildness, I suppose that most people today would say that wilderness is where wildness is, or that wildness is an aspect of the wilderness.

Wilderness is a place you go for a while, an escape to or from. It is a departure into a kind of therapeutic land management, a release from our crowded and overbuilt environment, an esthetic balm, healing to those who sense the presence of the disease but who may have confused its cause with the absence of the therapy. More importantly, we describe it to ourselves in a language invented by art critics, and we take souvenirs of our experience home as photographs. Typically, the lovers of wilderness surround themselves with pictures of mountains or forests or swamps which need not be named or even known, for they are types of scenery. But it is emphatically not scenery which is involved in either the ceremonies of Aborigines or the experience of the hermit saints. Something has intervened between them and the *zeitgeist* of the calendar picture. That something is the invention of landscape.

Wilderness remains for me a problematic theme, intimately associated in the modern mind with landscape. It is a scene through which spectators pass as they would the galleries of a museum. Art historians attribute the origins of landscape (in the Occident) to 16th century perspective painters, but I find a strange analogy to the descriptions of Mesolithic art, where "we are evidently approaching a historical sense.... The tiny size of these paintings is something of a shock after the Paleolithic. The immediate impression is of something happening at a great distance, watched from a vantage-point which may be a little above the scene of the action. This weakens the viewer's sense of participating in what is going forward. There is something of a paradox here, for in the graphic art of the paleolithic, though man was seldom shown, he was the invisible participant in everything portrayed, while now that he has moved into the canvas and become a principal, there is a quite new detachment and objectivity about his portrayal."[79] In other words, the first appearance of genre and perspective in pictorial art is Neolithic, and probably expresses a new sense of being outside nature. Something like modern landscape reappears later in Roman mosaics, prior to its rediscovery by Renaissance art, and I take this as evidence of renewed "distancing" and an expression of the Classical rationality that made possible the straight roads across Europe, based on survey rather than old trails.

I owe to David Lowenthall and Marshall McLuhan a debt for diverting me from writing and thinking about wilderness. Graduate work on the history of landscape, published as *Man in the Landscape*, left me susceptible to McLuhan's devastating analysis of 17th century science and art. Linear/mathematical thinking and the representation of places as esthetic objects distanced the observer from rather than connected him to his surroundings.[80] The place was framed. This was the esthetic origin of pictorial vision, of which wilderness is a subject matter.

Lowenthall did not describe so much as embody the humanist

position, in which the "love of nature" is understood as an esthetic experience, and any esthetic is a "congeries of feelings," a cultural ripple that can come and go in the dynamics of taste and fashion.[81] Lowenthall is wrong. He misunderstands the truly radical aspect of romanticism, misconstruing it as esthetic or iconographic rather than an effort to reintegrate cognition and feeling in an organic paradigm. But he may be right about landscape. It was the means of perceiving nature according to criteria established by art criticism, the avenue of "landscape" by which people "entered" nature as they did a picture gallery. As long as pictures were regarded as representations, the enthusiasm for landscape could still penetrate all areas of culture, in spite of the estrangement described by McLuhan. By the end of the 19th century the art world moved on to non-objectivity, leaving wilderness with the obsolescence and superficiality with which Lowenthall confused it.

The landscape cannot escape its origins as an objectifying perception, although it may be misused as a synonym for place, terrain, ecosystem, or environment. Photos of it are surrealistic in the sense that they empty the subject of intimate context. As pictures age they add layers of a cold impulse like growing crystals, making the subject increasingly abstract, subjecting real events to a drifting, decadent attention. When 19th century painters discovered photography they were freed, as Cezanne said, from literature and subject matter. Susan Sontag has it right about surrealism: disengagement and estrangement. It is, she says, a separation that enables us to examine dispassionately old photographs of suffering people.[82] It is a form of schizophrenia, a final effect of splitting art from its origins in religion. It becomes seeing for its own sake, what Bertram Lewin has called "neurotic scopophilia."[83] To this I add the photography of nature, which anti-hunters want to substitute for killing and eating. Pictures of nature exactly embody what is meant by wilderness as opposed to that wildness which I kill and eat because I, too, am wild.

## Wildness

Thank God Thoreau did not say, "In wilderness is the preservation of the world." Wildness, ever since Starker Leopold's research on heritable wildness in wild turkeys in the mid-1940s and Helen Spurway's "The Causes of Domestication,"[84] has for me an objective reality, or at least a degree of independence from arbitrary definitions.

Wildness occurs in many places. It includes not only eagles and moose and their environments but house sparrows, cockroaches, and probably human beings—any species whose sexual assortment and genealogy are not controlled by human design. Spurway, Konrad Lorenz's observation on the bodily and behavioral forms of domesticated animals, and the genetics of zoo animals provide substance to the concept. The loss of wildness that results in the heritable, blunted, monstrous surrogates for species, so misleading because the plants and animals which seem to be there have gone, are like sanity's mask in the benign visage of a demented friend.

What then is the wild human? Who is it? Savages? Why... it is us! says Claude Lévi-Strauss. The savage mind is our mind.[85] Along with our admirable companions and fellow omnivores, the brown rat, raccoon and crow, not yet deprived of the elegance of native biology by breeding management, it is us! Some among us may be deformed by our circumstances, like obese raccoons or crowded rats, but as a species we have in us the call of the wild.

It is a call corrupted not only by domestication but by the conventions of nature esthetics. The corporate world would destroy wildness in a trade for wilderness. Its intent is to restrict the play of free and selfish genes, to establish a dichotomy of places, to banish wild forms to enclaves where they may be encountered by audiences while the business of domesticating the planet proceeds. The savage DNA will be isolated and protected as esthetic relics, as are the vestiges of tribal peoples. This includes the religious insights of wild cultures, whose social organization represents exotic or vestigial stages in "our" history or "evolution," their ecological relations translated into museum specimens of primeval economics. My wildness according to this agenda is to be experienced on a reservation called a wilderness, where I can externalize it and look at it.

Instead my wildness should be experienced in the growing of a self that incorporates my identity in places. See Fred Myer, Roy Rappaport, D.H. Stanner, or Gary Snyder on the way the self exists in resonance with specific events in particular places among Australian peoples.[86]

The Australian outback is not a great two dimensional space, not a landscape, but a pattern of connections, lived out by walking, ritually linking the individual in critical passages to sacred places and occasions, so that they become part of an old story. To be so engaged is like a hunger for meat, irreducible to starches, the wild aspect of ourselves.

## Wild versus Domestic Metaphysics

The bones I sometimes think I have to pick with Gary Snyder are surely those remaining from a shared hunt and meal, pieces to be mulled over—to mull, from a root word meaning "to grind" or "to pulverize" which I take to mean that we are sitting at a fire together, breaking femurs to get at the marrow or the pith.

He has said that the intent of American Indian spiritual practice is not cosmopolitan. "Its content perhaps is universal, but you must be a Hopi to follow the Hopi way." A dictum that all of us in the rag-tag tribe of the "Wanta-bes" should remember. And he has said, "Otherworldly philosophies end up doing more damage to the planet (and human psyches) than the existential.conditions they seek to transcend."[87] But he also refers to Jainism and Buddhism as models, putting his hand into the cosmopolitan fire, for surely those are two of those great, placeless, portable, world religions whose ultimate concerns are not just universal but otherworldly. Yet, without quite understanding why, from what I have seen of his personal life, there is no contradiction. I suspect that Snyder in the Sierra Nevada, like Berry in Kentucky and Wes Jackson in the Kansas prairie, is not so much following tradition but doing what Joseph Campbell called "creative mythology."

When I am sometimes discouraged by the thought that Gary Snyder has already said everything that needs to be said, as in, for instance, "Good, Wild, Sacred,"[88] I reawaken my independence of spirit by thinking of his faith in agriculture and Buddhism, even though in reality he carefully qualifies both. No matter how benign small scale garden-horticulture may be, at its center is the degenerating process of domestication, the first form of genetic engineering. Domestication is the regulated alteration of the genomes of organisms, making them into slaves that cannot be liberated, like comatose patients hooked without reprieve to the economic machine.

As for coma, the excessive use of slave animals in experimental laboratories, their fecundating overspill as pets into city streets, and their debasement in factory farms has generated the "humane" movement, the dream of animal rights groups that by kindness or legislation you can liberate enslaved species. The clearest analogy is the self-satisfied, affectionate care of slaves by many pre-Civil War gentry. In our time, a huge, terrible yearning has come into the human heart for the Others, the animals who nurture us now as from our beginnings. Our gratitude to them is deep—so deep that it is subject to the pathologies of our crowded lives. In our wild hunger for the recovery of animal presence we have made and given names to pets, moulded their being after our cultural emphasis on individuals. Our hunting past tells us that the species is the "individual," each animal the occasion of the species' soul. Our humane movement personalizes them instead, losing sight of the species and its ecology. Worse, that self-proclaimed "kindness" marks the collapse of a metaphor central to human consciousness, replacing it with the metonymy of touch-comfort, hence the new jargon of "animal companion" for pet in the new wave of "animal facilitated therapy." It is a massive, industrial effort among an amalgam of health workers, veterinarians, pet food manufacturers, and institutions. The effects of the therapy are undoubtedly genuine, but its "cognitive style" connects at one end with the hair-splitting philosophical rationality of the animal ethicists and at the other with the maudlin neuropath keeping thirty cats in a three-room city house—an abyssal chaos of purposes and priorities.

The lack of ecological concern in almost all animal ethics is strangely similar to that "embarrassed silence" in anthropology—the posture of detached respect by which all ethnic rites are interpreted as serving social and symbolic functions for an erroneous religion. Animal ethics comes from the same Greek source as all our philosophy, passionately reasoning but grounded in detachment and skepticism. There seems to be no real feeling there for the living world. They simply do not ask whether the Holy Hunt might indeed be so.

As for killing animals to eat, in *The Tender Carnivore* I suggested, taste buds and tongue in check, that in an overpopulated world we could free the animals, including ourselves, make hunting possible, and terminate the domestication of multicellular life by eating oil-sucking microbes (which is entirely feasible). To my surprise I find that this is our direction, in our yogurt and cheese rush to avoid killing "higher" animals by substituting down a chain of being, killing asparagus instead of cows or yeasts instead of asparagus. But there is no escape from the reality that life feeds by death-dealing (and its lesson in death-receiving). The way "out" of the dilemma is into it, a way pioneered for us in the play of sacred trophism, the gamble of sacramental gastronomy, central myths of gifts, and chance, the religious context of eating in which the rules are knowing the wild forms who are the game. You cannot sit out the game, but must personally play or hide from it.

This brings us back to Buddhism. I remain a skeptical outsider, unnerved by the works of Gary Snyder and Alan Watts, whose combined efforts I consider to be a possible library on how to live. Still, the Hindus disdained Buddhism when they discovered how abstract and imageless it was, how shorn of group ceremony, the guiding insights of gifted visionaries and the demonstrable respect for life forms represented in their multitudinous pantheon. The Hindus at least saw personal existence as a good many slices of *dharma* in a variety of species before the individual finally escaped into the absolute, while the Buddhists argued that all you needed was the right discipline and you could exit pronto.

The Buddhists' contemporaries and fellow travelers, the Jains, famous for *ahimsa* (harmlessness), are familiarly portrayed moving insects from the footpath. But this is not because they love life or nature. The Jains are revolted by participation in the living stream and want as little as possible to do with the organic bodies, which are like tar pits, trapping and suffocating the soul. Historically, it would appear that both Buddhists and Jains got something from the Aryans who brought their high-flying earth-escaping gods from Middle East pastoralism. In the face of these invasions, the Hindus and their unzipped polytheism survived best in the far south of India where the Western monotheists penetrated least.

At a more practical level, everywhere the "world" religions have gone the sacred forests, springs, and other "places" and their wild inhabitants have vanished. The disappearance of respect for local earth-shrines is virtually a measure of the impact of the other-worldly beliefs.

Can there be a world religion of bioregions, a universal philosophy of place, an inhabitation of planet Earth with plural, local autonomy?

## Perception as the Dance of Congruity

Rene Dubos once observed that humans can adapt (via culture) to "starless skies, treeless avenues, shapeless buildings, tasteless bread, joyless celebrations, spiritualess pleasures—to a life without reverence for the past, love for the present, or poetical anticipations of the future. Yet it is questionable that man can retain his physical and mental health if he loses contact with the natural forces that have shaped his biological and mental nature."[89] But, unless these "forces" are the characteristics he mentions, what are they? His list is made up entirely of acts within a social and cultural milieu, by customary definition not "natural." Something "natural" looms behind all this, mediated by culture.

Dubos' statement is preceded by the observation that the human genetic makeup was stabilized 100,000 years ago. He quotes Lewis Mumford, "If man had originally inhabited a world as blankly uniform as a 'high rise' housing development, as featureless as a parking lot, as destitute of life as an automated factory, it is doubtful that he would have had a sufficiently varied experience to retain images, mold language or acquire ideas."[90]

What is this something natural necessary to become cultural? What is between culture and nature, betwixt the phenomenal or palpable world and the conceptual and ceremonial expressions of it? Connecting the cognition and the outer world is the event/structure, linking entity and environment. It is perception, the pre-cognitive act, mostly unconscious, which directs attention, favors preferences, governs sensory emphasis, gives infrastructure. Lee and Ong's distinctions between an "acoustical event world" and the "hypervisual culture" is just such a prior mode, giving primordial design to experience, limiting but not formulating the concepts and enactments by which events are represented.[91] Phonetic alphabet, pictorial space, and Euclidean theory are not only ideas and formulas, but frames supporting a kind of liminal foreknowledge of assumptions and inclinations.

Emphasis on perception does not mean that we shape our own worlds irrespective of a reality, or that one person's perceptual process

is as reliable as another's. Perception is not another word for taste. In
this, says Morris Berman, it transcends "the glaring blind spot of
Buddhist philosophy."[92] Its truest expression "by test" (my criteria:
quality of life; ecological integrity) in the world is the empirical effect
of its contiguity. It is the process of the first steps of directed attention
and vigilance. Perceptual habit is style in the sense that Margaret Mead
once used the term, to mean a pattern of movement and sensitivity, the
lively net of predisposition emerging from our early grounding, finally
affecting every aspect of one's expressive life. In our wild aspect such
unconscious presentations are centered in dance and narration, surrounded by innumerable and wonderfully varied moral and esthetic
presences. It presents us with an intuition of rich diversity whose
"forces" are purposeful and sentient. From Dubos' treeless avenues to
Mumford's parking lot, it is not a view that is absent, or things or
wilderness. It is a way of expecting and experiencing, encountering
inhabitance by a vast congregation of others unlike us, yet, like our
deepest selves, wild.

## 6. The Mosaic

We must now close the circle to that sweeping, four-word dictum
which is intended to close the door on access to the primitive: "You can't
go back."

### The Structural Dimension

The hereditary material is organized as a linked sequence of
separable genes and chromosomes. This genome is a mosaic of harmonious but distinct entities. This structure makes possible the mutation of
specific traits and the independent segregation of traits, the accumulation of multiple factors, and both the hiding and expression of genes.

The structure of the natural community, the ecosystem, is likewise
an integrated whole composed of distinct species populations and their
niches. The fundamental concept of modern biology is its primary
characteristic as a composite of linked and harmonious but separable
parts. The whole is neither the sum of its parts nor independent of any
of them. As with genes, substitutions occur. A given species can be

totally removed by extirpation or introduced into new communities. Witness for example the constitution of the prairie without the buffalo and the continuity of ecosystems after the successful introduction of the starling into North America.

Human culture, being genetically framed and ecologically adapted, is also an integrated conglomerate. Stories, dances, tools, and goods are sometimes completely lost from a society. At others they move from culture to culture, sometimes trailing bits of the context from which they come, sometimes arriving rough-edged and isolate, but being assimilated, modified or not, as a part of the new whole.

There is a common characteristic of each of the above examples from the genome of the individual, the material or expressive culture of a people, and the tapestry of the natural environment. The specific entity involves both a distinct portability and a working embeddedness. The reality is more complex but the principle is true: the capacity for a part to be transferred. It is then part of a new whole. The rest of the totality adjusts, the organism accommodates, the niche system stretches or contracts, the culture is modified.

Societies and cultures are mosaics. They are componential. Their various elements, like genes and persons, can be disengaged from the whole. Contemporary life is in fact just such an accumulation representing elements of different ages and origins, some of which will disappear, as they entered, at different times than others. The phrase "You cannot go back" can only mean that you cannot recreate an identical totality but it does not follow that you cannot incorporate components.

"You can't go back" is therefore a disguise for several assumptions, which in turn may hide ways of perceiving or preconstructing experience. One is the paradigm of uni-direction, the idea that time and circumstances are linear. Yet we "go back" with each cycle of the sun, each turning of the globe. Each new generation goes back to already existing genes, from which each individual comes forward in ontogeny, repeating the life cycle. While it is true that you may not run the ontogeny backwards, you cannot avoid its replays of an ancient genome, just as human embryology follows a pattern derived from an ancestral fish. Most of the "new" events in each individual life are only new within a certain genetic octave and only in their combinations. New genes do occur, but the tempo of their emergence is in the order of scores of thousands of years. The difference between the genomes of chimpanzees and humans is about one percent. Of the 146 amino acids of the Beta chain of blood hemoglobin the gorilla differs from humans at one site, the pig at ten, the horse at twenty-six.[93]

A paradox is evident: newness yet sameness; repetition and novelty, past and present. Recall that the historical consciousness of the West rejects this as illusory ambiguity. The rejection is a characteristic perceptual habit. In tribal life, such matters of identity ambiguity are addressed ritually in the use of animal masks and mimetic dances, on the grounds that we are both animal and human, a matter "understood" by certain animal guides. Genes are not only "how-to" information but are mnemonic, that is, memories. Ceremonies recall. The reconciliation of our own polythectic zoological selfhood is inherent in our ritualized, sensuous assent of multiple truth. It denies the contradiction, abolishes the either-or dichotomy in the simultaneous multitude that we are. Our primitive legacy is the resolution of contradiction by affirmation of multiplicity, plurality, and change.

In advocating the "primitive" we seem to be asking someone to give up everything, or to sacrifice something: sophistication, technology, the lessons and gains of History, personal freedom, and so on. But some of these are not "gains" so much as universal possessions, reified by a culture which denies its deeper heritage. "Going back" seems to require that a society reconstruct itself totally, especially that it strip its modern economy and reengage in village agriculture or foraging, hence is judged to be functionally impossible. But that assumption misconstrues the true mosaic of both society and nature, which are composed of elements that are eminently dissectible, portable through time and space, and available.

You can go out or back to a culture even if its peoples have vanished, to retrieve a mosaic component, just as you can transfer a species that has been regionally extirpated, or graft healthy skin to a burned spot from a healthy one. The argument that modern hunting-gathering societies are not identical to paleolithic peoples is beside the point. It may be true that white, ex-Europeans cannot become Hopis or Kalahari Bushmen or Magdalenian bison-hunters, but removable elements in those cultures can be recovered or recreated, which fit the predilection of the human genome everywhere.

### Three Important, Recoverable Components: The Affirmation of Death, Vernacular Gender, and Fulfilled Ontogeny

Our modern culture or "mosaic" is an otherworldly monotheism littered with the road kills of species. Road kills—such trivial death contrasts sharply to that other death in which circumspect humans kill animals in order to eat them as a way of worship.

This ancient, sacramental trophism is as fundamental to ourselves as to our ancestors and distant cousins. The great metaphysical discovery by the cynegetic world was cyclicity. It emerged in the context of the rites of death, both human and animal, as part of this flow. It is as old as the Neanderthal observation of hibernating bears as models of life given and recovered, and as new as Aldo Leopold's story "Odyssey" in *A Sand County Almanac* telling of an atom from a dead buffalo moving through the chain of photosynthesis, predation, decay, and mineralization. These concepts are about the nutritional value of meat in human metabolism as a reflection of a larger "metabolism," and about the gift of human consciousness in a sentient world in which food-giving symbolizes connectedness. Animals on the medicine wheel of the Plains Indians were said to be those that know how to "give away." "'Each dot I have made with my finger in the dirt is an animal,' said White Rabbit. 'There is no one of any of the animals in this world that can do without the next. Each whole tribe of animals is a Medicine Wheel, in that it is the One Mind. Each dot on the Great Wheel is a tribe of animals. And parts of these tribes must Give-Away in order that they all might grow. The animal tribes all know of this. It is only the tribes of People who are the ones who must learn it.'"[94]

William Arrowsmith, observing that in our time "we cannot abide the encounter with the 'other,'.... We do not teach children Hamlet or Lear because we want to spare them the brush with death.... A classicist would call this disease *hybris*.... The opposite of *hybris* is *sophrosyne*. This means 'the skill of mortality.'"[95] It is the obverse side of the "giving away" coin, the way of momentarily being White Rabbit, reminding the human hunter that he too once was a prey and, in terms of the cosmic circling-back, still is.

The difficult question of interspecies ethics centers on death-dealing. Death is the great bugaboo. How we resent its connection to food—and to life—and repress the figure of the dying animal. Gary Snyder's reply: "All of nature is a gift-exchange, a potluck banquet, and there is no death that is not somebody's food, no life that is not somebody's death. Is this a flaw in the universe? A sign of the sullied condition of being? 'Nature red in tooth and claw'? Some people read it this way, leading to a disgust with self, with humanity, and with life itself. They are on the wrong fork of the path."

Joseph Campbell has argued rightly that death was a great meta-physical problem for hunters, and concluded wrongly that it was solved by planters with their sacrifices to forces governing the annual sprouting of grain. But it was control, not acquiescence to this great round, that the agriculturalists sought. In the Neolithic, says Wilhelm Dupre, "The individual no longer stands as a whole vis-à-vis the life-community in the sense that the latter finds its realization through a total integration of the individual—as is the case by and large under the conditions of a gathering and hunting economy."[96]

"Hunters" is an appropriate term for a society in which meat, the best of foods, signifies the gift of life, the obtaining and preparation of which ritualizes the encounter of life and death, in which the human kinship with animals is faced in its ambiguity, and the quest of all elusive things is experienced as the hunt's most emphatic metaphor.

## Vernacular Gender

And so we bring to and from the mosaic of lifeways the hunt itself. Some feminists object that too much is made of it. But they misunderstand this killing of animals as an exercise of vanity, which they see as characteristic of patriarchy. They note that only a third of the diet is meat, the rest from plants, mostly gathered by women, as though there were a contest to see who really supports the society. In this they merely reverse the sexist view. Like so much of extremist feminism it is just a new "me first." They point out that in most hunting-gathering societies the women gather most of the food that is eaten. This view has the same myopia as that of the vegetarians—the tendency to quantify food value in calories. In any case they are wrong, as meat is so much higher in energy that the net energy gained from hunting is as great as that from gathering.[97]

While it is true that the large, dangerous mammals are usually hunted by men in hunting-gathering societies, it has never been claimed that women only pluck and men only kill. The centrality of meat, the sentient and spiritual beings from whom it comes, and the diverse activities in relationship to the movement of meat and the animal's numinous presence through the society, entail a wide range of roles, many of which are genderized. Insofar as the animal eaten is available because it has learned "to give away," there is no more virtue in the actual chase or killing than the transformation of its skin into a garment, the burying of its bones, the drumming that sustains the dancers of the mythical hunt, or the dandling of infants in such a society as the story of the hunt is told.

Meat, says Konner, is only thirty percent of the !Kung diet, but it equals the nutritional value of the plant foods and produces eighty percent of the excitement, not only during the hunt but in group life. The metaphysics of meat. The hunt itself is a continuum, from its first plan to its storied retelling, from the metaphors on food chains to prayers of apology, this carnivory takes nothing from woman, though it clarifies the very different meaning that different kinds of foods have in expressive culture. Broadly understood, the hunt refers to the larger quest for the way, the pursuit of meaning and contact with a sentient part of the environment, and the intuition that nature is a language. Hunting is a special case of gathering.

A critical dimension of the hunt is the confrontation with death and the incorporation of substance in new life, in all forms of sharing and giving away. Women are traditionally regarded as keepers of the mystery of death-as-the-genesis-of-life, hence the hunt is clearly connected with feminine secrets and powers, and we are not surprised to see Artemis and her other avatars, the archaic "Lady of the Beasts," and the Paleolithic female figurines in sanctuaries where the walls are painted with hunted game. More value is placed on men than women only as the hunt is perverted by sexism and war. Indeed, it is possible that sexism comes into being with the doting on fertility and fecundity in agriculture and the androgenous "reply" of nomadic, male dominated societies of pastoralism.

Hunting has never excluded women, whose lives are as absorbed in the encounter with animals, alive and dead, as those of men. If in some societies the practices of vernacular gender tend more often to relegate to the men the pursuit of large, dangerous game, it relegates to the women the role of singing the spirit of the animal a welcome, and to them the discourse at the hearth where she is the host. Roles and duties are divided, but not to make inequality. Among the Sharanahua of South America, the women, being sometimes meat-hungry, send off the men to hunt and sing the hunters to their task. They are commonly believed to transform men into hunters. Janet Siskind says, "The social pressure of the special hunt, the line of women painted and waiting, makes young men try hard to succeed." Women also hunt. Gathering, like hunting, is a light-hearted affair done by both men and women. The stable sexual politics of the Sharanahua, "based on mutual social and economic dependence, allows for the open expression of hostility," a combination of solidarity and antagonism that "prevents the households from becoming tightly closed units."[98]

Martin Whyte, comparing "cultural features in terms of their evolutionary sequence," concluded that as civilization evolves, "women tend to have less domestic authority, less independent solidarity with other women, more unequal sexual restrictions and perhaps receive more ritualized fear from men and have fewer property rights than is the case in the simpler cultures."[99]

All in all a far cry from the more strident views, whether of feminists, the obsolete social evolution of the neo-Marxists, or the flight from life of the humane animal protectionists. On the whole, plant foods are not shared as ceremoniously as meat. They do not signify the flow of obligations in the same degree. But this is not a statement about women as opposed to men.

## The Temporal Mosaic: The Episodic Character of Individual Life

Being individuals slow to reach maturity, we are among the most neotonic of species. This resiliency makes humans prime examples of

"K" type species evolution (education, few offspring, slow development). "Culture" constitutes the social contrivances that mitigate neotony. The transformation of the self through aging is inevitable, but whether we move through successive levels of maturity and the fullest realization of our genome's potential depends on the quality of the active embrace of society in all of the nurturance stages. Incomplete, ontogeny runs to the dead end of immaturity and a miasma of pathological limbos.[100]

The important nurturant occasions are like triggers in epigenesis. Neoteny, the many years of individual immaturity, depends on the hands of society to escape itself. This mitigation of our valuable retardation is in part episodic and social, a matching of the calendars of postnatal embryology by the inventions of caregivers. Occasions make the human adult. If culture in the form of society does not act in the ceremonial, tutoring, and testing response to the personal, epigenetic agenda, we slide into adult infantility—madness. This fantastic arrangement is foreshadowed in the nucleus of every cell. It is an expectancy of the genome, fostered by society, enacted in ecosystems.

Two of the transformative stages of human ontogeny have been studied in detail among living hunter-gatherers—infant/caregiver relationships and adolescent initiation. The archaeological record leaves little doubt that we see in them ancient patterns which may be incompletely addressed in ourselves. Foremost is the bonding/separation dynamic of the first two years. The interaction of infant and mother and infant and other caregivers emerges as a compelling necessity, perhaps the most powerful shaping force in the whole of individual experience. The "social skills" of the newborn and the mother's equally indigenous reciprocity create not only the primary social tie but the paradigm for existential attitudes. The lifelong perception of the world as a "counterplayer"—caring, nourishing, instructing, and protecting, or vindictive, mechanical, and distant—arises here.

The process arises in our earliest experience and is coupled to patterns of response. Hara Marano says, "Newborns come highly equipped for their first intense meetings with their parents, and in particular their mothers.... Biologically speaking, today's mothers and babies are two to three million years old.... When we put the body of a mother close to her baby, something is turned on that is part of her genetic makeup."[101] Details of the socially embedded rhythms of parenthood vary from culture to culture, but they can hardly improve on the basic style or primary forms found in hunter-gatherer groups. Studies of babies and parents in these societies reveal that the intense early attachment leads not to prolonged dependency but to a better functioning nervous system and greater success in the separation process.[102]

Something of the same can be said for the whole of ontogeny, especially those passage-markers by which the caregivers celebrate and energize movement across thresholds by the ripe and ready. Notable among these is adolescent initiation, a subject to which a vast body of science and scholarship has been devoted. Yet again it has fundamental forms for which individual psychology is endowed. Much of modern angst has its roots in the modern collapse of this crucial episode in personal development.

Early experience has this formative and episodic quality, with varying degrees of formality in its context. The hunt is one, bringing into play in the individual the most intense emotions and sense of the mysteries of our existence, to be given a catharsis and mediating transformation. The hunt is a pulse of social and personal preparation, address to presences unseen, skills and strategies, festive events and religious participation. We cannot become hunter-gatherers as a whole economy, but we can recover the ontogenetic moment. Can five billion people go hunting in a world where these dimensions of human existence were played out in a total population of perhaps one million? They can, because the value of the hunt is not in repeated trips but a single leap forward into the heart-structure of the world, the "game" played to rules that reveal ourselves. What is important is to have hunted. It is like having babies; a little of it goes a long way.

## Endemic Resources and the Design of a Lifeway—a Post-Historic Primitivism

In her book, *Prehistoric Art in Europe*, N. K. Sandars identifies four strands of the primordial human experience: (1) "The sense of diffused sacredness which may erupt into everyday life," (2) "an order of relationships the categories of which take no account of genetic barriers and which will lead to ideas of metamorphosis inside and outside this life," (3) "unhistorical time" and (4) "the character or position of the medicine man or shaman."[103]

These are not, of course, removable entities as such, but they constitute aspects of the Paleolithic genius, emergent gestalts from the separate and portable elements of a culture. As ideals not one of these is a regression to obsolescence but a forward step to Heidegger's *dasein*, Merleau-Ponty's and Whitehead's event world, Eliade's centrality of the rites of passage, Odum's redaction of ecological entities as process and relationship. It is not a matter of what ought to be done or how life could be, or even of greater meaning and understanding, but of the nature of experience. I would summarize these "experiences" as follows:

1. Therio-metaphysics. Animals as the language of nature, a great Semiosis. Reading the world as the hunter-gatherer reads tracks. The heuristic principle and hermeneutic act of nature and society as the basic metaphor. Eco-predicated logos.

2. The Voice of life. Sound, drum, song, voice, instrument, wind, the essential clue to the livingness of the world. It is internal and external at once, the game told as narrative, the play of chance. In story, Snyder has called it "the primacy of together-hearing."

3. The Fledging and Moulting principle. Epigenesis as the appropriate and sequential coupling of gene and environment, self and other. The ecology of ontogenesis as a resonance between bonding and separation that produces identity. Transitions marked by formal acts of public recognition.

4. Sacramental Trophism. The basic act of communion, transformation, and relatedness, incorporating death as life. It is centered on the act of bringing death and of giving to death as the central celebration of life.

5. The Fire Circle. All forms of social connection in relation to scale. Vernacular gender. Examples: Homeostatic demographic units. The dialectical tribe in Australia: family, band, and tribe affiliation. Sizes 25/500. "In terms of conscious dedication to human relationships that are both affective and effective, the primitive is ahead of us all the way," comments Colin Turnbull.[104]

6. Vocational Instruments. Dealing directly with the means of subsistence by hands-on approach. Tools are a gestural response to life, subordinate to thought, art, and religious forms. Marshack speaks of "the demands of fire culture" as one of lore and skills in which the tale is a "metaphysical gift" making the world "an object of contemplation."

7. Place instead of Space, Moments replace Time, Chance instead of Strategy. Place is at once an external and internal state in a journey home. The place is a process, not coordinates, yet a specific geology, climate and habitat.

8. Occasions of the Numinous in the relocation of the signs of sacred presence, the mystery of being, and the participatory role of human life, not as ruler or viceroy but as one species of many, in a mood not of guilt or conflict but of affirmation.

9. The escape from domestication, a liberation of nature into itself, including human nature, from the tyranny of the created blobs and the fuzzy goo of emotional—and epoxic glue of ethical—humanism.

Primitivism does not mean a simplified or more thoughtless way of life but a reciprocity with origins, a recovery misconstrued as inaccessible by the ideology of History. In the latter view one puts on costumes and enacts another culture as the French aristocracy imitated shepherds during the Renaissance or as middle class "dropouts" in the 1960s put on gingham gowns and bib overalls.

From the ahistoric perspective you cannot "go back" to recover "lost" realities, nor can you completely lose them. So long as there is a green earth and other species our wild genome can make and find its place. Like many difficult things the transformation cannot be made solely by acts of will. One can simulate the external features of a primitive life—for example, the limitation of possessions and the non-ownership of the land—but something precedes the outward form and its supporting ideology. That something is the way in which the sensuous apprehension is linked to the conceptual world, the establishment early in life of a mode by which experience and ideas interact, in perception.

It is, of course, a cyclic matter in which childhood experience leads to appropriate thought and custom, which in turn mentors individual genesis. Breaking into the circle is hard, as we urban moderns can only start with an idea of it. Rare are those who can make that leap from the idea to the mode without early shaping. As a result most of us get only glimpses of what we might be were we truer to our wildness, among them some of the anthropologists who study tribal peoples. Or, we get intimations from the archetypes arising in our dreams or given in visionary moments.

In sum it is an archetypal ecology, a paraprimitive solution, a Paleolithic counter-revolution, a new cynegetics, a venatory mentation. Whatever it may be called, our best guides, when we learn to acknowledge them, will be the living tribal peoples themselves.

# Ecocentrism, Wilderness, and Global Ecosystem Protection

by George Sessions

## I. Wilderness: From Muir to Pinchot to Muir

The ecophilosopher Holmes Rolston quotes disapprovingly from a 1978 United States Forest Service document on "wilderness management" which asserts that: "Wilderness is for people…. The preservation goals established for such areas are designed to provide values and benefits to society…. Wilderness is not set aside for the sake of its flora or fauna, but for people."[1]

It is disappointing to find Forest Service theorists in the late 1970s still promoting narrowly human-centered values of the function and values of wilderness. In so doing, they continue in the tradition of the founder of the United States Forest Service, Gifford Pinchot. As the ideological arch-rival of John Muir at the turn of the century, Pinchot promoted the anthropocentrism that has pervaded conservationist and land-use agency policy since the turn of the century: that, in Pinchot's words, "there are just people and resources."[2] On this view, natural ecosystems (including non-human wild species) have no inherent worth or value-in-themselves; their value consists rather in the instrumental contribution they make to rather narrowly conceived ideas of human welfare and enjoyment.

It is ironic that these Forest Service "resource" theorists continue in this unecological vein when we consider that Aldo Leopold (one of the founders of the Forest Service wilderness concept in the 1920s) published his justly famous ecocentric "land ethic" forty years ago. In fact it was Aldo Leopold, together with Rachel Carson, who deserves the main credit for initiating the wave of holistic ecological thinking which resulted in the "ecological revolution" (the "Age of Ecology") of the 1960s and 70s.[3] And it is precisely this understanding of the world which still fails to penetrate the consciousness of many professional "natural resource managers" and other contemporary anthropocentrists.

In the 19th century Henry David Thoreau and John Muir developed what we would call today an ecocentric or deep ecological consciousness. At the time of his famous 1851 Concord speech Thoreau had seen through the anthropocentric illusions of the modern world when he claimed that "in wildness is the preservation of the world.... In short, all good things are wild and free." Muir carried the ecocentric/wilderness philosophy to even greater heights and was primarily responsible for spearheading the drive for wilderness preservation in the late 1800s as the first president of the Sierra Club.[4] American policy makers, beginning with Theodore Roosevelt, adopted Pinchot's speciesist version of conservation resulting in a major setback for an ecological understanding of the human/Nature relationship, and an acceleration of the ecological destruction which has continued through the 20th century, now partly under the guise of "wise use and responsible conservation of *our* natural resources."

Pinchot's anti-ecological resourcist ideology (and the official policies toward wilderness and wild Nature which flowed from it) were, however, embedded in wider cultural patterns stemming from long-held Western philosophical and religious beliefs and assumptions that humans were destined to dominate Nature, and that Nature existed solely for the use and enjoyment of humans.[5]

As part of a series of contemporary "state of the art" papers on wilderness and ecosystem research, Thomas Fleischner points out that wilderness values during the Pinchot era (and as a result of the provisions of the Wilderness Act of 1964) have focused on two main types of use: resource extraction and recreation. He claims that "the former included such endeavors as mining, logging, and grazing, while the latter included fishing, hunting, and increasingly nonconsumptive activities such as hiking and nature study."[6]

It should be pointed out that the National Parks based on the "preservationist" ideal (although still preservation of the wildlife and the natural "esthetics" primarily for the enjoyment of this and future generations of humans) does not allow resource extraction within its boundaries including its newly designated wilderness areas. Many of the Parks however have been damaged by park policies over the years which have catered to dominant American values and lifestyles which see the Parks as essentially "natural scenery" and recreational escapes for city-dwellers. This has encouraged a Disneyland atmosphere of excessive tourism and overdeveloped facilities, upgraded high speed roads ("scenic drives") for automobiles and oversized motor homes, mechanized recreation such as snowmobiling which disturbs the serenity and the wildlife, and overcrowding: what Edward Abbey called "industrial tourism."[7] In short, there is a constant push from the commercially oriented to turn the non-wilderness portions of the Parks into resorts and international "roadside attractions."

The ecological revolution (which essentially overturned Pinchot's anthropocentrism and resurrected the ecocentrism of Thoreau, Muir, and Leopold) actually began shortly after World War II with the rise to prominence of the science of ecology. It also surfaced as an ideological battle within the conservation organizations, primarily the Sierra Club. After the death of Muir, the Club's vision had narrowed to that of a "hiking organization" concerned primarily with the protection of wilderness seen as providing high quality recreation, aesthetic experience, and "spiritual" renewal for its members. The general attitude among conservationists of this period seemed to be that, as long as fragments of wilderness (or birds, as in the case of the Audubon Society) were protected, the rest of the planet could, and should, rightfully be the arena of endless human growth, development, and resource extraction governed by economic criteria, free enterprise, and hopefully, principles of human justice and equality. The human domination of Nature meant the replacement of the "undeveloped" natural world by an artificial environment (a human spaceship) built to human specifications and piloted by humans. It was believed that scientists, technologists, and resource managers would devise ways to do this without damaging the planet (if damage to the Earth was even a concern for most people). And the whole modernist vision and world view was driven by a somewhat nebulous and ill-defined concept of inevitable human "progress."

The Wilderness Conferences, sponsored by the Sierra Club, beginning in 1949 and continuing for over twenty years, soon provided a public forum for ecologists to broach issues much wider than the value of wilderness as mere recreation. As executive director of the Club during these years, David Brower (referred to by historian Stephen Fox as "Muir reincarnate") in effect began to "ecologize" the Club and, by extension, the wider public, by encouraging professional ecologists and biologists to speak at these conferences on the ecological importance of wilderness and wild places, on the importance of old growth forest ecosystems which the Forest Service was beginning to clear-cut at an alarming rate, and on the dangers of human overpopulation and its ultimate impact on natural ecosystems, wilderness, and species diversity. Through his newly found ecological approach to conservation and his aggressive campaign tactics, Brower turned the Sierra Club into what the *New York Times* called in 1967 "the gangbusters of the conservation movement."[8]

By the spring of 1963, the National Park Service took steps to change its wilderness and wildlife management policies in an ecological direction when it implemented the suggestions of its Advisory Board on Wildlife Management (the "Leopold Report," so named for its chairman, Starker Leopold, a son of Aldo and a zoologist at UC-Berkeley). The Leopold Report proposed that the wilderness National Parks, using Yellowstone as a model, be managed as "biotic wholes"—as "original ecosystems." According to Alston Chase, Leopold claimed that his committee was proposing "a philosophy of management that could be applied universally" [e.g., to the African wildlife parks].[9]

But in his recent book on Yellowstone, Alston Chase argues that ecosystem management is a failure. He blames ecosystem wilderness management and deep ecological philosophy for the decline of wildlife and deterioration of the Park and urges a return to heavily manipulative and intrusive scientific wildlife management—a condition which would result in what some have described as a "natural zoo."[10] Environmentalists and ecologists have countered that, like most of the Parks when established, Yellowstone does not comprise a complete ecosystem. In order for ecosystem management to work, the boundaries of Yellowstone must be expanded to comprise a complete ecosystem—in effect, a

"Greater Yellowstone Ecosystem." To create such a system, the park's design must reflect ecological rather than only political and economic criteria. Predator-prey balances must be reestablished, such as reintroduction of the wolf. Some campgrounds and facilities must be relocated, such as Grant Village/Fishing Bridge, which unfortunately was built in prime grizzly habitat. To its credit, the Park Service, even after the recent major fires, continues to adhere to the overall concept of ecosystem management.

When Congress passed the Forest Reserve Act in 1891 enabling President Benjamin Harrison to set aside thirteen million acres in "forest reserves," John Muir had reason to believe they would be protected as wilderness. But the "forest reserves" later became National Forests and the U.S. Forest Service had other plans for them. As a branch of the Department of Agriculture, the Forest Service sees its primary function as anthropocentric commercial "resource" exploitation. Just as the early pioneers cleared wild forests to make way for agriculture, the Forest Service continues with its anti-ecological policies of destroying the last old growth (ancient) forest ecosystems in America to make way for agriculture in the form of monocultural "tree farms." Surely, the destruction of the last temperate zone forest ecosystems in America will go down as one of the great ecological crimes and blunders of this century, comparable to the present rainforest destruction in Central and South America and southeast Asia.

As the ecophilosopher John Rodman pointed out in 1976:

> The charges frequently made in recent years by Preserva-
> tionists and others—e.g., that the Forest Service is a
> captive (or willing agent) of corporate interests, that it
> allows ecologically-disruptive clear-cutting as well as
> cutting in excess of official quotas, while permitting
> grazing corporations to overgraze the land while paying
> fees far less than they would have to pay for the use of
> private land, etc.—represent less the latter-day capture of
> an agency by one or more of its constituencies than a
> maturation of the basic principles of the founder [Pinchot].
> The Forest Service is, in effect, a perennial government
> subsidy, in exchange for certain regulatory controls, to
> certain types of corporations.[11]

Forestry critic Chris Maser further points out that, contrary to the claims of timber corporations and forestry representatives, we are nowhere near attaining "sustainable forestry," even on those lands where tree harvesting is appropriate, because we are training "plantation managers" instead of foresters: "We are liquidating our forests and replacing them with short-rotation plantations. Everything Nature has done in designing forests adds to diversity, complexity, and stability through time. We decrease [this] by redesigning forests into plantations."[12]

The phenomenon of destroying forests and replacing them with unstable tree plantations is now occurring all over the world. Well-intentioned people who advocate just "planting trees" to help stem the "greenhouse effect" are proposing an ecologically simple-minded solution. What is needed is a massive global effort to *restore* diverse forest ecosystems. Fortunately, widescale public pressure is now being applied to modern forestry ideology and practices to bring about radical ecological reform including the protection of the last stands of old growth (ancient) forest. And even Pinchot-trained foresters are beginning to undergo an ecological change of heart. As Canadian forester Bob Nixon recently remarked:

> As a forester, I learned to view forests as a source of industrial fibre. Now, I know that forests are so much more than vertical assemblages of lumber, so very much more important than just a source of consumer products.... Natural forests, the new research tells us, are no longer something to move through, in the economic sense, in our quest for higher gains, but indeed are a key element in the balanced functioning of planetary life.[13]

John Muir would be pleased. But even as ecological consciousness increases throughout the world, the crucial issues are the amount of ecological damage we have inflicted upon the planet since Muir's day, together with the short time remaining to effect the needed changes.

## II. Reasons for Protecting Wilderness and Wild Species

One of the activities of ecophilosophers and ecologists since the 1970s has been to explore the arguments (reasons, justifications) which have been given to protect wilderness and wild species from industrial exploitation and other forms of destruction. One of the first attempts to do this was the classification scheme developed in 1979 by the Australian philosopher, William Godfrey-Smith (now William Grey).[14] Grey characterized "wilderness" as "any reasonably large tract of the Earth, together with its plant and animal communities, which is substantially unmodified by humans and in particular by human technology."

The Australian deep ecology theorist, Warwick Fox, has further expanded upon these arguments, agreeing with Grey that they ascribe only *instrumental* value to non-human Nature. I here use his description of Grey's analysis of the arguments.[15]

(1) THE SILO ARGUMENT. This argument claims that wilderness and the non-human world ought to be preserved as a stockpile of genetic diversity for agricultural, medical, and other purposes.

(2) THE LABORATORY ARGUMENT. Wilderness ought to be preserved for scientific study. (The biologist Daniel Janzen argued recently that scientists will have to *buy* the tropical forests so they will have a place to "biologize.")[16]

(3) THE GYMNASIUM ARGUMENT. Wilderness should be saved for human recreational purposes.

The image of a "gymnasium," however, is somewhat misleading. Some human activities and sports which are allowed in some wilderness areas, such as downhill skiing, hang gliding, the use of motorized vehicles (e.g., snowmobiles), trophy hunting, and competitive mountaineering and rock climbing, can have this characteristic. But some forms of backpacking, mountaineering, and other wilderness activities have led to spiritual development and identification with wild Nature: what Arne Naess calls "modesty in the mountains."[17] These latter kinds of activities have produced some of the world's most devoted ecocentric environmentalists, including John Muir, Dave Brower, Dolores LaChapelle, Edward Abbey, Gary Snyder, and Arne Naess, and should be the kinds of activities which are encouraged in wilderness areas.

(4) THE CATHEDRAL ARGUMENT. Wilderness should be preserved for its aesthetic pleasure and spiritual inspiration.

While Godfrey-Smith (Grey) holds that these instrumental value arguments are "powerful arguments for preservation which can be derived from the purely anthropocentric considerations of human self interest," like most contemporary ecophilosophers, he is more sympathetic to the "recognition that biological systems are items which possess intrinsic value, in Kant's terminology, that they are 'ends in themselves.'"

Warwick Fox claims that the *Cathedral argument* needs to be split into two separate arguments; one for aesthetic appreciation and the other for the spiritual or religious qualities. Hence we get:

(5) THE ART GALLERY ARGUMENT. Wilderness should be preserved for its aesthetic qualities (e.g., beauty, richness of texture or color). The *Cathedral argument* would now emphasize spiritual or religious qualities exclusively (e.g., enlivening of one's spirit, intimations of the presence of God). As Howard Zahniser of the Wilderness Society expressed this in 1952:

> We deeply need the humility to know ourselves as dependent members of a great community of life and this can indeed be one of the spiritual benefits of a wilderness experience. [In civilization man was surrounded and impressed by his own achievements. Removed from them it was possible] to recognize one's littleness, to sense dependence and interdependence, indebtedness, and responsibility.[18]

(6) THE MONUMENT ARGUMENT. Warwick Fox considers this an additional instrumental argument which is based upon the symbolic value of the nonhuman world to humans. He quotes Bryan Norton who argues that "other species, which struggle to survive in living, unmanaged ecosystems [ought to be preserved because they] are our most powerful symbols of human freedom."[19]

I would add several other instrumental arguments to this collection:

(7) THE MINDING ANIMALS ARGUMENT. In a detailed and fascinating discussion of the relation of animals to humans, Paul Shepard rejects the

economic, ecological, and ethical arguments as sufficient for protecting wild animals (except for the ethical argument which claims that their very existence justifies their protection). Shepard's argument is that "the human mind needs [wild animals in their unmanaged natural habitats] in order to develop and work. Human intelligence is bound to the presence of animals."[20]

(8) THE NATURAL HUMAN ONTOGENY ARGUMENT. This is related to the above argument yet I think it is distinct. Paul Shepard further generalizes his "minding animals argument" to claim that there is a natural psycho-genetic development for humans. This normal growth process includes a stage at which there must be an intimate bonding and relating to wilderness and wild animals. Thus, wilderness must be protected to allow the opportunity for humans to proceed through their normal developmental processes. This argument begins to merge with Arne Naess' claims about the nature of full human maturity. Naess' concept of the "ecological self" is a self which expands with increasing maturity through an identification with other people, to an identification with other species and nature, to the cosmos itself.[21]

(9) THE ANTI-TOTALITARIAN POLITICAL ARGUMENT. In a powerfully written paper, Wayland Drew examines the three major anti-utopian novels of this century (*We*, by the Russian, Eugene Zamiatin; Huxley's *Brave New World*; and Orwell's *1984*) and discovers that all three draw an invidious comparison between the coming totalitarian state and wilderness. Only in wilderness, these authors claim, is there the possibility of freedom and escape from the total control and mind conditioning of the totalitarian state. The argument is that wilderness is necessary for humans as a standard for freedom and autonomous behavior, and as a refuge from totalitarianism. This argument also relates to the concerns expressed by some ecophilosophers that, by cutting ourselves off from intimate contact with wild free Nature, we humans are, in fact, domesticating ourselves.[22]

Aldous Huxley carried this a step further by claiming, in 1958, that human overpopulation was the chief factor driving the world toward totalitarianism. He was no doubt influenced in this judgement by his brother, Sir Julian Huxley, who, as director general of UNESCO, was one of the first to warn of the overpopulation/resource crisis in 1948. By the early 1960s, Aldous Huxley had become an advocate of an ecological world view.[23]

Holmes Rolston has written a number of insightful papers which sort out the various kinds of arguments for, and values in, wilderness and wild species. He discusses many of the arguments mentioned above in value terms: (1) the economic value; (2) the scientific value; (3) the genetic diversity value; (4) the aesthetic value; (5) the recreational value; (6) the religious value; (7) the cultural symbolization value; and others, while adding (8) the life support value: another instrumental argument.[24]

(10) THE LIFE SUPPORT SYSTEM ARGUMENT. This argument was made famous during the "ecological revolution" of the 1960s and 70s by professional ecologists from Rachel Carson to Barry Commoner and Paul Ehrlich and led most people to begin considering themselves "environmentalists" and members of an "endangered species." The argument is that Nature and natural processes provide the life support system in which humans and all other species are embedded, and these processes are now seriously threatened by excessive human activities. This prudential argument for human survival was able, at least initially, to get the public's attention in a way in which none of the other arguments could. Anne and Paul Ehrlich have elaborated upon this argument in recent years claiming that wild species are like rivets in an airplane. If we continue "popping rivets" (driving more species into extinction), at some point which cannot be precisely determined, there will be structural failure and our life support system (the Earth) will fail. Further, they argue, natural processes perform many services for humans (nutrient recycling, the weather cycles) which cannot be duplicated by human technology and ingenuity.[25]

(11) THE GAIA HYPOTHESIS ARGUMENT. This "argument," that the Earth is itself a living organism, begins to dovetail with the life support system concept which holds that all the world's ecosystems are interrelated in the larger biosphere or ecosphere. It is a short step from Barry Commoner's "first law of ecology," that "everything is connected to everything else," to the realization that this huge biological bundle of interrelationships is alive. The scientific community is now beginning to take the Gaia hypothesis seriously. And the Gaia concept meshes

beautifully with an ecological understanding of the Earth while replacing, as it does, the 17th century Cartesian-Newtonian image of the Earth as a giant machine or clock-work, or more recently a spaceship: human artifacts subject to endless tinkering and improvement for human ends. But as an argument for the protection of wilderness or undeveloped natural areas, it is indeterminate.

For example, James Lovelock, a chemist under contract with NASA and the modern originator of the scientific Gaia hypothesis, had little sympathy with contemporary environmentalism. He held that Gaia has "vital organs" which need to be protected but, overall, the biosphere (Gaia) is very reliant and can withstand considerable abuse and continued development. He claimed that even nuclear war might not seriously damage Gaia. It has been reported that Lovelock's more recent views are closer to those of modern environmentalists. Other supporters of the Gaia hypothesis, inspired by the anthropocentric theology of Teilhard de Chardin, argue that we are the consciousness and intelligence of Gaia, the steering or control mechanism; thus our human destiny is to take charge of Gaia and alter the ecosystems to suit us. This understanding of the role of humans in Gaia differs little from the traditional Western views of human domination and control over the Earth. Thus, the Gaia concept needs to be combined with ecocentrism and ecosophy (ecological wisdom) in order for it to have an ecologically positive influence.[26]

The majority of contemporary ecophilosophers would allow various degrees of weight to the instrumental arguments or reasons for protecting wilderness, wild species, and natural ecosystems, but they would also argue that the *primary* and overriding reason for protecting them is that they have "intrinsic value" or value-in-themselves. As long as one attempts to remain within the terminology of technical modern ethical theory, I prefer to use the terms "inherent worth" or "inherent value" as defined by Paul Taylor and Tom Regan, respectively, in place of "intrinsic value." According to Taylor, something has intrinsic value when *humans* place a value on it for its own sake. By contrast, Regan claims that "the presence of inherent value in a natural object is independent of any awareness, interest, or appreciation of it by a conscious being." This meaning seems more faithful to the spirit of ecocentrism in that it moves beyond the pale of anthropocentrism.[27]

Aside from Thoreau and Muir and other early environmentalists and ecologists, one of the first clear statements that all individuals, species (including humans), and ecosystems have *equal* inherent value (the ecocentric position) was made by Arne Naess when he proposed the principle of *biospherical (ecological) egalitarianism* in 1972 (although Naess would say that "principle" is too strong a term for what he means). His formulation was that all individuals and species have "an equal right to live and blossom." Naess intended the word "right" to be used in a non-technical way.[28]

In his outstanding 1977 critique of Christopher Stone's legalistic defense of the "standing of trees" and of the animal liberationist theory of Peter Singer, John Rodman also made a very forceful statement of the inherent value of the non-human world. In taking a skeptical stance toward the "extension" of humanistic legal and moral theory to the non-human, he suggested that "to affirm that 'natural objects' have 'rights' is symbolically to affirm that *all natural entities (including humans) have intrinsic worth simply by virtue of being, and being what they are.*"[29]

In his 1978 critique of anthropocentric humanism, and of conservation as "resourcism," David Ehrenfeld examined the usual instrumental arguments for protecting species and then argued for what he calls the *Noah Principle*. He bases this upon a statement made in 1958 by one of the founders of ecology, Charles S. Elton: "The first [reason for conservation], which is not usually put first, is really religious. There are some millions of people in the world who think that animals have a right to exist and be left alone, or at any rate that they should not be persecuted or made extinct as species." Ehrenfeld interprets this to mean that:

> [Communities and species] *should be conserved because they exist and because this existence is itself but the present expression of a continuing historical process of immense antiquity and majesty.* Long standing existence in Nature is deemed to carry with it the unimpeachable right to continued existence.... This is, as mentioned, an ancient way of evaluating "conservability," and by rights ought to be named the "Noah Principle" after the person who was one of the first to put it into practice. For those who reject the humanistic basis of modern life, there is

simply no way to tell whether one arbitrarily chosen part
of Nature has more "value" than another part, so like Noah
we do not bother to make the effort.

Citing Ehrenfeld's Noah Principle, Anne and Paul Ehrlich agree that all
individuals and species have a right to exist which they claim is "the first
and foremost argument for the preservation of all nonhuman species."[30]

Contemporary Christian ecotheologians have also claimed that
wilderness and wild species have inherent worth because they are God's
creation. J. Baird Callicott points out that David Ehrenfeld's Noah
Principle is just such a defense based upon orthodox Judaism. Similarly,
John Muir frequently claimed, in a spirit reminiscent of St. Francis, that
all species have a "right to exist" because they are "God's creatures" and
part of "God's family," although modern scholarship tends to interpret
Muir's mature religious views as pantheist and even Taoist, rather than
theist.[31]

Two leading Christian thinkers who are developing ecocentric
positions are the Protestant theologian John B. Cobb, Jr. and the Catholic theologian Thomas Berry. Cobb's theological position is based
upon Whiteheadian process philosophy, combined with a sophisticated
knowledge of ecological principles. Cobb was also one of the first
theorists to analyze the anthropocentric development of Western academic philosophy. Cobb has however recently misrepresented deep
ecology philosophy by claiming that, for deep ecology, value resides in
the whole ecosystem but not in individual members of the whole. On the
contrary, deep ecology values both equally.[32]

Thomas Berry has reinterpreted the Christian evolutionary cosmology of Teilhard de Chardin along ecocentric lines. In his powerful essay
"The Viable Human," Berry claims that the present destructive non-
sustainable situation has arisen on Earth because,

> at the present time, the human community has such an
> exaggerated, even pathological, fixation on its own com-
> fort and convenience that it is willing to exhaust any and
> all of the earth's resources to satisfy its own cravings. The
> sense of reality and of value is strictly directed toward the
> indulgences of a consumer economy.... The naive as-
> sumption that the natural world exists solely to be pos-

sessed and used by humans for their unlimited advantage cannot be accepted. The earth belongs to itself and to all the component members of the community. The entire earth is a gorgeous celebration of existence in all its forms. Each living thing participates in the celebration as the proper fulfillment of its powers of expression.

While Aldo Leopold referred to humans as just "plain members" of the ecological community (to counteract the extreme anthropocentrism of the dominant Christian tradition), Berry claims that humans are a unique and special species as a result of our self-reflexive consciousness. We are the only species that can understand the overall outlines of the cosmological and biological evolutionary processes; our theoretical science can be used to understand and appreciate the world and to produce ecologically benign technology. Human self-reflexivity can result in a more objective cosmic and ecological perspective. And it can make us aware of the ecological crisis, so that we, as a species, can correct ourselves. Berry and the physicist Brian Swimme have proposed a new cosmological/evolutionary/ecological creation myth for modern societies.[33] Arne Naess also agrees that the ability to understand and identify with Life on Earth and with the cosmos makes humans "very special beings!"[34]

These views provide powerful alternatives (a) to recent Christian "stewardship" models which still portray humans as dominant over the Earth and its other creatures, and (b) to the misanthropy which some ecologists and environmentalists feel in moments of despair over thoughtless and greedy human destruction of the Earth. Further, they provide a long-range evolutionary perspective, similar to the perspective of Loren Eiseley, in which to envision humans as special *guests* in God's creation. Given this perspective, it is appropriate for humans to learn to behave more as humble and well-mannered guests, instead of as arrogant and self-centered conquerors, by "living lightly" and creatively on the Earth, and in harmony with God's other creatures: our co-guests.

In summary, some ecologists and ecophilosophers hold that the narrowly anthropocentric utilitarian arguments are sufficient to justify the protection of wilderness and the present proliferation of wild species on Earth. However, most hold that wilderness and wild species have intrinsic value or inherent worth equal to that of humans (or to put it in a less misleading fashion, that all individuals and species have an equal "right" to exist on the planet), and that this is the primary justification for their protection. Seen at another level, however, the anthropocentric/ecocentric distinction begins to break down. As Arne Naess points out, "The value of opposing two general attitudes within the ecological movement, the anthropocentric and the biocentric [ecocentric] is limited, as is every contrast."[35] The "resource extraction" (genetic bank) kinds of arguments are somewhat problematic especially when the function of the arguments is purportedly to *protect* wilderness and wild species from exploitation. The laboratory (scientific research) argument also has problems. Scientific research can be conducted from an attitude of genuine "respect for nature," or the scientific research argument can also be used to justify self-aggrandizing scientific research projects which often serve no positive function and result in the callous exploitation of plant and animal communities.[36] As we are beginning to realize, the life support system and Gaia arguments, dealing with the habitability of the Earth for all species, are of crucial importance. They call attention to the all-important ecological insight of the interrelatedness and interdependence of all Life on Earth at a time when biological systems are beginning to break down in dramatic ways.

Similarly, the arguments based upon individual and social/political freedom, human maturity and potential, and the freedom and autonomy of other species, begin to coalesce with spiritual, ethical, and religious arguments and considerations. At this point, these "deeper" human-oriented concerns shade off into a concern for the equal inherent worth of wilderness and other species. Humans and the rest of Nature are *truly and deeply* interconnected and interrelated in terms of their mutual long-term interests and welfare.

William Godfrey-Smith makes an intriguing comment concerning the question, "What is the *use* of wilderness?":

> But note how the very posing of this question about the *utility* of wilderness reflects an anthropocentric system of values. From a genuinely ecocentric point of view the question "What is the *use* of wilderness?" would be as absurd as the question "What is the *use* of happiness?"[37]

Viewed from a deep ecocentric/ecological self level, the flourishing of wild species in wilderness and the best interests of humans coincide. Wilderness and wild species have to be preserved not only for planetary health and the inherent worth of biodiversity, but equally as well for who we are as human beings.

## III. The Implications of Conservation Biology for Gaia's Evolutionary Processes

During the 1970s and 80s ecophilosophers have concentrated their efforts on issues such as (a) whether modern humanistic ethical theory can be "extended" to cover the concerns raised by ecology, or whether a "new" environmental ethic will be required; (b) whether or not nonhuman individuals, species, and ecosystems have inherent worth and, if so, how much inherent worth; (c) the cataloging and evaluation of the various arguments for the protection of wild species and wilderness; and (d) whether the existing anthropocentric technological/industrial society can be "reformed" in properly ecological ways, or whether ecological realities will require a new "post-modern" society based on an ecological metaphysics and world view.

These issues, in various forms, have a history which can be traced back to St. Francis, Spinoza, and the Romantic movement. And they have received intermittent but more or less continuous attention since the 19th century writings of George Perkins Marsh, John Stuart Mill, Henry David Thoreau and John Muir.[38] I would claim that an overall broad consensus has now emerged on these issues among most ecophilosophers and professional ecologists. That is, modern moral theory *cannot* be extended to cover adequately ecological situations; non-human individuals, species, and ecosystems have *equal inherent value or worth* along with humans; and a new post-modern non-consumerist ecological society *is* required based upon an ecocentric world-view. I submit that it is now time for a sizeable number of ecophilosophers to begin to devote their attention to a new set of more pressing practical issues and problems: helping to devise, to critically evaluate, and to advocate various strategies and plans for *protecting* wilderness, wild species, and humans, and for easing the transition to an ecocentric world view and society. While we ecophilosophers continue to debate the values of wilderness, wild species, and an ecological world, the possibilities for a rich diverse world will vanish irretrievably within the next ten to twenty years unless effective action is taken *now!*

Arne Naess has already recognized multiple functions and activities as existing with the "long-range deep ecological movement." There is a philosophical component (both in terms of pure theory such as Ecosophy T, and the "intellectual activism" involved in proposing and evaluating strategies), together with the more grass-roots activism involved with life-style, bioregional living, and non-violent "direct action" to resist further environmental destruction.

In the 1960s, professional biologists and ecologists, beginning with Rachel Carson, and including Garret Hardin, Lamont Cole, Raymond Dasmann, Barry Commoner, and Paul Ehrlich, stepped outside their narrow areas of professional scientific expertise and began to warn the public of the impending ecological disaster. They also proposed various public strategies to cope with these problems. The "intellectual activism" begun by these ecologists has now been institutionalized into a new branch of the science of ecology called "conservation biology." Mitch Friedman says that:

> Conservation biology considers the application of eco-
> logical theory and knowledge to conservation efforts. The
> development and utilization of this new discipline is a
> welcome advance in conservation, where ecological con-
> siderations tend to be overshadowed by political and
> economic forces, in part due to poor understanding of the
> effects of land-use decisions.[39]

Conservation biology has been spearheaded largely by the ecologist Michael Soulé, a former student of Paul Ehrlich, who has recently worked closely with Arne Naess. Soulé refers to conservation biology as a "crisis discipline" which has to apply its findings in the absence of certainty. This new field integrates ethical norms with the latest findings of scientific ecology.[40]

Soulé has provided scientific definitions for the terms "conservation" and "preservation." In his usage, "preservation" means "the maintenance of individuals or groups, but not for their evolutionary change." He proposes that "conservation" be taken to denote "policies and programs for the long-term retention of natural communities under conditions which provide for the potential for continuing evolution." Mitch Friedman carries this a step further by introducing the concept of "ecosystem conservation" which "involves the preservation of ecosystem wilderness: enough of the land area and functional components—the creatures and their habitat—to insure the continuation of processes which have co-evolved over immeasurable time."[41]

One can, of course, quibble over the choice of terms. "Conservation" has negative associations with Pinchot and the resource conservation and development approach. But "preservation" does not necessarily mean maintaining something in a static state such as "preserving jam" or "deep freezing" a wilderness. Perhaps "protection" would be a more neutral term. But the phrase "ecosystem conservation" means protecting the ongoing dynamic continuum of evolutionary processes which is the ecological planetary condition to be desired.

Friedman points out that Soulé recognizes "three objectives in the establishment of nature reserves. The protection of endangered, endemic, and other species of concern (this implies the protection of habitat for these species); the conservation of functioning communities; and the maintenance of biotic diversity, or the maximum number of species."[42] It is thus clear that the *primary goal* in setting aside and evaluating wilderness areas (nature reserves), from the standpoint of conservation biology and the present ecological crisis, is the conservation and protection of wild plants and animals (biotic diversity) and the possibilities for the continuation of evolutionary processes.

Based on these objectives, we need to look at present ecological realities. As Friedman points out:

> An element of panic is present with the literature of conservation biology, as well as among the conservation community at large. This panic originates predominantly from the present rate of species extinction, and the forecasts for impending mass extinction. We presently have scarcely a clue of even the total number of species on the planet, with estimates ranging between three and thirty-seven million. Yet, some researchers are predicting that

anthropogenic extinctions, at current rates (which do not
consider military disasters or other unpredictable events),
may eliminate as many as a third of the planet's species
over the next several decades (Meyers 1987). This is
shocking to anyone who treasures the intrinsic values of
Earth's natural diversity and fecundity, or who fears for
the fate of humanity and the planet as a whole. While most
of these extinctions are occurring as a result of tropical
rainforest deforestation, the same processes are occurring
in temperate areas, including the United States (Wilcove
et. al. 1986).[43]

Skipping over important discussions of island biogeography and
ecological concepts such as Minimal Viable Populations (MVP), we
come to the nub of the problem for ecosystem conservation:

It is not enough to preserve some habitat for each species
if we want to conserve ecosystems; the habitat must
remain in the conditions under which the resident species
evolved. For this reason, national forests, under present
"multiple use" management, may not be effective nature
reserves for many species.

Historically, national parks and other reserves have
been established according to political, or otherwise
nonbiological considerations.... To conserve species diversity, the legal boundaries of nature reserves should be
congruent with natural criteria (Newmark 1985). For
instance, a reserve may be large [e.g., Everglades National
Park] while still not protecting the ecological integrity of
the area.

Newmark (1985) suggests that reserves contain not
only entire watersheds, but at least the minimum area
necessary to maintain viable populations of those species
which have the largest home ranges. Others have stated
that complete, intact ecosystems should be preserved.
(Terbourgh and Winter 1980; Noss 1985).[44]

In the 1970s Michael Soulé examined 20 wildlife reserves in East
Africa, including the massive Tsavo and Serengeti National Parks. He
and his fellow researchers projected that:

all of the reserves will suffer extinctions in the near future. Their study predicts that a typical reserve, if it becomes a habitat island, will lose almost half of its large mammal species over the next 500 years.... When a habitat island, for instance a national park surrounded by national forest, is reduced in size (i.e., clear-cutting along the park boundaries), the number of species in that island will decrease. The empirical evidence for the relaxation effect is alarming, and reflects the urgency with which we must re-evaluate our conservation strategies and remedy the situation.[45]

As Edward Grumbine points out:

Newmark (1985) investigated eight parks and park assemblages and found that even the largest reserve was six times too small to support minimum viable populations of species such as grizzly bear, mountain lion, black bear, wolverine, and gray wolf. A recent study by Salwasser et. al. (1987) looked beyond park boundaries and included adjacent public lands as part of conservation networks. The results were the same. Only the largest area (81,000 square km) was sufficient to protect large vertebrate species over the long term.... Virtually every study of this type has reached similar conclusions: No park in the coterminous U.S. is capable of supporting minimum viable populations of large mammals over the long term. And the situation is worsening.[46]

Frankel and Soulé claim that "an area on the order of 600,000 square km. (approximately equal to all of Washington and Oregon) is necessary for speciation of birds and large mammals."[47]

The inescapable conclusion is this: for those ecophilosophers, ecologists, environmentalists, and other people around the world committed to wilderness, wild species protection, and the protection (conservation) of the ongoing evolutionary processes of Nature, there must be a recognition that current global wilderness/nature reserve protection policies are failing miserably. Past practices and policies have been based on inadequate ecological understanding. Humans have effectively clogged the evolutionary arteries of Mother Gaia. Policies and strategies must be devised to halt further ecological destruction and to bring about a realistic balance between humans and the non-human sphere. Along with protecting the ozone layer, minimizing the severity of the greenhouse effect, and stabilizing human population growth, the most crucial ecological task facing humanity at this time is to devise realistic nature reserve protection strategies, begin their implementation within the next five to ten years, and bring about an appropriate reorganization of human society consistent with these strategies. Ecophilosophers can play an important role in this process.

Narrowly "rational" scientistic approaches must give way to the wider approach which Naess calls "ecosophy" (ecological wisdom). For, as Paul Shepard has claimed: "there is an ecological instinct which probes deeper and more comprehensively than science, and which anticipates every scientific confirmation of the natural history of man." The overwhelming dimensions of the approaching human overpopulation/ecological/environmental crisis were seen intuitively and clearly by some of the ecologists and radical environmentalists of the 1940s, 50s, and 60s. They tried to sound the alarm at a time when their more conservative colleagues, and most of the public, were thoroughly immersed in the narrowly human-centered industrial/consumerist vision of reality. These visionaries include Aldo Leopold, William Vogt, Fairfield Osborn, Sir Julian Huxley, Aldous Huxley, Robinson Jeffers, Raymond F. Dasmann, Paul Ehrlich, Dave Brower, Loren Eiseley, Paul Shepard, Edward Abbey, and Gary Snyder, among others.[48]

For example, at the 1959 Sierra Club Wilderness Conference, ecologist Raymond Cowles warned that the exploding human populations of Africa would eventually overrun the nature preserves and wildlife of Africa unless a new direction was taken. In the 1960s, Dave Brower was saying that "I believe in wilderness for itself alone. I believe in the rights of creatures other than man." Estimating that roughly ten percent of the Earth was still wilderness at that time, he claimed that it should be protected: "We should touch nothing more in the lower forty-eight...whether it's an island, a river, a mountain wilderness—nothing more. What has been left alone until now should be left alone permanently." And we should go back over the ninety percent of the Earth humans have claimed as theirs "and do better, with ingenuity. Recycle things. Loop the system." In 1969, Gary Snyder outlined a post-industrial ecological vision in which "the unknown evolutionary destinies of other life forms are to be respected." He proposed a society in which a much smaller human population would live "harmoniously and dynamically by employing a sophisticated and unobtrusive technology" in a world environment which is "left natural."[49]

The scientific confirmation of their ecological intuitions and proposals came in the early 1970s with the Club of Rome reports based upon the computer studies of Jay Forrester, Donella and Dennis Meadows, and now the ecological findings of the conservation biologists.[50]

## IV. Approaches to Global Ecosystem Protection Zoning

It was David Brower who, in 1967, made perhaps the first world-wide zoning proposal to protect wilderness and wild species. Claiming that less than ten percent of the Earth had, at that time, escaped the technological exploitation of humans, he proposed protecting the remaining wilderness and "granting other life forms the right to coexist" in what Jerry Mander called an Earth International Park. Brower's increasingly radical ideas were a factor in his forced ouster from the Sierra Club, at which point he formed the more ecologically radical Friends of the Earth in 1969.[51]

Another major zoning proposal came from the ecologist Eugene Odum in 1971. The ecophilosopher John Phillips developed an argument (which has an uncanny resemblance to Naess' Ecosophy T) to support and expand upon Odum's proposal:

> Being is good. Increase of diversity is increase of being. Integration of diversities into systems is increase of being. Therefore diversity is good and integration of diversities is good. What is good should be protected. Therefore, Protect Diversity! Protect integrated systems of diversity!

The Odum/Phillips zoning policy recommendation is that:

*The biosphere as a whole should be zoned,* in order to
protect it from the human impact. We must strictly confine
the Urban-Industrial Zone and the Production Zone (agriculture, grazing, fishing), enlarge the Compromise zone,
and drastically expand the *Protection* Zone, i.e., wilderness, wild rivers. Great expanses of seacoasts and estuaries must be included in the Protection Zone, along with
forests and prairies and various habitat types. We must
learn that the multiple-use compromise Zone is no substitute, with its mining, lumbering, grazing, and recreation in
the national forests, for the scientific, aesthetic, and genetic pool values of the Protection Zone. Such zoning, if
carried out in time, may be the only way to limit the
destructive impact of our technological-industrial-agri-
business complex upon the earth.

Phillips concluded by saying that "to go so far as to zone the biosphere
and set aside an adequate Protection Zone would be a supreme act of
rationality by which the rational animal could protect the rest of life on
earth, and himself, from his own irrational temptations."[52]

The Odum/Phillips zoning proposal is basically sound in calling for
greatly expanded Protection Zones, and it is consistent with the Brower/
Mander concept of an Earth International Park. But it needs further
refinement.

Arne Naess distinguishes between wilderness protection zones or
parks (where people do not live and resource extraction is prohibited)
and *free nature*. Examples of "free nature" would be areas of relatively
sparse human inhabitation (such as the foothills of the Sierra, parts of
northern Europe, and much of the Third World) where the natural
processes are essentially intact. These areas should be zoned to protect
the natural processes and wildlife while encouraging non-exploitive
bioregional living.

Paul Shepard made a daring proposal for global ecosystem protection zoning in 1973. In order to allow for the huge expanses of
unmanaged wilderness needed "for ecological and evolutionary systems on a scale essential to their own requirements," he proposed
allowing the interiors of continents and islands to return to the wild.
Based on the assumption that human population would stabilize by the

year 2020 at eight billion, humans would live in cities strung in narrow ribbons along the edges of the continents. Hunting-gathering forays would be allowed into the wilderness, but there would be no permanent habitation.[53]

Based on his ecocentric orientation, Shepard, like Brower, foresaw the huge amounts of wilderness which would be required for the healthy ecological and evolutionary functioning of Gaia. But this proposal has a number of practical problems which include (1) the tremendous political/legal/economic issues, not to mention the actual physical task, involved in relocating humans to the edges of the continents and (2) the pressures which would be exerted by these concentrated human populations on the ocean shoreline ecosystems and estuaries. At this stage of history, it is probably more realistic to expand wilderness protection zones with the basically existing patterns of human settlement in mind.

Gary Snyder has traced the etymology of "wild" and "wilderness" to the concepts of free and autonomous; to the Tao (the "way of Nature," spontaneous and "generating its own rules from within"). He points out that pre-Columbian American wilderness was "all populated" with humans. In fact, "there has been no wilderness without some kind of human presence for several hundred thousand years."[54] It is important to provide a special kind of protection/wilderness/free nature zoning for those remaining tribal peoples, such as the Bushmen of the Kalahari desert and the tribes at the headwaters of the Amazon, who are still living in essentially traditional ways with minimal impact on wild ecosystems and the wildlife.

One of the central features of thinking about the "new ecological society" is the move toward decentralization and "bioregional" ways of life which involves reinhabiting and restoring damaged ecosystems.[55] But Roderick Nash, a major proponent of wilderness protection, has worried that a total movement toward bioregional reinhabitation of the Earth at this point (what he calls the "garden scenario") would be ecologically disastrous. He says: "The problem, of course, is numbers. There are simply too many people on the planet to decentralize into garden environments and still have significant amounts of wilderness."[56] Elsewhere, Nash characterizes "bioregionalism" as "the contemporary attempt to 'reinhabit' wilderness areas."[57]

Nash is entirely justified in calling attention to the limitations of an overly ambitious bioregional program at this point in history. It is not clear, however, that the intent of contemporary bioregionalists is to reinhabit wilderness areas. Certainly, leading bioregional theorists such as Gary Snyder, Raymond Dasmann, Thomas Berry, and Kirkpatrick Sale, are fully aware of the importance of establishing greatly expanded wilderness/protection zones. Bioregional ways of life are appropriate and necessary for areas zoned as "free nature" and for ecologically restructured cities suggested in such projects as Peter Berg's Green Cities.[58] Ecological cities should have wild and semi-wild areas interspersed with human inhabited areas, either by protecting and expanding upon wild areas that now exist, or by restoring such areas. Humans could continue to visit some wilderness/protection zones in limited numbers, as they do now, provided they follow minimum impact ways and do not disturb the ecosystems and wild species.

Other ecophilosophers have taken philosophical stands on, and discussed strategies for, protecting wilderness and wild species. As Roderick Nash points out, "In 1982 [Edward] Abbey expressed his basic belief that humans had no right to use more than a portion of the planet and that they had already passed that limit. Wild places must be left wild."[59] And Paul Taylor has promoted the idea of extensive wilderness protection as species habitat. He claims that we must

> constantly place constraints on ourselves so as to cause the least possible interference in natural ecosystems and their biota.... If [humans] have a sufficient concern for the natural world, they can control their own population growth, change their habits of consumption, and regulate their technology so as to save at least part of the Earth's surface as habitat for wild animals and plants.[60]

Taylor also finds it necessary to make a distinction between "basic" and "non-basic" interests of humans. In order to allow for sufficient amounts of species habitat humans need to curb their population growth and reduce their non-basic wants and consumption habits when these come in conflict with the basic needs of other species for survival and well-being. Here, Taylor's analysis coincides with the function of Naess' distinction between vital and non-vital needs which is incorporated into the Deep Ecology Platform.[61]

Taylor also makes another important contribution with his discussion of the *bioculture*. He defines "bioculture" as "that aspect of any human culture in which humans create and regulate the environment of living things and systematically exploit them for human benefit."[62] Agriculture, pets, domestic animal and plant breeding, and "tree plantations" all belong to the human bioculture. Establishing wilderness protection zones would, in effect, separate the world of the wild from the exploitive human activities of the bioculture. "Free nature" would be a sort of hybrid buffer zone between protection zones and the bioculture with ecological processes predominating.

Many groups which consider themselves ecological are, in effect, primarily involved with an ecological "reform" of the bioculture. The organic farm movement, inspired by Wendell Berry and Wes Jackson, is an example of this. The concern of the animal rights movement with the "rights" of all animals, often fails to distinguish the conditions of domestic animals in the bioculture from the conditions of wild animals in wilderness with sometimes alarming and anti-ecological results. The goals of the Forest Service, and similar efforts world-wide, to clear-cut natural forest ecosystems and replace them with "tree plantations," can now be seen as an attempt to continually extend the bioculture at the expense of the wild. As Taylor points out, the "ethics" of the bioculture differs from the basically "non-interference" ethics of the wilderness. Perhaps some ecologically enlightened version of the "stewardship" model is appropriate for the bioculture. Other problems arise when wild animals stray from the protection zones into the biocultural zones, and when there are "mixed communities" of wild and domestic as in "free nature."[63] It is important for those primarily concerned with biocultural "ecological" reform to expand their outlook to encompass a genuinely ecocentric perspective to insure that their policies and programs are compatible with, and supportive of, the overall long-range ecological health of Gaia.

Arne Naess has already begun the critical ecophilosophical task of providing analyses of policies and proposals designed to protect wilderness and evolutionary processes, and to bring about a "sustainable" society. One such major proposal is the "World Conservation Strategy" developed in 1980 and backed by the United Nations Environmental Program. Naess finds much to agree with in this proposal, but claims that it is narrow in perspective in that it lacks an ecocentric perspective. He has also examined the Brundtland report and finds that it lacks an adequate emphasis upon human population stabilization and reduction.[64]

Two other strategies for protecting natural ecosystems and wildlife are the Biosphere Reserve concept (part of UNESCO's Man and the Biosphere Program) and the World Heritage Site system. According to Edward Grumbine,

> A model biosphere reserve consists of four integrated zones: a large protected core; a buffer zone; a restoration zone; and a stable cultural area where "indigenous people live in harmony with the environment".... The National Park Service has informally adopted the biosphere reserve model as a guide to regional land planning [and] after eighteen years, 41 biosphere reserves exist in the U.S. many of which occupy both national park and forest lands.[65]

Grumbine sees some possibilities with World Heritage Site designations, but claims that there are serious problems with the Biosphere Reserve concept; the zones are not properly interrelated and the "self-sustaining" core is not large enough to allow for speciation. He suggests that the biosphere reserve model be replaced by a national system of biological reserves. This needs to be supplemented by a major program of *ecological restoration*:

> Restoration of damaged lands must be married with the goal of native diversity. This follows the *wilderness recovery* strategy of Noss (1986a) and would include large scale restoration of natural fire cycles, recovery of threatened, endangered, and extirpated species, road closures and reforestation projects, stream rehabilitation to increase native anadromous fisheries, and much more (see Berger 1985). Once an area was restored, nature would take its course with minimal interference from managers.

> The amount of work to be done would likely offset the loss
> of jobs in exploitive industries.[66]

The concept of ecological restoration is a crucial one for all of the zones, but some have tried to use it as an excuse for "mitigation" procedures: in this case claiming that we can continue to develop (i.e. destroy) natural ecosystems/wildlife habitat and then compensate these losses by "restoring" an equivalent area elsewhere. This is a short-sighted foolhardy approach which I see as part of the overall "Disneyland syndrome" in that it neglects the difficult and expensive process of restoration together with the very real likelihood that restoration projects may only be partially successful and will not duplicate the genuine article. The "mitigation" of wildlife habitat painfully recalls the similar process of Native American "resettlement." Europeans dispossessed native peoples of their best tribal lands and moved them to the marginal edges (only to find, to their chagrin, that these "useless lands" were sitting on huge deposits of coal, oil, and uranium!). Similarly, proposals are under consideration to drill for oil and gas being discovered under existing wildlife refuges and sanctuaries!

The question still remains: How much of the Earth should be protected in wilderness/protection zones? The main answer to this question has already been given by conservation biologists based upon the most recent ecological research: enough wildlife habitat to protect species diversity and the ecological health of Gaia and to allow for continued speciation and evolutionary change. Arne Naess has provided a future ecological vision towards which we can progress:

> I am not saying that we should have preserved the primor-
> dial forest as a whole, but looking back we can imagine a
> development such that, let us say, one third was preserved
> as wilderness, one third as free Nature with mixed com-
> munities, which leaves one third for cities, paved roads,
> etc. This would probably be enough, and I guess most
> people with influence in matters of the environment would
> agree. But of course, it is a wild fantasy, which is, inciden-
> tally, an important kind of wilderness![67]

Holmes Rolston recently claimed that a shift of focus should be made from individuals or species to ecosystems, and proposes an Endangered Ecosystems Act to accompany the Endangered Species Act.[68] The need for legal protection of entire ecosystems is certainly felt when environmentalists have to argue for the protection of the endangered spotted owl in order to protect the old growth forest ecosystems which comprise their habitat. But while that would be an important step, in terms of the present state of global ecological destruction, we are actually way beyond that point. The entire Earth or Gaia is endangered and needs to be protected as a whole through the process of immediate global protection zoning. To wait until it can be established that some specific component (this or that species or ecosystem) is in danger is ultimately to miss this point! This is not to say, however, that determining which species are or ecosystem is near extinction should not be used in setting priorities for and choosing strategies of protection efforts.

It should be clear that the first step in protecting Gaia is to halt any further development and destruction of wild species habitat. We have seen that ecologists and environmentalists, from Dave Brower to Paul Shepard, Gary Snyder, Eugene Odum, and Ed Abbey, have been proposing such a step since the 1960s. Paul Ehrlich once again reiterated this proposal in 1985 when he said that "in a country like the United States, there is not the slightest excuse for developing one more square inch of undisturbed land." In their 1987 summary of world environmental problems and proposed solutions, the Ehrlichs generalized their stand to the whole Earth: "The prime step [is] to *permit no development of any more virgin lands*.... Whatever remaining relatively undisturbed land exists that supports a biotic community of any significance should be set aside and fiercely defended against encroachment."[69]

As Thomas Fleischner points out, "Over ninety-five percent of the contiguous United States has been altered from its original wilderness state. Only two percent is legally protected from exploitive uses."[70] And even that two percent lacks adequate protection. Forest Service "designated wilderness" actually allows exploitive mining, sport hunting, and domestic animal grazing. Legislative efforts are now being made to revise the existing mining laws which have been the cause of much public land abuse. But some have claimed that, apart from forest destruction, the greatest cause of ecological destruction on public lands (both wilderness and non-wilderness lands) has been cattle and sheep grazing. Domestic animal grazing on public lands destroys the natural plant and grass communities, damages streams and other water supplies, competes with wildlife, encourages predator elimination, and should, therefore, be phased out.[71]

Once the ecologically destructive uses of now-existing Forest Service wilderness have been eliminated, the additional three percent of *de facto* wilderness should be placed in protection zones which would bring the contiguous United States total to five percent protected habitat. That would still leave the contiguous United States about thirty percent short of a 1/3 wilderness—1/3 free nature—1/3 bioculture ratio (disregarding for the present the zoning of "free nature"). Under the provisions of the Wilderness Act of 1964, the battles over the classification of wilderness lands in the National Parks and Forests have already been fought and environmentalists have had to compromise severely in both cases, particularly the latter. Now the battle to zone lands as wilderness is occurring over the 250 million acres administered by the Bureau of Land Management (the BLM). The BLM is studying only ten percent of its land (twenty-five million acres) for possible wilderness designation, and the likelihood is that, after political wrangling and compromise, only ten to fifteen million acres will be protected. This decision is to be made in 1991.

The Wilderness Act of 1964 is essentially a pre-ecological document and, accordingly, its provisions and stated purposes do not reflect the huge tracts of wilderness (protection zones) required for species and ecosystem protection, large mammal speciation, and planetary health. A recent newsmagazine article discussing the Wilderness Act and the upcoming BLM wilderness fight still discusses the issues largely in terms of anthropocentric special interest compromise politics: of wilderness recreation versus "motorized-recreation and commercial interests." The ecological issues are all but ignored. In order to boost the protection zone percentages toward the thirty percent figure, it would probably be necessary to place almost all Forest Service and BLM lands in protection zones and restore them to wildlife habitat. The recent proposals by Deborah and Frank Popper of Rutgers University to return the Great

Plains to buffalo habitat would also greatly increase ecosystem protection areas. Earth First! has been carefully studying the American land situation for some time and has developed ecologically realistic detailed plans for greatly increasing wilderness areas in the United States.[72]

This brings us to the issue of global human overpopulation. The pressure of the existing five to six billion people on Earth, magnified by the incredible levels of consumption, industrialization, and toxic waste and other pollution occurring in the industrial countries, is already exerting intolerable pressure on the global biosphere and wildlife habitat. As Edward Grumbine points out, the success of ecosystem conservation requires "decreases in human population and industrial activity."[73] Many professional ecologists and environmentalists hold that a maximum viable global human population living comfortably at a "basic needs" consumption level (and allowing for the evolutionary and ecological requirements of the planet) would be no more than one to two billion (certainly much less than the current number), with the United States at about 100 million.[74]

While demographers and population biologists had recently predicted and hoped that the global human population would level off at about ten billion by the end of the next century, the latest United Nations projections are that, at current rates of increase, the population will soar to fourteen billion before leveling off at that time. If this figure is reached, in all likelihood it will prove to be a total disaster for Gaia and its wildlife, while also resulting in unimaginably degraded living conditions for humans globally. This prospect, together with the present and projected rates of species extinction over the next ten to fifteen years underscores the necessity for establishing global protection zones immediately and for mounting an all-out world-wide campaign to bring population growth to a halt by humane means as quickly as possible. As someone once said, "Trend is not destiny!"

Aside from Third World countries, with their high rates of population increase together with its negative effects on natural ecosystems and wildlife, there are special areas of environmental concern within the industrial world. Raymond Dasmann once made a distinction between "ecosystem people" and "biosphere people":

> Ecosystem people live within a single ecosystem, or at
> most two or three adjacent and closely related ecosystems.
> They are dependent upon that ecosystem for their sur-
> vival....Biosphere people draw their support, not from the
> resources of any one ecosystem, but from the biosphere....
> Biosphere people can exert incredible pressure upon an
> ecosystem they wish to exploit, and create great devasta-
> tion—something that would be impossible or unthinkable
> for people who were dependent upon that particular eco-
> system....I propose that the future belongs to...[ecosystem
> people].[75]

Japan, with its total dedication to industrialization and international markets, large industrial population and consumption patterns, current buying power, lack of natural resources, and import policies, has become the most obvious example of a "biosphere people," currently surpassing even the United States and Europe as the world's leading destroyer of global natural ecosystems. It is exploiting the last of the old-growth forests from Brazil and Peru, to the United States (and Alaska), and southeast Asia, in addition to its immense negative impact on the oceans.

In a series of public lectures in Japan in 1989, Arne Naess pointed to the current total indifference of the Japanese government and people towards the world-wide concern for the environment. Japan's direction for the last thirty years has been a "blind, ruthless, increasingly destructive policy of economic growth at any cost." And citing the World Conservation Strategy, the United Nations Charter for Nature, the Brundtland Report, and Worldwatch's State of the World reports, Naess challenged Japan to become a leader of world conservation strategy.[76]

Another special area of environmental concern is my home state of California. As the sixth largest economic power in the world, and with heady images of "Pacific Rim" international trade dominating current economic talk, California is, in many ways, trying to emulate Japan. And, like Japan, California is headed for a disastrous social/environmental future. For example, the present population growth rate is out-of-control with 600,000 people moving to California each year (the main factor preventing California, and the rest of the United States, from stabilizing its population is immigration: both legal and illegal). Some California cities and counties, especially in the central and northern foothill areas, are growing at rates in excess of the fastest growing countries in Africa. The great agricultural areas of the Central Valley have been drenching the soil with pesticides and chemical fertilizer for over forty years and groundwater supplies are now condemned as contaminated. Further, as one of the most diverse biotic areas in the world, California is now experiencing a wildlife crisis as a result of habitat loss from increasing development, commercial poaching, and agricultural waste water and selenium poisoning of waterfowl sanctuaries. Cities cannot meet federal standards for clean air, and this air pollution is contributing to the death of forests near Los Angeles and along the west slope of the Sierra.

A recent poll showed that Californians now realize that the quality of life is declining in the state: gridlock is common in the cities, there is overall overcrowding, and the amenities of life are fast slipping away. And yet the talk is still of adding more freeway lanes and, among the more enlightened, more mass transit together with slow-growth initiatives. What is conspicuously absent is an awareness of the necessity to stop further development and to stabilize and reduce population.

California is a land of strange contradictions, where anthropocentric Disneyland economic growth fantasies and New Age religious cults exist side-by-side with the birthplace of world ecocentric environmentalism (John Muir, David Brower, and the Sierra Club) and, not surprisingly, where the citizens alternate between electing such arch anti-environmental pro-development governors as Ronald Reagan and George Deukmejian, and the environmental visionary, Jerry Brown.

In 1965, the year after California applauded itself for surpassing New York as the most populous state in the Union, ecologist Raymond Dasmann outlined the population/environmental situation for Californians. He proposed that population growth be stopped through a strategy of "not planning for growth."[77] Perhaps the only strategy at this point that will allow California an ecologically viable future would be the passage of a state-wide no-growth initiative together with the implementation of protection zones around all remaining wild and semi-wild areas. Development and population growth must be stopped. Viewed ecologically, immigration only adds to the cumulative impact of existing human overpopulation on the affected ecosystems. As Gary Snyder points out, one of the first principles of bioregional living is to "Quit moving—stay where you are!"[78]

After the promising environmental awakening of the 1960s and 70s, we in the United States have experienced a decade-long environmental hiatus during which a conservative Republican president did everything in his power to obstruct environmental protection: from neutralizing the efforts of the EPA and suppressing acid rain studies, appointing anti-ecological pro-developmental people as heads of the Department of Interior and the Forest Service, promoting the "Sagebrush Rebellion" aimed at selling off public lands in the West to private developers for commercial exploitation, to refusing to allow money to be spent for the acquisition of additional parks and protected wildlife habitat. Further, the Reagan administration shocked the world community in 1985-86 by cutting off United States funding for the United Nations population control programs presumably on the grounds that these organizations provide abortions. The Reagan administration chose to ignore totally the environmentally comprehensive 1980 Global 2000 report to the U.S. President: a report which caused a considerable stir among government heads in other parts of the world.[79]

## V. The Need for Ecologically Structured Governments

The 1990s hold every promise of being like the 1960s in at least one major respect: it will be an era of global environmental reawakening and concern. This is the result largely of such massive biospherical environmental disruption as the greenhouse effect, ozone layer depletion, and tropical rainforest destruction along with the rising rates of species extinctions. Major news media (*Time, National Geographic, U.S. News and World Report*) have already announced this event: *Time* called for a "universal crusade to save the planet."

One main factor which stands in the way of "saving the planet" is a fundamental difference of emphasis among those in the "environmental movement": this is a problem which emerged in the 1960s and has been with us ever since. The environmental decade of the 60s began with

Rachel Carson's warnings about pesticides which negatively affected *both* the health of humans in the bioculture *and* the natural ecosystems and wildlife. She was a trained biologist/ecologist and ecocentrist who cared deeply about the natural world (the oceans and the birds). Her source of inspiration was Albert Schweitzer and his "reverence for life" principle.

But soon a new breed of environmentalist arrived on the scene, exemplified by Barry Commoner, Ralph Nader, and the Environmental Defense Fund. The concerns of these "newer man-centered leaders," according to Stephen Fox, centered on industrial pollution as the essence of the environmental problem. Commoner, who was not trained as an ecologist, became known as the "pollution man," and he soon took a stand against the warnings of Paul Ehrlich and other ecologists on human overpopulation. (As of 1988, Commoner was still denying that there is a global human overpopulation problem.) According to Fox,

> What united all these new crusaders against pollution...was a certain distaste for the traditional conservation movement. It was hard to imagine Nader even stepping outdoors; he once lamented that more people were drawn to birdwatching than to Congress watching. Commoner likewise had little interest in the outdoor pastimes of most conservationists. Disapproving of "a passive admiration for nature," he made his priorities clear: "I happen to think that people are more important than whooping cranes".... The conservation movement, after dealing with pollution for years in its own way, no doubt had mixed feelings about these upstarts.[80]

David Brower worried that, in the public rush to embrace the new narrowly anthropocentric survival environmentalism of the late 1960s, the deeper message of ecological interrelatedness and ecocentric concerns such as wilderness, natural ecosystem and wild species protection, would be lost in the shuffle. And he was right.

The environmental legislation passed in the late 1960s and early 70s in the United States reflected the anthropocentric/biocultural "pollution" orientation of this new and more narrowly focused version of "man-centered conservation." In 1968, Paul Ehrlich had proposed an ecologically oriented governmental Department of Population and Environment. He also called for an "international policy research program [to] set optimum population-environmental goals for the world and to devise methods for reaching these goals" (incidentally, he also mentioned a problem called the "greenhouse effect").[81] While various population control and ecologically oriented environmental programs were established in the United Nations, the United States settled for the Environmental Protection Agency (EPA) and the Clean Air and Water Acts. The EPA essentially reflects the Commoner/Nader "pollution" approach to the environment; it is charged with enforcing the provisions of the Clean Air and Water Acts, and monitoring pesticides, chemicals in our foods, and toxic wastes: the "externalities" of industrialization. The only legislation concerned with wider ecological issues, the Endangered Species Act, came much later (1976) almost as an afterthought and, from an ecological perspective, it is a minimally effective token piece of legislation. Environmental legislation during the 1970s was reformist in nature, and designed essentially to allow industrial growth and development (and the GNP) to continue at accelerating rates, while making minimal concessions to "cleaning up the environment" and protecting ecological integrity.

National opinion polls taken over recent years have consistently shown an unusually high level of support for environmental issues. The degree of this high level support is unprecedented among political issues: most Americans consider themselves to be "environmentalists." But an analysis of these polls reveals that much of this support consists of "pollution" consciousness; concern for ecocentric issues such as human overpopulation, and wilderness and wild species protection, although still relatively high, ranks much lower than concern for air and water pollution, toxic waste disposal, etc. And these issues are often seen, or at least are presented, as separate and isolated problems. A reasonably sophisticated awareness of the ecologically interrelated nature of these issues, and of the overall threat to the continued integrity and viability of the planet, appears not to be very widespread among the general public. Efforts to educate the public, by the news media, to a comprehensive and interrelated ecological understanding of the current environmental crisis is virtually non-existent. Comprehensive ecological/environmental education in the schools, although mandated by law in states such as California, is also pitifully inadequate or non-existent.

After the unprecedented anti-environmentalism of the Reagan administration, George Bush now claims to be an "environmental president." Bush appears to be encouraging a more vigorous approach to some forms of pollution control by the EPA, but he also continues to pursue the Reagan-type exploitation of public lands by appointing pro-development people as heads of the Department of Interior and other public land agencies, and by promoting oil drilling in marine mammal sanctuaries off the California coast and in the Arctic National Wildlife Refuge. Although Bush and his advisers have had access to the environmentally comprehensive annual Worldwatch *State of the World* reports since 1984, his concept of environmentalism has advanced little beyond the "pollution control" level—what the leading environmental textbook writer, G. Tyler Miller, refers to as the first and most simplistic level of environmental awareness.[82] This kind of environmental leadership, either at the national or international levels, is not going to even begin to deal effectively with the current dimensions of the environmental crisis.

The leaders and citizens of industrial countries have yet to face realistically the fundamental incompatibility of continued human expansion and economic growth and development on the planet (the economic/consumerist vision of reality), and the viable ecological functioning of the Earth. They continue to equate economic growth and development with "progress" when, at this point in the history of the Earth, just the opposite is the case. Years ago, Karl Polanyi pointed to the radical nature of the modern consumer society which involves "no less a transformation than that of the natural and human substance of society into commodities."[83]

Thomas Berry claims that:

> We have the ecologist standing against industrial enterprise in defense of a viable mode of human functioning.... This opposition between the industrial entrepreneur and the ecologist has been both the central human issue and the central earth issue of the late 20th century.... The efforts of the entrepreneur to create a wonderworld are, in fact,

> creating a wasteworld, a nonviable environment for the human species.... The ecologist is offering a way of moving toward a new expression of the true wonderworld of nature as the context for a viable human situation.... Ecologists recognize that reducing the planet to a resource base for consumer use in an industrial society is already a spiritual and psychic degradation.[84]

If Americans (and the rest of the world) are to become really serious about dealing effectively with our ecological problems, then there must be a decisive shift from the economic/consumerist vision of reality to a genuinely ecological/ecocentric world view. Gary Snyder has claimed that "economics must be seen as a small sub-branch of ecology."[85] As a way of making the transition to an ecologically sustainable society, Anne and Paul Ehrlich have promoted Herman Daly's steady-state economic theories.[86] In short, the "business" of America (and the rest of the world) must become ecological protection.

The structures of governments must be reorganized from their present narrowly human-centered fragmented pro-developmental orientation to reflect this new emphasis upon ecocentrism and ecological protection. A new overarching agency (possibly called the Ecosystem Protection Agency) needs to be established to coordinate and oversee the efforts of protecting *both* the biocultural zones and the ecological protection zones. This agency should be heavily staffed with professional ecologists and conservation biologists.

The EPA would become a branch of the Ecosystem Protection Agency and could continue as the primary agency concerned with the ecological reform of the bioculture (air and water pollution, toxic waste, pesticides, organic agricultural reform). The Forest Service would be removed from the Department of Agriculture and, together with the Department of the Interior (including the National Park Service), and the Bureau of Land Management (all of the public land agencies), and the Fish and Game Commission, should be overseen by the Ecosystem Protection Agency. Their function would shift primarily to ecosystem conservation and protection. The functions of the Bureau of Reclamation and the Army Corps of Engineers would be primarily ecosystem restoration. An agency concerned with encouraging the stabilization and reduction of population would also be an integral part of this approach.

With the "cold war" presumably winding down, a large portion of the defense budget must be allocated to environmental protection: acquiring land for the protection zones, adequately staffing environmental protection agencies, providing for educational programs to educate the public to current environmental/ecological realities and proposed solutions. An honest sophisticated national debate on these issues should be promoted by the government. The effects of ozone layer depletion, the greenhouse effect, acid rain, air and water pollution, pesticides, human overpopulation, and current economic/consumerist growth policies freely cross the boundaries between the bioculture, "free nature," and ecosystem protection zones. Only such an integrated holistic approach to environmental problems has a chance of succeeding in the decades to come.

Increasingly, our environmental problems are being recognized as global in scope and, as such, require effective international cooperation. Noel Brown, director of the U.N. Environment Program, indicated that an "ecological council" comparable to the Security Council could soon be a reality.[87] The United Nations also needs to reorganize its population control agencies and environmental protection programs to reflect a unified integrated Ecosystem Protection approach. Ecosystem and environmental protection must be given a very high priority on its agenda. The United Nations General Assembly in effect adopted an ecocentric approach when it approved the World Charter for Nature in 1982. The Charter asserts that:

> Every form of life is unique, warranting respect regardless of its worth to man, and to accord other organisms such recognition, man must be guided by a moral code of action.... Nature shall be respected and its essential processes shall not be disrupted.

The Reagan administration gained further international notoriety by opposing the Charter for Nature. As Arne Naess points out, "the General Assembly voted in favor of the charter by a vote of 111 to 1. The United States of America cast the sole dissenting vote!"[88]

The urgency of our current environmental situation suggests that the United Nations must agree to step up efforts by several orders of magnitude to help stabilize the human population in the shortest time possible while, at the same time, protecting human dignity, ideals of justice, and the freedom of individual choice. The United Nations must continue to help feed the hungry and improve basic living conditions in Third World countries; to help and encourage nations establish ecosystem protection zones and protect wildlife; and to help police these zones (as in Africa where large mammal populations have suffered precipitous declines in the 1980s at the hands of poachers); and to discourage industrial consumerism as part of an overall program of ecological and economic sustainability. The United Nations also needs to further develop major educational programs to "ecologize" the peoples of the world.

In pursuing these goals and strategies, it should be recognized that the situations of First World and Third World countries are very different. As Arne Naess has pointed out, unlike First World countries which are already overdeveloped and ecologically unsustainable, Third World countries will need to continue to develop, although hopefully along ecologically sustainable paths. It is important for environmentalists and ecologists in First World countries to be sensitive to the unique human problems in Third World countries. It is unrealistic and unjust to expect Third World countries to turn to the protection of their natural ecosystems and wildlife *at the expense* of the vital needs of their human populations. At the same time, the magnitude and severity of the global environmental crisis must be fully appreciated. Third World countries should be encouraged to place as high a priority on the expansion of ecosystem protection zones, and the protection of large areas of "free nature" as is possible, given their other pressing needs and concerns. Rich industrial nations will, of necessity, have to pay most of the costs of global environmental protection and restoration. The 1988 Worldwatch Institute's *State of the World* report estimates these costs at $150 billion a year. Naess claims that a step in the right direction would be for industrial nations to forgive Third World debts and loans.[89]

As Paul Taylor points out:

A world of harmony between human civilization and

nature is a distinct empirical possibility.... It should be evident from my discussion of the biocentric outlook and the attitude of respect for nature that an *inner* change in our moral beliefs and commitments is the first, indispensable step. And this inner change is itself a psychological possibility. Some people have actually made such a change, exercising their autonomy in the decision to adopt new moral principles regarding their treatment of the natural environment and its living inhabitants.[90]

The shift from an economically exploitive world view and society to an ecological "green" society expressing deeply satisfying human goals and values should be seen as a joyous positive gain for humanity and Gaia rather than as a self-denial for individuals and humanity as a whole. To experience the transition in this way no doubt requires a conversion to an ecological consciousness, an "ecological self," or what Paul Taylor refers to as an "inner change." While this inner change is occurring among people at an increasing rate throughout the world, and while people are adopting ecologically compatible bioregional ways of living, ecological destruction and species extinction is also accelerating at a terrifying rate. Interim legalistic strategies such as ecosystem protection zoning and the ecological restructuring of governments to help cope with the magnitude and severity of the environmental crisis seem indispensable as this point. Ultimately, we must work at all the levels of ecological protection and restoration, social justice, and human spiritual renewal and ecological sanity simultaneously.

# The Utility of Preservation and the Preservation of Utility: Leopold's Fine Line

by Curt Meine

"Conservation, without a keen realization of its vital conflicts, fails to rate as authentic human drama; it falls to the level of a mere Utopian dream."

*Aldo Leopold, 1937*[1]

"There is nothing more practical in the end than the preservation of beauty, than the preservation of anything that appeals to the higher emotions of mankind."

*Theodore Roosevelt, 1903*[2]

A short-tailed weasel, poking its curious nose out from under a pile of old boards, provides a fitting introduction.

Last summer my brother and I rebuilt an aging footbridge, piling the old planks randomly along the creek bank. The crossboards, green with moss, brown with fungal rot, grey with the weather of the northwoods, were an eyesore. Rusty tenpenny nails, four to each rotten remnant, protruded dangerously. We were intent on getting the bridge finished, so we left the clean-up for later.

This summer I put aside time to correct the mess. As I set about dismantling the main pile, I noticed a scurrying amidst the alders, ferns, and blue flags every time I removed an offending board. The scurryer always returned—until I returned for more boards. Curiosity finally overcame both of us. It seems my brother and I, in our act of destruction, had created ideal short-tailed weasel cover, and with every piece of old bridge I removed, the weasel's roof diminished. I ceased and desisted, and moved on to another pile. The weasel, constitutionally incapable of apathy, sat on his front porch to watch me work, his dark summer pelage gleaming like graphite in the morning light.

It seems that the most tempting metaphors from nature must always be read with the most caution. I recall a chilly spring morning spent foraging about a junk yard for a steering column mount for my old Plymouth Valiant, quite impressed by the quality of habitat the old wrecks provided for black-capped chickadees. Were they better off there than in the prairie edge oak grove that the junk yard now occupied, a hundred and fifty years later? Then there was the sparrow who, with no sentimental qualms whatsoever, built a nest in the hybrid maple outside the college chemistry building. The nest consisted wholly of shredded computer paper. Neither the sparrow, the tree, nor the paper were native to the scene. Who can measure that practical adaptation against the enormous changes wrought by the computer?

In the case of the northwoods weasel, the point is a modest one: just that the line between utility and beauty is not always as sharp as we sometimes suppose. Depending on definitions and circumstance, waste can contain wonder, preservation can lead to destruction, and one generation's "wise use" can become the height of folly to those who follow. The formula for conservation, like the formula for human wisdom itself, is rarely as simple as it appears.


Likewise, individual lives are never as simple as the convenient isms in which we attempt to confine them. If they sometimes come close, it is testament less to the profundity of the ism than to the ardor with which some individuals adhere, whether consciously or not, to self-imposed constraints. The biographical approach to history has many shortcomings, but one special saving grace: it allows us to understand historical forces not as monolithic isms, but as influences in the experience of fellow human beings. It restores soul to the skeleton of facts and the sinew of ideas.

The impetus for this discussion was provided by a reviewer of Aldo Leopold's life, who noted that "he had a practical understanding of conservation as wise use as well as a sense of the aesthetic and idealistic;

his evolving ecological reasoning bound those strains. His life is evidence that the traditional division between aesthetic preservationists and utilitarian conservationists is not as clear as historians have often portrayed it."[3] An accurate judgment. Leopold, with good reason, has become a significant figure in our search for environmental sanity, and his reputation continues to grow as the aptness of his thought reveals itself. Yet, like any seminal figure, Leopold was a complicated individual whose growth, inner tensions, and external influences have often been overlooked in the rush to embrace, apply, analyze, modify, reject, and otherwise employ his "Big Conclusions." Details may be the dross of great social movements, but they are the very stuff of historical biography.

Environmental scholars have seized upon the split between utilitarians and preservationists—or between the utilitarian and aesthetic; the distinction bears examination—as a primary organizing device in interpreting the human-environment relationship. It has become a standard approach to understanding American conservation and environmental history: the natural world is either a commodity to be controlled and used (albeit "wisely"), or it is a beautiful unity, possessed of inherent value and right, to be preserved. The prototypical joiners of the battle, Gifford Pinchot and John Muir, continue through their ideological successors to vie for the consciences of the committed.

That the "traditional division" is real, and the approach valid, is attested to by its usefulness as an historical tool. With it we have begun to understand the unfolding of conservation as a human idea and as a force on the landscape. It has helped us to identify the social response to rates of environmental change unprecedented in the human experience, and to gauge, we hope, adequate responses. This is not the place for a full critique of the utilitarian/preservationist dichotomy in the historiography of conservation. My intent is only to show, through the words and experiences of Aldo Leopold, that there are chinks in this scholarly edifice, and that if we take them into account we may thereby gain some better handholds with which to scale it.

What follows is a sort of informal extended discussion with Leopold. Once one recognizes that Leopold does not so easily fit the standard pattern in the utility vs. beauty debate, certain selections from his writings over the years stand out as particularly provocative. I have gathered some of these together and discuss them as they pertain to three areas of vital concern to Leopold throughout his varied career: wildlife management, wilderness preservation, and general conservation philosophy. Leopold speaks for himself in the quotations; I provide context and interpretation.

Before that, however, it may be worthwhile to identify several problems with the strict reliance on the utilitarian/preservationist dichotomy in interpreting conservation history. This is not an exhaustive roster; only a list of some of the most obvious.

Conservation is a big thing, and this historical tool is better suited for certain tasks than others. The use vs. preservation argument has different shades of meaning when applied to minerals, oceans, the atmosphere, groundwater, surface waters, wetlands, soils, vegetation, forests, biodiversity, game, fish, wildlife, range, parks, wilderness, recreation, and agriculture. We yearn for a uniform approach to these components of the whole, and the need for holistic consideration grows only more apparent with every pollution, extinction, and erosion; yet is it wise to ignore the particular needs and dynamics of each? The traditional split in conservation is most applicable, conversely, in considering the most extreme—and usually important—cases: the initial status of the early national forests, the damming of Hetch Hetchy, the demise of the old growth forests, the fate of remaining roadless areas, the management of wild lands, the oil fields of Alaska. Yet, conservation has always involved quieter, more pervasive skirmishes, away from the major battlegrounds. Those conflicts continue—on tired acres of farmland, in poisoned plumes of well water, on the outskirts of cities whose fringes sprawl as their guts decay, in ecosystems whose deaths are too slow to draw cameras or concern. These demand *conserving* as much as do our most spectacular sceneries, but the traditional divisions speak less directly to the issues here.

The division also tends to oversimplify. "Use" can connote a wide range of relationships—depending on *its* use—from outright slavery and exploitation to communion and symbiosis. It is an unalterable fact that all living creatures "use" their environments. But to those who would convert this biological truth into economic dogma, and so prescribe the intensive use of all natural objects and processes, we hastily add that preservation does not imply a lack of purpose or utility. To the contrary: if we breathe, it is because the oxygen-producing capacity of our photosynthetic planet-mates still persists; if we drink, it is because the world's hydro-logic still pertains; if we eat, it is because wild nature gave us seed stock, and because a sufficiency of soil fertility and solvent farmers still endure; and if we dream, it is because there still exists wilderness large enough to contain our wonder.

The use/preservation division is limited to a certain degree by the fact that it is largely an outgrowth of the American experience. The conquest of this continent was so fast, so destructive, and remains so vital a part of American myth and symbolism that the environmental rationale it left in its wake can be applied only with caution to other settings. Conservation now knows no boundaries—as if it ever really did—and the American experience, crucial though it be in the development of an environmental ethic, cannot be its sole source.

A more subtle fault with the dichotomy is that it too easily glosses over some basic historical connections: that, for instance, but for the progressive utilitarian conservation movement of the early 1900s, we would be arguing over ashes, stumps, and gullies; that the movement for wilderness protection and preservation emerged in large part from the utilitarian confines of the U.S. Forest Service; that the programs for endangered species preservation had their "test runs" on game species; that before ecology got "deep" it had to get *born*. Perhaps these boil down to this: solid thematic analysis requires that the historical homework be done first, and not as an afterthought. In our explorations, we have favored the end over the sequence—the intellectual climax over the successional sere. In crisis situations, this may be a necessary sacrifice; in charting long-term strategies, though, more attention must be given to the historical record.

A further fault of the dichotomy is that it all but automatically opposes the active development or use of "resources" with the passive preservation of same. It allows little space for alternative activity. This concern will emerge strongly in Leopold's words. Leopold was impatient with those who believed that the cure-all to conservation dilemmas was negative, compulsory restriction; he constantly pressed for positive actions, for taking necessary steps to create new initiatives that engaged people in the landscape, and engaged their aesthetic sense in doing so. Again, this applied variously from case to case, most strongly in farm wildlife programs and education, less so in the obvious instance of wilderness preservation. The point: "hands off" is not always, or necessarily, the wisest conservation strategy.

Finally, there is the concern that a strict adherence to the worldview implied by the utility vs. beauty split can only reinforce the alienation that afflicts the human/nature relationship. Others are far better qualified than I to discuss the philosophical implications of this.

Again, these cracks in the edifice, and others I have overlooked, should serve not to diminish the real value of the utility vs. preservation approach, but to lead in to a discussion of how Leopold confronted it in his own work, and how he used it to address ultimate conservation questions. I hope to show how Leopold endeavored to be a healing presence, how he tried to staunch the flow from this psychic split in American conservation—perhaps even in the American mind—by emphasizing common ground wherever it existed, by employing history's synoptic view as a unifying force, by appealing to abstractions but always being comfortable with and conversant in the technical details of conservation, and by reaching outside conservation proper for insight and reinforcement. It may be that the harmony he sought came to exist more in his own soul than in the exterior landscape, but it is equally true that that landscape, what he once termed "that great biota we call America," was nudged closer toward a richer and more enduring balance through his efforts.

The place to begin is at the end, when Leopold was in the fullness of his creative powers, when his vision of conservation was equal to his profound concern, when he himself recognized his role as a statesman in the movement, when he was secure in both his emotion and intellect. He is standing before a class of post-World War II undergraduates, pausing mid-way in his wildlife ecology course to discuss, briefly, his aims and motives:

...I am interested in the thing called "conservation." For this I have two reasons: (1) without it our economy will ultimately fall apart; (2) without it many plants, animals, and places of entrancing interest to me as an explorer will cease to exist. I do not like to think of economic bankruptcy, nor do I see much object in continuing the human enterprise in a habitat stripped of what interests me most.

If the individual has a warm personal understanding of land, he will perceive of his own accord that it is something more than a breadbasket. He will see land as a community of which he is only a member, albeit now the dominant one. He will see the beauty, as well as the utility, of the whole, and know the two cannot be separated. We love (and make intelligent use of) what we have learned to understand.

...Once you learn to read the land, I have no fear of what you will do to it, or with it. And I know many pleasant things it will do to you.[4]

This is the statement, not just of a teacher who has learned the value of reason and enthusiasm in communicating lessons that mere propaganda cannot, but of a human being who has wrestled with his own vision through a lifetime of constant inquiry, and emerged with full confidence in his values, his message, and his approach. There is circumspection here—the ability to understand other points of view while staying centered on one's own—and there is tolerance. There is concern, but not despair. Leopold was under intense pressure during these final years of his life, yet here, as in many of his final statements, we sense a balance that allows him to spread the pressure evenly and so maintain his equilibrium.

The reference to the "utility and beauty of the whole," and the emphasis on their essential coherence, is not anomalous. It is, in fact, rather typical, a clear expression of a theme that had existed in Leopold's thinking and writing since his earliest days as a working conservationist. That theme changed in its make-up and application through Leopold's forty-year career, growing along with his emerging ecological worldview, and culminating in his framing of the land ethic.

Leopold is a delightfully difficult figure to pigeonhole. He is a fountainhead for the deep ecologists, but his utilitarian roots are deep, important, and unavoidable. He was the most practical of field men, and a pioneer in several branches of conservation, but his grasp of conservation's breadth and historical context continues to challenge the field forces, not to mention the general public, to seek new levels of understanding. Because Leopold became best known as an active defender of wilderness and as a popularizer, through *A Sand County Almanac*, of ecological reasoning, and because that fame first came in a big way during the environmental awakening of the 1960s and early '70s, when the subtleties of personality and history were easily overwhelmed, we have begun to understand only recently the quiet, personal, ongoing dialectic that produced Leopold's mature philosophy.[5]

One cannot divorce Leopold the hunter from Leopold the preservationist, although that divorce in the wildlife conservation community has often been loud and bitter. As a naturalist and a sportsman, Aldo from his earliest excursions practiced his pastimes with self-discipline, self-criticism, and enthusiasm. They were paths to a common destination: the natural world. Leopold came to maturity during the heyday of Teddy Roosevelt as national leader, conservationist, and cult figure, and Aldo absorbed the flavor of the era. In many ways, he resembled Roosevelt more than either of the rivals for Roosevelt's conservationist heart, soul, and ear—Muir and Pinchot. Roosevelt could live the extremes of their positions, and did, bringing the same brio to the creation of national parks, forests, and refuges that he brought to the buffalo plains of North Dakota and the safari savannas of Africa. If ever the split between Muir and Pinchot was manifest, it was in the person and era of Roosevelt.

Leopold's training at the Yale Forest School placed him in the *avant-garde* of conservation, and if Pinchot's utilitarian doctrine ran free in the classrooms, it did not necessarily in the minds of the individual students. Leopold learned his proper forestry, but he had other classes, and experiences, and realizations on which to draw during his years out east. Forestry had become a profession, and was quickly gaining all the statistical, technical, and bureaucratic accoutrements of a profession. Nonetheless, it was still a romantic profession, and in the end the fire that fueled most young foresters was not board feet figures, or sales potentials, or tensile strengths, but *forests*, especially *western* forests. So with Leopold. In 1909, he arrived on the Apache National Forest in the Arizona Territory ready to impose forestry, but just as eager to explore the new country. In the age of Roosevelt, both were still possible.

Leopold left wildlife behind during these early years in the U.S. Forest Service, but never completely. And his vision of forestry even at that time was broad enough to foreshadow his later devotion. An important early expression of this was in a letter to his fellow officers on New Mexico's Carson Forest in 1913, when during his protracted bout with nephritis he took the opportunity to discuss the mandate of the Forest Service:

> ...We are entrusted with the protection and development, through wise use and constructive study, of the timber, water, forage, farm, recreative, game, fish, and aesthetic resources of the areas under our jurisdiction. I will call those resources, for short, "The Forest".... And it...follows that the sole measure of our success is the *effect* which [our efforts] have on *the Forest*.[6]

Already he was thinking of the forest as something more than trees, and of trees as something more than timber. As a forester, Leopold made his share of mistakes during the early years of his career, but he had not made the more serious blunder of putting economic function above functional integrity.

It was shortly after his recuperation from nephritis that Leopold began to devote more of his energies to the cause of wildlife conservation—or, more properly, *game protection*. The wildlife movement had not yet broadened enough to take into account, except in a few special cases, non-game species or habitats, or to think in terms of positive conservation measures. Yet, Leopold as leader of the local movement in the Southwest plainly recognized both the ethical and political import of a balanced approach to wildlife protection. He said to a group of Albuquerque Rotarians in 1917:

> ...We conceive of these wild things as an integral part of our national environment, and are striving to promote,

restore, and develop them not as so many pounds of meat,
nor as so many things to shoot at, but as a tremendous
social asset, as a source of democratic and healthful
recreation to the millions of today, and the tens of millions
of tomorrow....

...It is our task to educate the moral nature of each and
every one of New Mexico's half million citizens to look
upon our beneficial birds and animals, not as so much gun
fodder to satisfy his instinctive love of killing, but as
irreplaceable works of art, done in life by the Great Artist.
They are to be seen and used and enjoyed, to be sure, but
never destroyed or wasted.[7]

Leopold conveyed this attitude through his leadership, and his notable
successes during the years 1916-1921, in so *western* a state as New
Mexico, were due in large part to this combination of boldness and
balance.

Of course, anyone familiar with Leopold's personal progression
toward an ecological view knows that at this time not all species
qualified as "irreplaceable works of art." Leopold's ethical boundaries,
and one might say his aesthetic boundaries, still failed to include species
not deemed "beneficial," i.e. varmints. The issue of predator persecution
would not even present itself *as* an issue for another decade. That story
has been told elsewhere, by Leopold and others. The important point is
that while other leaders in the field were still fighting for restrictive
measures and negative actions—lower bag limits, predator control, a
strict system of refuges—Leopold was beginning to lay the foundations
for a positive management approach. He would continue to fight,
differentially, for these other measures, but he would increasingly
concern himself with outlining the real biological needs that would
produce self-sustaining wildlife populations.

This priority would allow Leopold to rise above, or at least to
sidestep, the rancorous debates, often centered on the utility/ preservation line, that rocked the wildlife conservation movement from 1924-
1934. That line cut somewhat differently than it had earlier in the
century, when the debates concerned parks, forests, waters, and range-
land. The use and/or preservation of those features involved large
institutions, industries, and agencies, and most of the action took place

in the halls of government. With wildlife, the institutions were often more local, the habits and attitudes more personal, the sociological baggage heavier. Leopold was familiar with all the players, and as well versed in the issues—refuge policy, predator control, bag limits and other hunting regulations—as anyone in the country. He spoke out when necessary, but devoted most of his energies, especially after he left the Forest Service in 1928 to go full-time into wildlife work, to the background field work necessary to get game management off the ground.

Although Leopold preferred such background work, his prominence in the wildlife conservation community would not allow him to avoid the divisions within it. He was in a precarious position. Under the employ, at first, of the arms and ammunitions industry, he was naturally the object of suspicion from the vocal opponents of hunting, especially the bird-watching contingent (this despite Leopold's credentials as an ornithologist). Hunters were skeptical of, if not openly hostile to, Leopold's management theories, especially as they affected the traditional American freedom to hunt on private lands. Game farmers did not share Leopold's enthusiasm for the preservation and creation of habitat. Zoologists rarely concerned themselves with the applied arts of conservation, and those who did viewed Leopold as an agent of the hunting fraternity. Few appreciated Leopold's vision of a wildlife profession skilled in the recognition, study, preservation, and careful manipulation of habitat to enhance the survival of game in the wild, under conditions as close to natural as possible.

Difficult though his position was, Leopold benefitted from the direct confrontation with antagonistic attitudes that came with his prominence. As he prepared to chair the American Game Policy Committee, the purpose of which was to forge a new policy to guide wildlife conservation across the country, Leopold began to think more deeply about the motives behind conservation, and his own aims as promoter of an alternative vision. In an unfinished manuscript, he assessed the relationship between the "wild lifers" and the "gunpowder faction":

> The devotees of each [faction] like to consider it the antithesis of the other. The nature student is at small pains

to conceal [the belief] that he is superior to mere atavistic blood-letting, while the sportsman sees a lack of Rooseveltian robustness in hunting with field glass or camera. This mutual intolerance would be amusing if it were merely personal. The fact is, however, that each side is nationally organized, and that a state of deadlock between the opposing factions has more than once prevented action on measures obviously advantageous to both, not to mention the development of ideas which might lessen the apparent conflict of interest.[8]

To Leopold, at this point, those who appreciated the utility of a wild creature and those who appreciated its beauty were both missing the boat; whether useful, or beautiful, or both, no creature could survive and perpetuate itself if its basic habitat needs were not met. Leopold, of course, was in Wisconsin by this time, and the impact of intensifying agriculture on wild game in the midwest in the 1920s made this a point of utmost concern.

Leopold was a superb diplomat. When his game policy statement came out, it explicitly called for greater cooperation among the factions as well as greater attention to habitat needs, including those of non-game species. This did not guarantee its success, however, and Leopold took the initiative in selling the policy, in speeches, articles, conferences, and editorials, in cornfields and classrooms and offices. He met continued resistance, again from all sides of the issue. In a reply to one preservationist broadside against the new policy and Leopold's work, Leopold laid his feelings on the line:

>...Does anyone still believe that restrictive game laws alone will halt the wave of destruction which sweeps majestically across the continent, regardless of closed seasons, paper refuges, bird-books-for-school-children, game farms, Izaak Walton Leagues, Audubon Societies, or the other feeble palliatives which we protectionists and sportsmen, jointly or separately, have so far erected as barriers in its path?
>
>...I have tried to build a mechanism whereby the sportsmen and the ammunitions industry could contribute financially to the solution of this problem without dictat-

ing the answer themselves.... These things I have done, and I make no apology for them.[9]

In making this reply, Leopold gave evidence that, on this issue at least, pragmatism took precedence over purism. Never one to be content with platitudes, Leopold may have begun his conservation career with romantic visions, but he had long since tempered them with practical realizations. To achieve conservation, attitude change was essential, but not enough; success required sound science, political will, and much education. Realism and self-criticism were also crucial:

> I realize that every time I turn on an electric light, or ride on a Pullman, or pocket the unearned increment on a stock, or a bond, or a piece of real estate, I am "selling out" to the enemies of conservation. When I submit these thoughts to a printing press, I am helping cut down the woods. When I pour cream in my coffee, I am helping to drain a marsh for cows to graze, and to exterminate the birds of Brazil. When I go birding or hunting in my Ford, I am devastating an oil field, and re-electing an imperialist to get me rubber. Nay more: when I father more than two children I am creating an insatiable need for more printing presses, more coffee, more oil, and more rubber, to supply which more birds, more trees, and more flowers will either be killed, or what is just as destructive, evicted from their several environments.
>
> What to do? I see only two courses open to the likes of us. One is to go live on locusts in the wilderness, if there is any wilderness left. The other is surreptitiously to set up within the economic juggernaut certain new cogs and wheels whereby the residual love of nature, inherent even in "Rotarians," may be made to recreate at least a fraction of those values which their love of "progress" is destroying. A briefer way of putting it: if we want Mr. Babbitt to rebuild outdoor America, we must let him use the same tools wherewith he destroyed it. He knows no other.[10]

In his classic essay "Nature," Ralph Waldo Emerson discussed the role of nature as "Commodity"; Henry David Thoreau opened *Walden* with a chapter on "Economy," and an accounting of his bean business;

John Muir was raised on a farm and in his later years managed a California orchard; Ansel Adams needed silver nitrate to produce his images. Even these most potent sources of the aesthetic and preservationist impulse have accounted for their utilitarian debts to the natural world—and have done so far more openly, one somehow feels, than the utilitarian leaders have accounted for their physical, cultural, and spiritual debts. Leopold had no illusions about the task of conservation, and in a time when the conservation of wildlife existed more on paper than it did on land, in the air, or under water, he was prepared to strip down to realities in order to begin the process of real advance. "It takes all kinds of motives to make a world," he wrote in 1933.

> If all of us were capable of beholding the burning bush, there would be none left to grow bushes to burn. Doers and dreamers are the reciprocal parts of the body politic: each gives meaning and significance to the other. So also in conservation. Just now, conservation is short of doers. We need plants and birds and trees restored to ten thousand farms, not merely to a few paltry reservations.[11]

Progress may not always require balance, but in this time and place it did. Leopold saw the need in the outer world because his inner constitution reflected it. The environment is a dynamic entity, as is civilization. In the past, conservation has changed to reflect different adjustments necessary between them. Leopold's work succeeded, and still endures, because he anticipated needs and responded to them with astounding accuracy.

In 1933, *Game Management*, several years in the writing, finally came out. Leopold, in its opening line—a statement guaranteed to raise the hackles of many preservationists, then and now—made his intention plain: "Game management is the art of making land produce sustained annual crops of wild game for recreational use."[12] It requires some reading beyond the opening line, however, to sense Leopold's full vision, and to realize that Pinchot never saw forests the way Leopold saw game.

At the end of his opening chapter, Leopold answered those who (paradoxically?) for aesthetic reasons might be reluctant to hunt, or perhaps even watch, game tainted by the taste of management, even the faint taste advocated by Leopold:

> There are still those who shy at this prospect of a man-made game crop as at something artificial and therefore repugnant. This attitude shows good taste but poor insight. Every head of wild life still alive in this country is already artificialized, in that its existence is conditioned by economic forces. Game management merely proposes that their impact shall not remain wholly fortuitous. The hope of the future lies not in curbing the influence of human occupancy—it is already too late for that—but in creating a better understanding of the extent of that influence and a new ethic for its governance.[13]

Even as Leopold was writing these words, words that would go on to train several generations of wildlife managers, his understanding of the promise of ecologically informed management was increasing, and filtering through his conservation philosophy as a whole. Some of this emerged in later portions of the book in chapters entitled "Economics and Esthetics" and "Game as a Profession." In the former, Leopold devoted several pages to "Management of Other Wild Life," a clear indication that he recognized management for the consumptive use of animals through hunting as only a necessary first step toward a broader commitment to "wild life" conservation:

> The objective of a conservation program for non-game wild life should be exactly parallel [to that for game]: to retain for the average citizen the opportunity to see, admire and enjoy, and the challenge to understand, the varied forms of birds and mammals indigenous to his state. It implies not only that these forms be kept in existence, *but that the greatest possible variety of them exist in each community.*[14]

Leopold's impatience with factionalism and grandstanding again emerged in this same section. He described the advances that even the most basic research into wildlife habitat needs could promote, to the benefit of all factions (including the oft-forgotten animals):

Measured by its effectiveness, this [research] is worth
10,000 platitudes about forests and wild life. The crying
need at this stage of the conservation movement is *specific
definitions* of the environment needed by each species.
...There is, in short, a fundamental unity of purpose
and method between bird-lovers and sportsmen. Their
common task of teaching the public how to modify economic activities for conservation purposes is of infinitely
greater importance, and difficulty, than their current differences of opinion over details of legislative and admin-
istrative policy. Unless and until the common task is
accomplished, the detailed manipulation of laws is in the
long run irrelevant.[15]

In the final passage of *Game Management*, under the label "Social
Significance of Game Management," Leopold expanded fully on his
expectations for the profession he was creating. In the process, he
explicitly alluded to the pure utilitarian and aesthetic attitudes toward
land that he was trying to unify through his overarching intentions. The
passage is long, but worth quoting in full, for in it one reads Leopold
defining himself:

The game manager manipulates animals and vegetation to
produce a game crop. This, however, is only a superficial
indication of his social significance. What he really labors
for is to bring about a new attitude toward land.
The economic determinist regards the land as a food-
factory. Though he sings "America" with patriotic gusto,
he concedes any factory the right to be as ugly as need be,
provided only it be efficient.
There is another faction which regards economic
productivity as an unpleasant necessity, to be kept, like a
kitchen, out of sight. Any encroachment on the "parlor" of
scenic beauty is quickly resented, sometimes in the name
of conservation.
There is a third, and still smaller, minority with which
game management, by its very essence, is inevitably
aligned. It denies that kitchens or factories need be ugly,
or farms lifeless, in order to be efficient.
That ugliness which the first faction welcomes as the
inevitable concomitant of progress, and which the second

> regretfully accepts as a necessary compromise, the third
> rejects as the clumsy result of poor technique, bunglingly
> applied by a human community which is morally and
> intellectually unequal to the consequences of its own
> success.
>
> These are simply three differing conceptions of man's
> proper relation to the fruitfulness of the earth: three
> different ideas of productivity. Any practical citizen can
> understand the first conception, and any esthete the sec-
> ond, but the third demands a combination of the economic,
> esthetic, and biological competence which is somehow
> still scarce.
>
> It would be, of course, absurd to say that the first two
> attitudes are devoid of truth. It seems to be an historical
> fact, however, that such few "adjustments" as they have
> accomplished have not kept pace with the accelerating
> disharmony between material progress and natural beauty.
> Even the noble indignation of the second school has been
> largely barren of any positive progress toward a worthier
> land-use.
>
> ...Examples of harmonious land-use are the need of
> the hour.[16]

Even in his most technical discussions of game management, Leopold
never failed to include or imply the aesthetic component of the work;
conversely, even in his most preservation-oriented wilderness papers,
he never dismissed or ignored the human economic response, but always
faced it head on. His conviction was that the two "conceptions" of
conservation were not necessarily incompatible; that, in most cases,
they could be harmonized to the benefit of both man and nature.

By the late 1930s, Leopold was one of many who saw that they *had
to be* harmonized. This is a complicated period, and the influences on
Leopold were manifold—his university appointment, his travels in
Germany and Mexico, the Dust Bowl (especially), the economic up-
heaval of the Depression, the intellectual syntheses in biology, the
purchase of his worn-out farm along the Wisconsin River. Even as the
ink on *Game Management* dried, Leopold began to confront the essential question that ecology now raised: where does utility lie? By now
Leopold had completed his full conversion on the question of predator
control, but this was only the most conspicuous expression of his

expanded vision. In terms of wildlife management, ecology had brought all species into the intellectual fold (if not yet the field operations) of conservation. Species, until then, had gained attention if they were useful, helpful or harmful, beautiful or ugly, from the human perspective; henceforth, they were worthy of attention by the basic fact of their existence. And to the intellectual leaders in ecology, the old message of George Perkins Marsh reemerged with new clarity and immediacy: *ultimate* utility lay in the overall stability of the environment. Ecology and evolution had emerged as intersecting axes on which we might gauge that relative stability.

Leopold's published and unpublished work is rich in the early exploration of this new world of understanding. Reflecting the synthesis of the times, his topics overlap and connect, and their foci are just that—specific points where Leopold grounds broader discussions. Wildlife management—soon to become wildlife ecology—was no longer an obscure, isolated, foundling field; it was now a crucial nexus of inquiry. Hardly used to its legs, it was being asked to run. Leopold summarized the changes, and his own comprehension of ecology's full meaning, in his 1939 paper, "A Biotic View of Land":

> ...The emergence of ecology has placed the economic biologist in a peculiar dilemma: with one hand he points out the accumulated findings of his search for utility in this or that species; with the other he lifts the veil from a biota so complex, so conditioned by interwoven cooperations and competitions, that no man can say where utility begins or ends. No species can be "rated" without the tongue in the cheek; the old categories of "useful" and "harmful" have validity only as conditioned by time, place, and circumstance. The only sure conclusion is that the biota as a whole is useful, and [the] biota includes not only plants and animals, but soils and waters as well.[17]

With that, all bets were off in terms of indiscriminate environmental manipulation. The game of conservation now had new rules, and the "line" had to redefined.

Leopold's wildlife work would concentrate during his final decade on the refinement of management techniques, but not at the expense of the broader vision. On the contrary, part of the responsibility of wildlife ecology, to Leopold's thinking, was to "help rewrite the objectives of science."[18] Increasingly frustrated with the excesses of what he began to call "power science," Leopold saw clearly the potential of his field to counter the increasing intensity of land-use technologies. Although at work in that bastion of utility, the land grant college of agriculture, he was not reluctant to state his case. In a 1946 review, "The Outlook for Farm Wildlife," he explicitly depicted rural development as a competition between two different attitudes toward farm life:

> 1) The farm is a food factory, and the criterion of success is salable products.
> 2) The farm is a place to live. The criterion of success is a harmonious balance between plants, animals, and people; between the domestic and the wild; between utility and beauty.
> Wildlife has no place in the food factory farm, except as an accidental relic of pioneer days. The trend of the landscape is toward a monotype, in which only the least exacting species exist.
> On the other hand, wildlife is an integral part of the farm-as-a-place-to-live. While it must be subordinated to economic needs, there is a deliberate effort to keep as rich a flora and fauna as possible, because it is "nice to have around."
> It was inevitable and no doubt desirable that the tremendous momentum of industrialization should have spread to farm life. It is clear to me, however, that it has overshot the mark, in the sense that it is generating new insecurities, economic and ecological, in place of those it was meant to abolish. In its extreme form, it is humanly desolate and economically unstable. These extremes will someday die of their own too-much, not because they are bad for wildlife, but because they are bad for farmers.[19]

The very meaning of utility in conservation, Leopold now realized, had changed as a consequence of economic pressures and technological advance. The unnecessary disappearance of cultural amenities from the landscape was not merely unfortunate; it was a warning sign.

Through these years, Leopold had simultaneously, and importantly, begun to buttress his science with aesthetic appreciation, based on the insights of ecology. This intermixture would, like estuarine waters, prove uncommonly rich and productive. Leopold composed most of the *Sand County* essays at this time. His teaching began to stress perception, as opposed to manipulation, as the first priority. As he reached the end of his days, his philosophy came into final focus, culminating in his bequest of "The Land Ethic."

Ironically, ecology had wrought a revolution in the old debate: perception now had *survival value*; aesthetic sensitivity, as partially redefined by the new science, was *useful*. Leopold, typically, made the point more wryly. In *Sand County*, he described the "educated lady, banded by Phi Beta Kappa," who had never seen or heard the Canada geese "that twice a year proclaim the revolving seasons to her well-insulated roof." He then wondered: "Is education possibly a trading of awareness for things of lesser worth? The goose who trades his is soon a pile of feathers."[20]

In 1921, Leopold opened the debate over the fate of wildlands within the National Forests with these words:

> Very evidently we have here the old conflict between preservation and use, long since an issue with respect to timber, water power, and other purely economic resources, but just now coming to be an issue with respect to recreation. It is the fundamental function of foresters to reconcile these conflicts, and to give constructive direction to these issues as they arise.[21]

Appearing in the *Journal of Forestry*, Leopold's words were certain to provoke consternation among many of his professional colleagues. Sensitive to bureaucratic politics and tradition, he made his case by invoking the holy writ: "The argument for such wilderness areas is premised wholly on highest recreational use."[22] As a case example, Leopold suggested granting protected status to the headwaters of the Gila River in western New Mexico. In so doing, he provided a wonderful oxymoron for foresters and future environmental historians to ponder. "Highest use," he insisted, "demands its preservation."[23]

With that statement, Leopold signalled a breakaway from the utilitarian fold. That romantic yearning that had led him into forestry and lured him west still guided him, even after ten years of ascension through the Forest Service hierarchy. His break was not total. He remained in every way a loyal, dedicated, and innovative Forest Service officer. His move, though, would force the Forest Service to recognize and, even more important, *protect* values not easily quantified. And it would force Leopold, as (in Bob Marshall's words) "the Commanding General of the Wilderness Battle," to justify a bureaucratic innovation not easily understood.

Wilderness protection was different from wildlife protection. The grain of attitudes ran in a different direction. The economics were less amenable. A sense of history was more essential. The constituency was harder to define—if indeed it existed at all. And after all, the country already had a National Park Service devoted to preserving wild wonders; wasn't that enough?

It was not enough for Leopold and a small circle of his Southwestern colleagues. The parks were closed to hunting, and in any case were being riddled with roads and tourist accommodations. And scenery was not enough. Leopold wanted a *functional* wilderness, "big enough to absorb a two weeks' pack trip," yet accessible to those not wealthy enough to travel to the ends of the earth. His 1921 call for preserving wilderness areas was *very* utilitarian, but launched him on a career of advocacy which would, again, show that the line dividing utility and preservation was neither simple nor immutable.

That Leopold's initial interest in wilderness was more than aesthetic, i.e. involved more than scenery, became plain when the subject of wilderness status for national forest lands first arose in his discussion with Arthur Carhart in 1919.[24] Carhart, a landscape architect, argued for preservation of the scenic value of Colorado's Trapper's Lake through protection of its immediate shoreline. Leopold had something bigger in mind. Leopold's seminal 1921 article, "The Wilderness and Its Place in Forest Recreation Policy," stressed exclusively the recreational value of wilderness areas, with no word given to the scenic, biological, or ecological values, and only an implication of the social, cultural, and historical values. And yet, of course, there was an aesthetic aspect to the form of recreation that Leopold was seeking to perpetuate: the sort of travel and hunting Leopold most enjoyed demanded a large and wild environment.

By 1924, when the Gila Wilderness Area was designated, Leopold was beginning to express deeper reasons for preserving wilderness. But even in doing so, there was always a practical tack to his argument. Preservation, evidently, had higher "uses":

> What I am trying to picture is the tragic absurdity of trying to whip the March of Empire into a gallop. Very specifically, I am pointing out that in the headlong stampede for speed and ciphers, we are crushing the last remnants of something that ought to be preserved for the spiritual and physical welfare of future Americans, even at the cost of acquiring a few less millions of wealth or population in the long run. Something that has helped build the race for such innumerable centuries that we may logically suppose it will help preserve it in the centuries to come.[25]

In essence, Leopold was asking the Forest Service to commit itself, in a real way, to forest uses other than those most readily translatable into the "ciphers" of economics; simultaneously—and in a fashion similar to his later wildlife work—he was emphasizing the very practical value of wilderness preservation. The line between utility and preservation had become very thin indeed.

In the years immediately following Leopold's move to Wisconsin in 1924, he produced a series of articles advocating preservation and rounding out the reasoning behind "the wilderness idea." Directed to different audiences, and appearing in a broad range of publications, these articles nonetheless spoke from common themes: wilderness as a complement to civilization; the limits of standard economic arguments; the need for balance in the nation's vision of land use; the important role of wilderness in American history. Fighting a rearguard battle, Leopold chose not to deny the possible economic value of the lands in question, but used this as a starting point. Realism was his hallmark:

> The Forest Service will naturally select for wilderness playgrounds the roughest areas and those poorest from the

economic standpoint. But it will be physically impossible to find any area which does not embrace some economic values. Sooner or later some private interest will wish to develop these values, at which time those who are thinking in terms of the national development in the broad sense and those who are thinking of local development in the narrow sense will come to grips. And forthwith the private interests will invoke the aid of the steamroller. They always do. And unless the wilderness idea represents the mandate of an organized, fighting, and voting body of far-seeing Americans, the steam roller will win.[26]

In order to build such a mandate, Leopold resorted less to Muir-like evocations of wild beauty and sublime majesty than to building an appreciation of the contrast value of wilderness. This called for a sense of history and cultural wholeness that stood out against the prevailing mood of America in the 1920s:

...the measure of civilization is in its contrasts. A modern city is a national asset, not because the citizen has planted his iron heel on the breast of nature, but because of the different kinds of man his control over nature has enabled him to be. Saturday morning he stands like a God, directing the wheels of industry that have dominion over the earth. Saturday afternoon he is playing golf on a kindly greensward. Saturday evening he may till a homely garden or he may turn a button and direct the mysteries of the firmament to bring him the words and songs and deeds of all the nations. And if, once in a while, he has the opportunity to flee the city, throw a diamond hitch upon a pack-mule, and disappear into the wilderness of the Covered Wagon Days, he is just that more civilized than he would be without the opportunity. It makes him one more kind of man—a pioneer.[27]

In a time when wilderness preservation was still only a distant dream, Leopold recognized the *realpolitik* need, not to alienate potential supporters, but rather to gather them in through an expanded vision of the national saga, and the national landscape. Building on the ideas of Muir and Frederick Jackson Turner, drawing on such writers and poets as Whitman, Stephen Vincent Benet, and Sinclair Lewis, Leopold argued for wilderness protection, not as a denial of the American myth of progress, but as a fulfillment of it.

The economic dragon had to be faced, however. "Economic development," then as now, was roughly synonymous with "roads." To Leopold, it was a matter of scale and balance: roads were not good nor evil in and of themselves; their utility, or lack thereof, was simply a function of time, place, and density. Viewed on a national scale, and in historical context, the rise of the automobile culture demanded a parallel commitment to wilderness preservation:

> Roads and wilderness are merely a case of the pig in the parlor. We now recognize that the pig is all right—for bacon, which we all eat. But there was no doubt a time, soon after the discovery that many pigs meant much bacon, when our ancestors assumed that because the pig was so useful an institution he should be welcomed at all times and places. And I suppose that the first "enthusiast" who raised the question of limiting his distribution was construed to be uneconomic, visionary, and anti-pig.[28]

In an article tellingly titled "Wilderness as a Form of Land Use," Leopold made the same point more broadly:

> Our system of land use is full of phenomena which are sound as tendencies but become unsound as ultimates.... The question, in brief, is whether the benefits of wilderness-conquest will extend to ultimate wilderness-elimination.
>
> ...To preserve any land in a wild condition is, of course, a reversal of economic tendency, but that fact alone should not condemn the proposal. A study of the history of land utilization shows that good use is largely a matter of good balance—of wise adjustment between opposing tendencies.[29]

Leopold did not expand here on what he thought the "benefits" of wilderness "conquest" had been, and it is difficult to know the degree to which he was holding his tongue in his cheek in order to make his argument. It is plain that the ecological value of wilderness was not yet a substantial component of "the wilderness idea." But if one of the benefits of wilderness conquest had been a heightened appreciation of the *remaining* wilderness, then use and preservation were inevitably and closely coupled, and the history of that coupling became a crucial part of the consideration of any future development aspiring to the adjective "wise."

Leopold put this latter point into a particularly American context. Scorning the superficial definitions of utility and Americanism that marked the "Babbittian" decade, he presented the wilderness not as a source just of use *or* beauty, but as *the* source of a still incomplete, evolving nation:

> Wilderness as a form of land use is, of course, premised on a qualitative conception of progress. It is premised on the assumption that enlarging the range of individual experience is as important as enlarging the number of individuals; that the expansion of commerce is a means, not an end; that the environment of the American pioneers had values of its own, and was not merely a punishment which they endured in order that we might ride in motors. It is premised on the assumption that the rocks and rills and templed hills of this America are something more than economic materials, and should not be dedicated exclusively to economic use.[30]

The American experience of wilderness would come to be overshadowed in Leopold's wilderness philosophy by the globally applicable lessons of ecological stability and land degradation. Through the 1920s, however, it was a principal, and effective, part of his argument. The "real" wilderness, as white Americans had known it, was forever gone—that too was a lesson of history—but its cultural significance was as potent as ever. By forcing those who patriotically invoked the symbol to confront the stark reality of dwindling wild spaces, Leopold implicitly and explicitly invited action. To those who questioned whether there was any place for wilderness in an America whose business was business, Leopold could simply ask, "Shall we now exterminate this thing that made us American?"[31]

After this pulse of wilderness advocacy papers in the mid-1920s,

Leopold devoted most of his energies to laying the foundations of
wildlife management. When he returned to an active role in the wilderness preservation movement in the mid-1930s, he did so with all the
additional insight that his intellectual evolution could bring to the cause.
The significance of wild lands was no longer just aesthetic, recreational,
cultural, historical, or social, but scientific and ecological.

With this came an intensified sense of the practical benefits to be
gained by preserving wilderness. Leopold stressed these in his contribution to the inaugural issue of *The Living Wilderness*, the journal of the
new Wilderness Society:

> I suspect...that the scientific values [of wilderness] are
> still scantily appreciated, even by members of the Soci-
> ety....
>         The long and the short of the matter is that all land-use
> technologies—agriculture, forestry, watersheds, erosion,
> game, and range management—are encountering unex-
> pected and baffling obstacles which show clearly that
> despite the superficial advances in technique, *we do not
> yet understand and cannot yet control* the long-time
> interrelations of animals, plants, and mother earth.[32]

The logical corollary? We needed the dynamic of wilderness as an
alternative to the dynamic of civilization. Leopold had said as much in
the 1920s, but his emphasis then was on the benefits to society; now he
emphasized the benefits to the whole culture-nature community.

Focussed in thought by journeys to Germany and the Sierra Madre,
humbled by his increasing appreciation of the complexity of population
ecology (the "yet" would fade from his statement above), and tempered
by the harsh lessons of the Dust Bowl years, Leopold would henceforth
emphasize this argument for wilderness above all others. It would take
its place at one end of the spectrum of his overall conservation philosophy, inseparable from his other conservation interests. As Leopold
endeavored to integrate the "biotic view of land" into conservation
strategies, wilderness became the vital check:

> Every region should retain representative samples of its
> original or wilderness condition, to serve science as a

sample of normality. Just as doctors must study healthy
people to understand disease, so must the land sciences
study the wilderness to understand disorders of the land-
mechanism.[33]

Leopold employed this "land health" analogy regularly during these
years—the late 1930s and early 1940s—as he worked to communicate
the ecological message. The preservationist sounded very practical at
this point:

The most important characteristic of an organism is that
capacity for internal self-renewal known as health....
　　In general, the trend of the evidence indicates that in
land, just as in the human body, the symptom may lie in
one organ and the cause in another. The practices we now
call conservation are, to a large extent, local alleviations of
biotic pain. They are necessary, but they must not be
confused with cures. The art of land-doctoring is being
practiced with vigor, but the science of land-health is a job
for the future....
　　A science of land health needs, first of all, a base-
datum of normality, a picture of how healthy land maintains itself as an organism....
　　All wilderness areas, no matter how small or imperfect, have a large value to land-science. The important
thing is to realize that recreation is not their only or even
their principal utility. In fact, the boundary between recreation and science, like the boundaries between park and
forest, animal and plant, tame and wild, exists only in the
imperfections of the human mind.[34]

This is a utilitarian rationale on an expanded, even global, scale. It is one
we, five decades later, can only appreciate all the more; what a wonder
if we could have an intact expanse of buffalo range, or a great unbroken
stand of mixed hardwood-white pine forest, or a county or two of Iowa
tallgrass prairie, or a virgin salmon fishery, or a cylinder of pre-
industrial atmosphere. One suspects that even the most avaricious and
manipulative of modern utilitarians would cry to see them.
　　Although Leopold regularly emphasized the practical benefits to be
gained through preservation, we need to recall that his aesthetic re-

sponse, too, remained profound. One has only to read his *Sand County*
accounts of Arizona and Mexico, of Manitoba and Baja, California, of
the less monumental but still enriching wilds of Wisconsin. These were
written in the 1940s, when his wilderness philosophy was fully mature
and gave context to his memories. At the same time, he remained an
active defender, in print and person, for threatened wild lands from the
Arctic to the Wisconsin cutover. His art, his activism, his science, and
his concern, were of a piece.

There will always be those unable to think of wilderness as anything
but a "locking up of resources." If there be so unthinking a counterpart
on the preservationist side, is it the activist working to save such wild
remnants as remain, or the individual who will profess a love of
wilderness—the "parlor of scenic beauty" that Leopold referred to in
*Game Management*—but not allow that love to filter through to the other
compartments of his or her life? The latter, I think. If there is ever to be
a reconciliation of the utilitarian and preservationist traditions on this
issue, it will come only when enough individuals have come to understand the historical and geographical context of wilderness, and have
allowed that understanding to be translated, in as many ways as there are
modes of living, into personal commitment. This is the point to which
the evolution of Leopold's wilderness philosophy—and of his full land
ethic—finally led. In his final essay on wilderness, he wrote:

> Wilderness is the raw material out of which man has
> hammered the artifact called civilization....
> To the laborer in the sweat of his labor, the raw stuff
> on his anvil is an adversary to be conquered. So was
> wilderness an adversary to the pioneer.
> But to the laborer in repose, able for the moment to
> cast a philosophical eye on his world, that same raw stuff
> is something to be loved and cherished, because it gives
> definition and meaning to his life.[35]

Wilderness has given "definition and meaning" even to the lives of those
most removed from, ignorant of, hostile towards, and oblivious to, its
existence. This is a "use" so all-encompassing that even we who choose
to defend wilderness have difficulty comprehending it. It has challenged
the most creative of our race's minds. It goes beyond "use" to the very

essence of our existence, diving to the depths of our evolutionary origins, asking profound questions of human intentions, calling us to creation's brink. Though not always pleasant or comfortable, the human experience of the wild has *made us* human. This is true ontogenetically as well as phylogenetically—for each of us as individuals and all of us as human beings. Lose the wild, and we lose the human. That would be very impractical. And very ugly.

The same independent thinking that fueled Leopold's innovations in wildlife conservation and wilderness preservation also led him to expand the scope of his overall conservation philosophy. The "ratio" of utility to other values shifted as Leopold matured in his understanding of conservation; the usefulness of nature was never forgotten or ignored, but was finally placed within the broader context provided by ecology.

Leopold was never wholly comfortable with the utilitarian party line. Even in his headiest days as a young forester in Pinchot's Forest Service, he saw forestry as far more than the mere securing or refining of timber. "I have no ambition to be a timber-tester or tie-pickler," he declared when contemplating his professional path. Among his first self-initiated chores on the Apache: setting boundaries for a proposed game refuge. The Pinchot influence was pervasive in those early days of the Forest Service, but that is not to say that it was monolithic.

By the time Leopold first began to express himself on the broader meaning of conservation, he had clearly formulated a personal set of premises and conclusions far more comprehensive than those he was trained on. In an early effort to explain his views, 1924's "Some Fundamentals of Conservation in the Southwest," Leopold revealed his frustration with the limits of pure utilitarianism:

> Most religions, in so far as I know, are premised squarely on the assumption that man is the end and purpose of creation, and that not only the dead earth, but all creatures thereon, exist solely for his use. The mechanistic or scientific philosophy does not start with this as a premise, but ends with it as a conclusion.[36]

At the same time, however, his alternative was not to deny outright the utilitarian attitude, but to harness it to a decency guided by respect:

> ...the privilege of possessing the earth entails the responsibility of passing it on, the better for our use, not only to immediate posterity, but to the Unknown Future, the nature of which is not given us to know. It is possible that Ezekiel respected the soil, not only as a craftsman respects his material, but as a moral being respects a living thing.[37]

The use of the earth by humans was a given. But Utility, if it disregards past and future considerations, becomes ultimately self-defeating.

It is important to remember that Leopold arrived at this attitude as much through his reading of the landscape as his reading of books. He already had fifteen years of Forest Service field work under his belt, and his studies of overgrazing, vegetation change, soil erosion, and fire ecology in the Southwest had shown him the long-term results of short-sighted land use. Without that experience it is doubtful that the words he was reading would have resonated so deeply. Now, when Leopold called out Ouspensky and Bryant and Whitman against the rimrock of the Colorado Plateau, along the extrusions of the Mogollon Rim, and up the side canyons of the Gila, his own voice, enriched and invigorated by contact with the country, echoed back:

> Possibly, in our intuitive perceptions, which may be truer than our science and less impeded by words than our philosophies, we realize the indivisibility of the earth—its soil, mountains, rivers, forests, climate, plants, and animals, and respect it collectively not only as a useful servant but as a living being, vastly less alive than ourselves in degree, but vastly greater than ourselves in time and space—a being that was old when the morning stars sang together, and when the last of us has been gathered unto his father, will still be young.[38]

Aware of the limits of scientific truth, undaunted by the commands of formal philosophy, Leopold allowed his intuitive sense of the living, indivisible earth to inform, though not to dictate, his conservation stance. He recognized the earth "not only as a useful servant but as a living being." The understated tension in that phrase would be a constant goad to Leopold, changing along with his perspective and his priorities, leading him on to the synthesis of "The Land Ethic." Were Leopold not so inherently forward-looking, or so possessed of the naturalist's ingrained respect for the natural world, or so plain stubborn in his concern for the fate of the earth, the tension might have been depressive and draining. Instead, it became a creative tension, driving him to comprehend the rapidly changing relationship between humankind and the earth in a newly industrialized world.

That that relationship need not be destructive, and that we need not acquiesce in the impoverishment of the landscape, were fundamental premises behind Leopold's push to establish game management. Others could cry loudly over the loss of wildlife and righteously deplore its most obvious (if not most important) causes; Leopold, as noted above, was determined to counteract the broader trends by beginning the painstaking study of actual, on-the-ground needs of species. A conceptual revolution! When in 1933 Leopold next tried to summarize his conservation philosophy in "The Conservation Ethic," it bore the mark of his detailed habitat studies. The measure of success in conservation was not merely efficiency, even long-term efficiency. "The real end [of conservation]," he wrote, "is a universal symbiosis with land, economic and esthetic, public and private."[39]

Flush with the promise of this new aim and new methods with which to attain it, Leopold had high hopes for the infant field of game management, and was already anticipating the extension of its principles to other areas. Through active husbandry of the wild environment, people might make some progress toward that "universal symbiosis":

> ...[The] idea of controlled wild culture or "management"
> can be applied not only to quail and trout, but to *any living
> thing* from bloodroots to Bell's vireos. Within the limits
> imposed by the plant succession, the soil, the size of the
> property, and the gamut of the season, the landholder can
> "raise" any wild plant, fish, bird, or mammal he wants to.
> A rare bird or flower need remain no rarer than the people

willing to venture their skill in *building it a habitat*. Nor
need we visualize this as a new diversion for the idle rich.
The average dolled-up estate merely proves what we will
someday learn to acknowledge: that bread and beauty
grow best together. Their harmonious integration can
make farming not only a business but an art; the land not
only a food factory but an instrument for self-expression,
on which each can play music of his own choosing.[40]

The influence of the midwestern landscape is apparent in this statement.
Wildlife management learned many of its early lessons working with
small game on midwestern farmsteads, where "bread and beauty" could,
with effort, grow together. Leopold well understood, however, that
certain environmental values would be sacrificed if this formula were
applied consistently across the landscape; in the Gila Wilderness, for
instance, where bread—or beef, or timber—took a back seat to beauty.
The explanation of this seeming contradiction lay in an appreciation of
scale: bread and beauty grew best together on a continent—on a
planet—as well as on the back forty. The trick was to strike some balance
in a world so increasingly blind to beauty and hungry for bread that the
environmental conditions necessary for balance were threatened.

   This would be the moral lesson of the Dust Bowl and of other
environmental conundrums of the 1930s. Leopold saw aesthetic appreciation of the environment not as a luxury for the elect, but as an absolute
social necessity. Only by increasing the general sensitivity to environmental health, to the processes and functions that determined stability,
could that health be maintained for the common good. We have already
seen how important a role wilderness came to play in the framing of this
idea. But the land ethic toward which Leopold was moving applied
"across the board."

   Leopold was quickly coming to understand the full impact that the
science of ecology bore for conservation. An important milestone along
the path—in fact the place where the phrase "land ethic" first appeared
in Leopold's writing—was a 1935 address entitled "Land Pathology."
Leopold began by discussing the problems of applying the profit motive
in conservation, and then went on to consider the social, cultural, and
historical reasons for conservation's too limited success. The divorce of

utility and beauty, virtually institutionalized in the conservation bureaus, and in society at large, played prominently in Leopold's argument:

> Conservation is a protest against destructive land use. It seeks to preserve both the utility and beauty of the landscape. It now invokes the aid of science as a means to this end. Science has never before been asked to write a prescription for an esthetic ailment of the body politic. The effort may benefit scientists as well as laymen and [the] land.
>
> Conservationists are sharply divided into groups, interested respectively in soil fertility, soil erosion, forests, parks, ranges, water flows, game, fish, fur, non-game animals, landscape, wildflowers, etc.
>
> These divergent foci of interest clearly arise from individual limitations of taste, knowledge, and experience. They also reflect the age-old conflict between utility and beauty. Some believe the two can be integrated, on the same land, to mutual advantage. Others believe their opposing claims must be fought out and settled by exclusive dedication of each parcel of land to either one use or the other.
>
> This paper proceeds on two assumptions. The first is that there is only one soil, one flora, one fauna, one people, and hence only one conservation problem. Each acre should produce what it is good for, and no two are alike. Hence a certain acre may serve one, or several, or all of the conservation groups.
>
> The second is that economic and esthetic land uses can and must be integrated, usually on the same acre. To segregate them wastes land, and is unsound social philosophy. The ultimate issue is whether good taste and technical skill can both exist in the same landowner.[41]

Leopold again displays his unifying instincts here—but with even greater conviction as a consequence of several field assignments that showed him that such integration was not only desirable and possible, but necessary.

Such integration, however, was increasingly difficult in an urbanizing society whose legacy of conservation was nothing to brag about in the first place. Leopold saw clearly that the forces were pulling in precisely the wrong directions:

> The unprecedented velocity of land subjugation in America involved much hardship, which in turn created traditions which ignore esthetic land uses. The subsequent growth of cities has permitted a rebirth of esthetic culture, but in landless people who have no opportunity to apply it to the soil. The large volume and low utility of conservation legislation may be attributed largely to this maladjustment; also the dissentious character of the conservation movement.[42]

And when such tastes and traditions became compartmentalized in a nation's collective mind, they sooner or later became compartmentalized on the nation's landscape. In the worst case scenario, the segregation of ethics, aesthetics, and economics would work to the detriment of each. Leopold did not shy from the most disturbing lessons of this condition:

> ...Parks are over-crowded hospitals trying to cope with an epidemic of esthetic rickets; the remedy lies not in hospitals, but in daily dietaries. The vast bulk of land beauty and land life, dispersed as it is over a thousand hills, continues to waste away under the same forces as are undermining land utility. The private land owner who today undertakes to conserve beauty on his land does so in defiance of all man-made economic forces from taxes down—or up.[43]

Writing in the midst of the Depression, Leopold was particularly sensitive to the limits of "man-made economic forces" in bringing about the good life. In conservation he saw an important corrective, a balance-weight, a force that could begin again to meet the inseparable spiritual, material, and psychic needs of human beings:

> Every American has tattooed on his left breast the basic premise that manifestations of economic energy are inherently beneficent. Yet here is one which to me seems

malignant, not inherently, but because a good thing has
outrun its limits of goodness. We learn, in ecology at least,
that all truths hold only within limits. Here is a good
thing—the improvement of economic tools. It has exceeded the speed, or degree, within which it was good.
Equipped with this excess of tools, society has developed
an unstable adjustment to its environment, from which
both must eventually suffer damage or even ruin. Regarding society and land collectively as an organism, that
organism has suddenly developed pathological symptoms, i.e. self-accelerating rather than self-compensating
departures from normal functioning. The tools cannot be
dropped, hence the brains which created them, and which
are now mostly dedicated to creating still more, must at
least in part be diverted to controlling those already in
hand. Granted that science can invent more and more
tools, which might be capable of squeezing a living even
out of a ruined countryside, yet who wants to be a cell in
that kind of a body politic? I for one do not.[44]

In showing us the potential of conservation, Leopold has conversely
presented us with the ultimate result of the divorce of utility and beauty:
a compulsive, inescapable devotion to increases in efficiency and
productivity on such land as is left after the original stability and
productivity are vanquished, all to support an urbanized population that
finds its aesthetic desires satisfied mainly by urban pleasures, content
that it has "conserved nature" by setting aside a few scenic parks. It was
a dark vision from a dark time.

Leopold, on rare occasion, hinted at the full personal cost of being
ecologically literate, at the sadness that came from "living alone in a
world of wounds." Yet, he was not by nature a pessimist or a cynic. It
was always his style to realistically assess a situation, weigh the options,
make the best informed choice, and press forward. As the lessons of
ecology seeped ever more deeply into his work through the late 1930s,
his response continued to be positive and self-critical. The deeper his
realizations, the calmer but more confident his convictions. Without
letting down his wilderness guard—wilderness protection remained the
*initial* priority—Leopold began to explore in greater detail the questions
of humanity's proper interaction with the wild and the semi-wild.

Wrestling time and again with the very definition of "conservation," he regularly came up against the "age-old conflict between utility and beauty." If the roots of that conflict lay in the historically parallel rise of romanticism on the one hand and economics and engineering technologies on the other, then conservation had to be redefined as the biological sciences led the way toward a new, unified understanding of the natural world.

In unpublished "notes on a new theory of conservation," Leopold wrote:

> Conservation...has been presented to us as a threat of *deficit* in natural resources (timber famine, soil rape, extinction of wildlife). Certainly we are running deficits, but the cry of "wolf wolf" is negative and incomplete, like frightening children.... Ecological conservation is a *positive* proposal to learn the act of skillful burling.[45]

"Skillful burling" implied that the human race was, inevitably, a part of the great river, a force in its movement, a part of its flow, though not the sole reason for, nor object of, its progression. And to Leopold, the first step in achieving the kind of use truly worthy of the term "wise" was to overcome routine paeans to utilitarianism itself. In one of many articles Leopold wrote for farmers at this time, he made the point:

> Sometimes I think that ideas, like men, can become dictators. We Americans have so far escaped regimentation by our rulers, but have we escaped regimentation by our own ideas? I doubt if there exists today a more complete regimentation of the human mind than that accomplished by our self-imposed doctrine of ruthless utilitarianism. The saving grace of democracy is that we fastened this yoke on our own necks, and we can cast it off when we want to, without severing the neck. Conservation is perhaps one of the many squirmings which foreshadow this act of self-liberation.[46]

Once utility was thus brought under rein, or at least defined with greater care, Leopold had no qualms about admitting it a place in conservation.

At times, he anticipated our latest discussions of sustainable development:

> This science of relationships is called ecology, but what we call it matters nothing. The question is, does the educated citizen know he is only a cog in an ecological mechanism? That if he will work with that mechanism his mental wealth and his material wealth can expand indefinitely. But that if he refuses to work with it, it will ultimately grind him to dust. If education does not teach us these things, then what is education for?[47]

To say that utility was not the only or ultimate guide in conservation, but was yet a part of the human-environment relationship, was to suggest that better ethical guidelines were needed for its exercise. Another decade would pass before urgency prompted the compilation of "The Land Ethic," though Leopold now saw fully the extent of the conservation dilemma. In a lecture to a group of engineering students, he spoke directly:

> We end, I think, at what might be called the standard paradox of the twentieth century: our tools are better than we are, and grow better faster than we do. They suffice to crack the atom, to command the tides. But they do not suffice for the oldest task in human history: to live on a piece of land without spoiling it.[48]

After 1939, utility and beauty grew ever closer in Leopold's thinking. In formulating a "land aesthetic" that celebrated not merely the outward appearance of natural objects, but their evolutionary history and ecological role, he broadened traditional criteria of natural beauty to a point where they overlapped his sense of long-term utility based on ecosystem stability.[49] The converse interpretation is equally valid: in exploring the practical value of long-term ecosystem stability, Leopold came to appreciate the subtleties of an entity too beautiful in its workings merely to *study*. In either case, perception was the key. In some 1938 notes on "Economics, Philosophy, and Land," he wrote under the label "Esthetics":

We may postulate that the most complex biota is the most beautiful. I think there is much evidence that it is also the most useful. Certainly it is the most permanent, i.e. durable. Hence there is little or no distinction between esthetics and utility in respect of biotic objective.

Esthetics is an aspect of argument about land, not of land. It is part of the package system. We segregate esthetics so as to give farmers none and women's clubs a lot. In actual practice, esthetics and utility are completely interwoven. To say we do a thing for either reason alone is prima facie evidence that we do not understand what we are doing, or are doing it wrong.[50]

The dividing line for Leopold had become obscured when ecology itself proffered a vision of the whole too great to be contained by conservation as he had known it. And in applying that vision back to conservation, the old definitions of utility and beauty were altered. In an oft-quoted passage, Leopold wrote:

The last word in ignorance is the man who says of an animal or plant: "What good is it?" If the land mechanism as a whole is good, then every part is good, whether we understand it or not. If the biota, in the course of aeons, has built something we like but do not understand, then who but a fool would discard seemingly useless parts? To keep every cog and wheel is the first precaution of intelligent tinkering.[51]

Burling and tinkering, then, were legitimate human activities, but to undertake them with twentieth-century tools, without full environmental awareness, was foolhardy, wasteful, disrespectful, and dangerous.

The fallout—literal and figurative—of World War II added immediacy to the conservation cause. Leopold was only one, albeit leading, framer of the now globalized imperative. In his case, we find richer postwar shades of concern over the role of science, the primacy of indiscriminate utility, and the neglect of aesthetic values. One may read *A Sand County Almanac* as, in part, his personal reaction to these trends. There is also, however, a difficult-to-define fullness to Leopold's work at this point, a mellowness hard-won in the struggle to *comprehend* conservation. There is, too, a degree of clarity remarkable even by Leopold's standards:

> The citizen who aspires to something more than milk-and-water conservation must first of all be aware of land and all its parts. He must feel for soil, water, plants, and animals the same affectionate solicitude as he feels for family and friends. Family and friends are often useful, but affection based on utility alone leads to the same pitfalls and contradictions in land as in people.[52]

In the original foreword to *Sand County*, Leopold reiterated this provisional reconciliation of utility and beauty, and extrapolated to suggest the cultural potential of their symbiosis:

> We regard land as an economic resource, and science as a tool for extracting bigger and better livings from it. Both are obvious facts, but they are not truths, because they tell only half the story.
>
> There is a basic distinction between the fact that land yields us a living, and the inference that it exists for this purpose. The latter is about as true as to infer that I fathered three sons in order to replenish the wood pile.
>
> Science is, or should be, much more than a lever for easier livings. Scientific discovery is nutriment for our sense of wonder, a much more important matter than thicker steaks or bigger bathtubs.
>
> Art and letters, ethics and religion, law and folklore, still regard the wild things of the land either as enemies, or as food, or as dolls to be kept "for pretty."[53]

"The wild things of the land" still have instrumental value for humans, and always will. And perhaps the realization of their long-term instrumental value provides rationale enough for those reluctant to invoke inherent value. For real-world conservation purposes, the bottom line is that human demands and impacts are now pervasive, unavoidable, and unlikely to diminish. Under such circumstances, the preservation of beauty can be difficult to rationalize, much less achieve. But along these lines, it is instructive to note that, in one of the *least*-quoted passages from "The Land Ethic," Leopold combined an open "admission" of the utility of the natural world with his strongest defense of the preservationist position:

> ...A land ethic cannot prevent the alteration, management, and use of these "resources," but it does affirm their right to continued existence, and, at least in spots, their continued existence in a natural state.[54]

Can the two positions, in fact, be reconciled? Not easily, and not quickly. But Leopold, finally, was practical enough to see that they *had* to be, and idealistic enough to believe that they *could* and *should* be.

The distinction between preservation and utility will not go away in discussions of environmental history, strategy, and philosophy. Nor should it. But the question arises: by focussing on it exclusively, do we diminish the opportunity for more effective consensus not only on sustainable use, but on preservation as well? We may predict that a mature conservation/environmental movement will work across the full spectrum of land types, from the wild to the semi-wild to the cultivated to the settled to the urbanized, and will recognize the relevance of each to all the others. Leopold's statement of a land ethic has been criticized for being so broad, even poetic, as to lose meaning in application. I would defend Leopold here: only a broad and poetic statement could span the spectrum, and call every land-use and land-user to attention. Only such a statement could encompass the common ground between Pinchot and Muir. And although Leopold himself might look askance at such tributes, it is an encouraging sign that his name has now been bestowed on both a federal Wilderness Area (in his one-time home state, New Mexico) and a university Center for Sustainable Agriculture (in his native state of Iowa).

For we who particularly value wildness, there is hesitancy to admit that utility is a legitimate and, in any case, inevitable, aspect of humanity's relationship with the natural world. As David Ehrenfeld has recently written, "In an ideological war it is always dangerous to adopt the rationale of the enemy."[55] How do we avoid a potential weakening of the case for wilderness protection? By being on guard. By emphasizing the historical and geographical context. By being bigger, in numbers and character, than the blind economic determinists. By working and voting for the changes which will take pressures off of wilderness. By protecting our backyards. By spreading the word, until the ears of our rival brethren fall off if need be, that the attainment of a continuing, perpetual, self-renewing harmony with land—Leopold's definition of conservation—requires the fullest protection of wilderness and wild places that we can achieve. Ehrenfeld answers his own concern: "Neither of the classical varieties of conservation, protection or management, can by itself save the world's fauna and flora. Nevertheless, they are both vitally necessary as a holding action to save what can be saved until such a time, soon or far off, when humanity adopts a way of life in harmony with other life forms on the planet."[56]

From this vantage point, the environmental movement, back through the entire roster of protean conservationists and pitched battles, is seen as a process of continual evolution of the human race's ability to perceive and, hence, to anticipate and react to, new self-generated environmental conditions. The utilitarian cast of the early American conservation movement was probably necessary, but ultimately its finest contribution was its own subduction. Critics from other cultures and other regions of the world are likely, and perhaps even obligated, to point out, as did one recent commentator, that we in America sometimes "tend to equate environmental protection with the protection of wilderness," and that "this is a distinctively American notion borne out of a unique social and environmental history."[57] But if that tendency can continue, as it always has, to feed a broader, global environmental vision, then it may yet be our own greatest contribution to a new and improved worldview.

Edward Abbey described wilderness as a complement to, and a compliment to, civilization. An underlying assumption to this paper is that unless we understand, accept, and incorporate this complementarity, both our civilization and our remaining remnants of wild country face inevitable degradation. Leopold saw in conservation the possibility of a positive interaction of the complements. If conservation remains a state of harmony between people and land, we are now able to consider several corollaries: that that harmony cannot be static, for it eternally shifts and changes, creates and grows; that dissonance is not bad *in and of itself*—it allows us to appreciate harmony all the more—but only when it threatens the integrity of the whole composition; that even the most wonderful harmony can be enhanced by the solo expression. Allow cities. Allow wilderness. Allow everything in between. Allow people and land to sing separately, so that their recombinations may sing together even more wonderfully. But allow the song—*all songs*—to continue.

# Perceiving the Good

## by Erazim Kohák

The purpose of this paper is to examine the interaction between the ways we *perceive* value and the ways we conceive of it. I am persuaded that the ability to formulate an adequate and efficacious conception of value is contingent on a prior, prereflective *perception* of value—and that were we to *perceive* the world as devoid of value any *conception* of value that reason might formulate would remain as formal and unpersuasive as Max Scheler accuses Kant's conception of being.[1] Or, in a slightly different idiom, I shall claim that in order to *think the Good*, we need first to *see the Good*. The painful flaws in our conception of value which are letting us drift to an ecological apocalypse, I believe, reflect far more a perceptual than a conceptual failure, and so call less for a new conception of the good than for a new way of *seeing* the good.

That assertion, to be sure, might sound rather strange to a modern audience. We do not commonly associate the language of perception with the experience of value. Though upon reflection we might wish to deny it, in our habitual usage we tend to assume that it is discrete entities in space-time that are the objects of perception. At most, if we accept *Wesenschau* or eidetic intuition as a form of perception, we might add ideal objects to the class of the perceived. Value, however, tends to appear to us not as an object of perception but as a mode of perceiving an object that is value-free of itself. Hans-Georg Gadamer traced this subjectivisation of value with respect to aesthetic experience. The experience of beauty, he shows, appears to us as a subject's way of perceiving rather than as a perception of the beautiful.[2] *Mutatis mutandis*, his analysis describes the way we interpret all value experience. Though Plato spoke of *seeing* the Good and though the mediaevals had no problem with the idea of a beatific *vision*, our age has come to assume

173

that value is something that is not perceived but only conceived, constituted reflectively in the act of valuing.

And yet I wish to claim that in spite of our habitual interpretation of the experience we do indeed first perceive, only then conceive of value. Lest there be misunderstanding, when I speak of perception, I am using the term in the broadest sense of what Husserl describes as *gebende Anschauung* or, in Boyce Gibson's neologism, *dator intuition*.[3] It refers to any experience in which a subject encounters a content that presents itself as autonomously given, not a product of the act of perceiving only. For our present purposes, we can set aside the much debated question of whether or not such experience is or can be pre-linguistic and whether or not it can escape cultural conditioning, since we do not wish to advance any absolute foundational claims in its behalf. What matters to us is only that, in our ordinary experiencing and for the experiencing subject, such perception is indeed prereflective and prepredicative: that we see before we think, even if our thought can and does affect the way we see. By contrast, by conception I mean any reflective attempt, on however low a level, to order, interpret and categorise the givens of dator intuition, as for instance William James describes the process, in a close parallel with Husserl, in his *Principles of Psychology*.[4]

It is in this sense that I wish to claim that value is something we perceive, that is, encounter already on the prereflective, prepredicative level of dator intuition, or simply, of seeing. The stream of our lived experience, as James, too, recognizes in the work cited earlier, is not really a "buzzing, blooming confusion." It presents itself as a meaning-fully ordered context for which Husserl introduces the term *Lebenswelt*[5] and which Jan Patočka shows convincingly to be "a world of good and evil," that is, *ab initio*, prereflectively value laden.[6]

Considered in terms of ordinary experience, this is not actually a very radical claim. It might be that, were we in fact the pure intellects contemplating a self-contained reality from a standpoint three feet above existence for which we have too often mistaken ourselves. Yet we are surely not that. We are radically incarnate, active beings set in reality's midst. Quite apart from any cultural conditioning, simply in virtue of the fact that we are phototropic organisms that inhale oxygen and exhale carbon dioxide, light and airy places for us really are, not only appear, endowed with positive value *ab initio*. Certainly, upon reflection we might choose to invert that initial perception. When there is an air raid in progress, dark, dank dungeons become highly desirable. The point, though, is not our freedom to modify our initial value perception in reflection, but rather that initial value perception itself. Here we can invoke Merleau-Ponty: our prereflective experience is neither meaningless nor value-neutral but rather meaningfully ordered and value laden already in our initial perception.[7]

For our purposes, however, the interesting point is not simply that we in fact perceive value but rather *how* we perceive it. The example we cited earlier brings out one dimension of value perception. Because we are present in the world as agents, even in acts as elementary as breathing and moving, the world presents itself to us initially as either facilitating or as hindering our action. The impressive explanatory power of utilitarianism, I am convinced, stems less from the ingenious arguments of Jeremy Bentham, James and John Stuart Mill and their spiritual heirs than from the fact that utilitarianism articulates the fundamental, prereflective perception of value as utility. For, as Heidegger carefully points out in a wholly different idiom, the world amid which we find ourselves is in the first instance a world of *Zeuge*, of "gear" that either facilitates or hinders our action.[8]

There is, however, also another, altogether different but no less primary perception of value which stands out clearly within the transcendental-phenomenological brackets that Husserl proposes in his *Ideen*. In the second volume of that work,[9] Husserl cites an example that has since become quite familiar—that of touching my own body, as in the case of one hand touching the other. Here there is a dual experience. On the one hand, the hand I touch is there simply as an object in the world, facilitating or hindering my right hand's projects, there to be moved and manipulated much like any other object and like it endowed with a utilitarian value. At the same time, though, the left hand that my right hand encounters is also I. It is not simply an other object: that object is the presence of I, with its own inner autonomy, calling not only for manipulation but for respect. This is a person present in and as an object. I am aware of my presence in the left hand, and so with my right hand encounter it not simply as there-for-me, as I might a tool, but as there in its own right, endowed with its own integrity, the presence of an I that calls for respect.

This experience carries over into the experience of touching the body of another person and serves as the key to recognizing that body as the presence of a person. To speak of the body as "besouled," as Husserl at times does, does not do justice to his insight. The point is, rather, that an Other is embodied and present in and as a body. Thus that body, too, is not only there-for-me, one of the objects that aid or hinder me, but as there-for-itself, an appresented Other, with an integrity of its own and calling for respect.

Nor is that the case with the encounters with other humans merely. Though there is a scale of descending intensity, the same experience is repeated for instance in stroking a dog. Though the canine body is also one of the objects of my world, there for me, it is also very much the presence of an other, with his own thoughts, will, integrity and autonomy, appresented through the body I touch, and so calling for respect. That experience of the otherness of the Other may be much less intense in dealing with cold-blooded animals, yet it is never altogether absent. Nor is it absent in interaction with plants—a tree so clearly has a life and an agenda of its own—or, for that matter, in the cool feel of the boulder resisting the pressure of my palms as if, in a hopelessly inadequate anthropomorphic imagery, that boulder "wanted" to remain in its accustomed place. In contrast with the experience of the other as mine, one of the manipulanda of my life-world and perceived as useful or useless, we are here dealing with the experience of the other as Other, as having his intrinsic worth. When Whitehead spoke of the "subjective aim," he was not merely speculating but also, perhaps unwittingly, giving expression to a very fundamental dimension of prereflective experience.

For the sake of terminological clarity, let us designate this second aspect of prereflective value-perception as the perception of *intrinsic worth*, saving the term value for the value of utility. Terminology apart, though, I would suggest that these two dimensions of value/worth perception are equiprimordial and that their interplay provides the matrix for our further reflective efforts. Working still within our transcendental-phenomenological brackets, we can say that it is the primordial experience of the Other in his otherness, as appresenting me with an integrity and an intrinsic worth, calling for respect—Saint Augustine's *esse qua esse bonum est*—that makes moral categories relevant to our interaction with the world. Amid a world of objects-for-me, categories of use would suffice. When the world is perceived as endowed with an intrinsic worth, however, then such categories are not enough. There is a more basic level of consideration, that of respect. If anything, it is the perception of intrinsic worth of being that lends a derivative moral significance to what would otherwise be purely functional categories of use. The useful is also the good insofar as that which it serves is itself endowed with an intrinsic worth.

Within the matrix of our perception of intrinsic worth, the perception of the relative value of utility has the significance of making action possible. Action inevitably entails the augmentation of one being at the cost of a diminution of another, as in the notorious case of omelettes and eggs. Had we no basis for making preferential choices, were we only aware of the intrinsic worth of all being and of the grief of its passing, we should remain paralysed by what Paul Ricoeur once called *la tristesse du fini*.[10] It is the ability to discriminate on the basis of differential utility that makes preferential choices possible.

Yet there is a third moment as well. The perception of relative value of utility accompanies and modifies the perception of intrinsic worth, but it does not displace it. The lingering sense of the intrinsic worth of the being we have sacrificed to the greater utility of another remains, placing a limit on the self-righteous arrogance of utility. Saint Paul's claim, "Ye are not your own, we are bought with a price," applies to all our utilitarian preferences. Thus utility must tread humbly, with gratitude, ever conscious of the cost it exacts—and willing, at times, to be overridden by it. Utility can justify use, never abuse or heedless waste. Though justified by a relative utility in service of the intrinsic worth it augments, at a price, our action remains circumscribed by respect for what it uses.

Or so it appears, as long as our transcendental-phenomenological brackets remain firmly in place, shielding our analysis from the realities of our age, our culture and its life-world. With the brackets removed, the picture changes drastically, revealing a devastated world and a devastating culture, seemingly incapable of perceiving worth or value and able to conceive only of the value of utility. In its name, it appears committed to ever expanding consumption and largely unmoved by the most excellent and rigorously constructed arguments of environmental ethicists.[11]

Among ecological writers, it has become almost an article of conventional wisdom that what we are seeing around us is a product of the disenchantment of the world at the hands of Descartes, Galileo, and their scientific heirs.[12] As long as humans thought of the world as the cherished handiwork of a loving God, the thesis runs, they treated it with respect for its intrinsic worth. Once, though, they came to conceive of it as an accidental conglomeration of matter in motion, they came to treat it as no more than raw material for satisfying their needs. On this reading, the way we perceive the world is a function of the way we conceive of it—and if the practice that follows from it is flawed, the solution is to be sought in a more adequate conception. Philosophers, sensing themselves on familiar ground, have fallen to the task and offered an entire range of proposals for reconceptualizing the world.

If, however, our earlier argument is sound, then such an approach reverses the true priorities. On our reading, the reason why a civilization has lost all conception of intrinsic worth is to be sought in a more basic loss of all *perception* of intrinsic worth. Or perhaps not altogether all such perception: though we are all too ready to use our fellow humans as materials in social engineering, we still draw back short of full commercial utilization, at least when confronted with the actual pain of a concrete individual. Similarly, while willing to ignore the plight of animals in experimentation and in commercial farming, we are capable of responding with empathy to the suffering of an individual animal or the destruction of a particular tract of woodland, even if we do justify it in terms of recreational utility.

By and large, however, such residual glimpses of intrinsic worth appear to us a matter of sentimentality—and, on our reading, not because of the way we conceive of the world but because of the world we perceive. Milan Machovec sums it up with a poetic flair: "We walk on asphalt, not on the good earth, we look up at smog and neon, not at the starry heavens above."[13] For the vast majority of humans, the effective life-world of their daily acts consists almost entirely of artifacts, neither created nor grown but manufactured by humans for human use. A tree has its own life, its own agenda, a telephone pole exists solely for its human use—and in our effective life-world telephone poles vastly outnumber trees. Perceiving a world of artifacts whose sole meaning is their utility, humans might very well gain the impression that they are the source of all value, and that all reality is a reality-for-them. The reason is not that that is how humans of our age conceive of reality: many of them, after all, are religious believers and, intellectually at least, conceive of the world as God's creation. The reason is the way they perceive reality and the reality they perceive. Thus the most urgent task for our time might not be the one at which philosophers are most adept, that of constructing a more adequate conception of reality, but rather the premetaphysical, Socratic task of learning to *see the good*, to perceive the intrinsic worth of being.

How, though, can we regain the clear vision of the good, of the intrinsic worth of being, when there seem to be so many reasons for concluding that it is simply not there to be seen? Our effective life-world is one of artifacts whose being appears reducible to the use for which they were designed, in the service of their human makers. If that is so, then the quest is vain, and the resources of philosophy would be far better applied to the task of providing at least a framework of relative human value that would provide a context for a life worth living in a world devoid of absolute worth.[14] Even those thinkers who are not prepared to give up the hope of grounding the relative value of utility in the intrinsic worth of being admit that the unproblematic sense of the intrinsic worth of being that marked precritical metaphysics with its unquestioned assumption that *esse est unum, bonum et verum* has become effectively unavailable.[15] Today the task is less one of simple seeing than a Socratic one of anamnesis, of recalling an awareness long buried under a heavy layer of forgetting.

Here it may be useful to take a fresh look at contemporary philosophical efforts, not in terms of the overt tasks they set for themselves but in terms of the way they approach the underlying task of anamnesis. Here labels are not readily available, though I would suggest three broad strategies do emerge which we might label speculative, contemplative, and practical, respectively. And, while each category would include philosophies we normally do not associate with each other, in each case we can select an instance that can serve as a paradigm for this or that strategy.

Among the strategies I would label speculative perhaps none is as grand as Edmund Husserl's attempt to anchor the relativity of the subject in the absolute matrix of transcendental subjectivity. The *locus classicus* is the work of Husserl's final years, especially Part III.B. of his *Crisis of European Sciences*, the sentence which sums up the attempt most clearly comes in § 72—"...*die absolut fungierende Subjektivität zu entdecken, nicht als die menschliche sondern als die in der menschlichen, oder zunächst in der menschlichen, sich selbst objektivierende.*" Or, in David Carr's English translation, "...to discover the absolutely functioning subjectivity, not as human subjectivity, but as the subjectivity which objectifies itself, (at least) at first, in human subjectivity."[16]

The significance of that sentence, reinforced by the "monadological" reflections in Husserl's *Nachlaß*,[17] tends to be obscured by Husserl's own earlier, still significantly Cartesian conception of phenomenology as a quest for a higher order of objectivity, reducing the relativity of objectivity of fact to the absolute being of consciousness. For all of Husserl's strenuous efforts, that attempt remained tainted by a suspicion of psychologism, and, in spite of the dramatic shift from a static to a dynamic conception of phenomenology, the early portions of the *Crisis* can appear continuous with it.

Yet what Husserl is attempting to do in the last part of the *Crisis* as well as, *anticipando*, in *Ideen II* and again in several posthumous texts[18] appears to me as something of a different order. As the sentence we cited suggests, here the aim is no longer one of reducing the contingency of fact to the constituting subject but rather one of locating the subject within the matrix of absolute subjectivity. And, as I read the text, subjectivity here is not a code name for the subject, or for the subject writ large, not even for the subject read collectively, as it appeared in a number of earlier passages. Rather, it points to a characteristic of reality, of being itself, now presented not as merely objectively—causally and extensionally—ordered, nor as meaningfully ordered for the constituting subject, but rather as meaningfully ordered in itself, intrinsically and as such. Just as objectivity is not an object but a mode of being, so subjectivity is not a subject but a mode of being. The absolutely functioning subjectivity, Husserl stresses, is not a human subjectivity, though it objectifies itself therein. Rather, it is prior to human subjectivity, the meaningful ordering with which humans endow the world in constituting it. It is the intrinsic meaningful ordering of being itself, not a human perception of reality but its intrinsic meaningful order.

There are, of course, problems that need to be worked out and on which Husserl barely touches even in his posthumous texts. Still, in principle, were the task merely a conceptual one, we might well rest content with the solution Husserl proposes. That, after all, was what was wanting: a conception of being which would recognize it as primordially and intrinsically meaningfully ordered and value laden. And yet there is something disturbing in the very neatness of Husserl's proposal. Though Husserl insists that the absolutely functioning subjectivity is not a human one, only objectifies itself as such, it is, as Jan Patočka never tired of pointing out,[19] yet somehow familiar, continuous with human subjectivity. Reality as we experience it, however, is seldom only that. That familiar, meaningfully ordered reality, is always set within a prevenient horizon that is radically other—Husserl speaks of it as the dark horizon. That dimension of otherness, the hardness of reality, is there even in the most familiar objects, a radical otherness within the easy familiarity.[20] In fact, as we noted earlier, it is precisely the sense of otherness of reality, its own integrity irreducible to other-for-me, demanding respect, that evokes the lived experience of intrinsic worth. That is the experimental dimension that is lost in the conception of reality as meaningfully ordered by transcendental subjectivity. That is not a problem of error in argument. It is a structural problem. Speculative strategies, seeking to reach out to the otherness of the other, inevitably enclose the other within the reach of human subjectivity and so mask the dimension of its otherness—a criticism raised already by Kierkegaard against Hegel. In fact, the strategy may be stronger than the criticism: once humans have encountered being in its awesome, radical otherness, Husserl's strategy may well be optimal for coming to terms with it. The experience, though, must come first, and remain as an irreducible reminder.

The common trait of the approaches we would label *contemplative*, however inadequate that term may be, is precisely the attempt to assure the availability of that experience, suspending the constitutive activity of critical reason and standing in mute awe before the wonder of being. To be sure, philosophers being what they are, that mute awe produces a flood of words. Romantics—and, as Neil Evernden argues persuasively,[21] that is *not* a derogatory term—whether Henry David Thoreau or Richard Taylor or others,[22] have ever sought to use words not to construct an edifice of arguments but rather to evoke the experience of a world that is not yet "Man's" but God's, a world from which being has not withdrawn. Martin Heidegger's late writings, especially his meditations on Hölderlin's poetry, represent one of the clearest philosophical examples of that strategy.[23] Heidegger's use of language here is not referential. It is evocative, seeking to render words transparent to the experience of being in its majestic otherness, untamed by concepts. Even describing such texts as hermeneutic is inaccurate. Far more, it is an attempt at premetaphysical philosophizing, reaching back to the original philosophic wonder. Heidegger's immense impact on his contemporaries testifies to its effectiveness. Whether philosophic or poetic, the strategy of contemplation does confront the reader who is willing to suspend critical reason with the presence of being in its intrinsic worth.

To be sure, the contemplative strategy is even more vulnerable to criticism than the speculative. I would not attach particularly great weight to the most usual one, that the retreat to Walden Pond, to a New Hampshire clearing or the vastnesses of Schwarzwald is the privilege of a chosen few, not a viable social strategy. Of course it is true. Even if such a retreat were desirable and possible, there are not enough Waldens to go around. But the contemplative strategy had never been intended as a viable social option. It has always represented a solitary quest for personal insight, for a rediscovery of the wonder and worth of being which the seeker offers not as a model but as the source of an insight that can be shared. The perennial popularity of such personal testimonies suggests that the insight can be communicated and evoke a new vision even in those who share in it only vicariously.

The real weakness of the contemplative strategy, I would suggest, lies elsewhere. Heidegger unwittingly points to it when he writes that "any kind of polemics fails from the outset to assume the attitude of thinking."[24] His assertion is straightforward enough: contemplation in truth calls for a suspension of critical reason. With that suspension, however, goes also the ability to make those differential judgements of which we had spoken at the start. As a number of writers have pointed out recently, it is not Heidegger's unfortunate brush with national socialism that casts the gravest doubts on his thought. When an age is out of joint, humans err. Rather, it is his inability, years after the war, to condemn clearly and unambiguously the monstrous evil of which he had been a witness.[25] Certainly, as Heidegger pointed out in his *Letter Concerning Humanism*,[26] even the best intended categories of human utility (in our terminology) degenerate into self-blinding human *hybris* when they are not grounded in a sense of wonder and intrinsic worth of being. Yet Heidegger's own inability to provide the grounds for a clear moral judgement when such judgement was most called for attests that the reverse is no less true. A sense of wonder, stripped of all categories of utility, sinks into a poetic impotence. Both are needed.

It is this dual need that makes the strategies we had labeled, none too happily, the *practical* strategies so appealing. It is a broad category, intended to include all who, in doing what is useful, encounter what is good. To pick once more a philosophical example, Hans Jonas' analysis in *The Imperative of Responsibility*[27] is set in the intensely practical context of the recognition that first emerged with the detonation of the first nuclear bomb[28]—that humanity can no longer take either itself or its world for granted. Ethical decisions have acquired not only a personal but also a cosmic significance. Thus ethics requires not only a perspective of utility but also a matrix within which to set it, one that religion once provided but which we now must seek in recognition of the intrinsic worth of being as such.

Can we attribute intrinsic worth to being? Hans Jonas' answer is unambiguously affirmative, though the way he arrives at it inverts the traditional derivation. The tradition had always derived the value of particular acts from the intrinsic worth of being which those acts augment or diminish. Jonas' claim is the opposite—that being has intrinsic worth because it makes value possible. Though the end of an act may be intrinsically worthless, the fact that it is the end of a purposive action constitutes it as valuable—the desired becomes the desirable. Thus for Jonas it is in the context of praxis, of positing aims and so values, that being acquires its worth as the potential locus of value. Being, we could say, is good, intrinsically good, because it is capable of goodness, capable of being the bearer of value.

Such a one-paragraph summary, to be sure, fails to do anything like justice to Hans Jonas' rich and captivating argument, addressed directly to an age whose sense of value has been reduced to the appreciation of utility. The experience of utilitarian value is one we all share. By showing that the continuation of that acknowledged value is directly contingent on the survival of being—of humankind and the world—Jonas leads his reader to a concrete recognition of the intrinsic worth of being, a seeing of the good.

And yet here, too, doubts emerge. Can such a derivative conception of intrinsic worth, derived from utility, serve the dual function of Augustine's long forgotten *esse qua esse bonum est*—that of bestowing a moral significance on utility and that of putting an absolute limit on utility's claims? Jonas argues, painstakingly and persuasively, that it can—and, in any case, it is the best we have in face of a potential nuclear and ecological apocalypse. And yet a lingering doubt remains: does an act, worthless and pointless in itself, really bestow value on its object simply because it is purposive? Or does such a "value," internal to a value-less act, remain itself value-less? Do we need something more—that direct vision of the good, not as derivative from value but as autonomously given, a worth a priori analogous to Rudolf Otto's religious a priori?

In spite of such doubts, I believe that Hans Jonas' reflections may be the most valuable, if we interpret them in perceptual rather than conceptual terms. Here a minor phenomenological detour might prove helpful. Simply in terms of our actual experience, as we live rather than as we reflect upon it, when is it that humans experience the intrinsic worth of the other most clearly? It is not in loving, enjoying or needing the other. Though we may not live such situations as mourning a lost love in crassly utilitarian terms, they are still situations in which the other functions as being-for-us. Nor, contrary to popular lore, is it in losing what we once cherished, though such loss does graphically bring home what we once took for granted, how much the other meant *for us*. Still, it is the experience of our loss, and so presents the other as being-for-us. Rather, it is in the experience of active concern, active caring for the other, in giving of ourselves to the other that we encounter the other in his intrinsic worth. At one end of the scale, it is the doll the child most cared for, or the child the parent nursed most self-lessly, that will appear as most precious. At the other end, it is the person who most sacrificed for his country to whom that country will be most precious. Here a generalizable principle emerges: it is not in taking, in receiving from the other, but precisely in giving to him and giving up for him, in caring for him in his need and weakness, sometimes in spite of his use-lessness for us, that we come to perceive him in his intrinsic worth. Not in taking, but in giving of ourselves to the other do we learn to *see the good*.

That, again, is hardly earth shaking news—far more, it is a recognition as old as humankind, though overlaid with forgetting. For millennia, we have been told that it is more blessed to give than to receive, that whosoever would save his life must lose it, that we should seek first the Kingdom. The behavioral impact of such injunctions, however, has been less than overwhelming. Is there any reason to believe that they would carry greater weight today?

It is here that the considerations with which Hans Jonas begins his work become relevant. Though we have little noted it, the place of humans in the cosmos has been radically transformed within our life time, perhaps the first such transformation since *homo erectus* first walked upon this earth. Through all the millennia of human presence, humans have always been able to take nature for granted. It was eternally there, the dark ground and the expanding horizon of human life, the vast cyclical presence enfolding us in its rhythm, at times threatening, at times comforting, but always there. We could assume that we could slash and burn, we could kill and destroy, we could fight our wars, confident that when our tumults stilled nature would restore the devastated forests and purify the tainted seas, that there would always be the vast sky above and the fertile earth beneath our feet. There would always be more whales, there would always be more elephants.

Today we are confronting the day when there will be no more whales, when there will be no more elephants, wise and majestic in their gentleness. Literally as well as metaphorically, nature has become an endangered species. We have exceeded its ability to renew itself. Nature is no longer something we can take for granted, it has become our responsibility. We can no longer be content to care for humanity and use the rest of reality for that purpose. If we are to survive at all, we need to learn to care for the other.

Yet that precisely is the shift of which we spoke earlier as the key to learning to see the good, the shift from a basic posture of taking to giving, caring, from nature as being-for-us to nature as being-for-itself, or, in the jargon of the trade, from an anthropocentric to an ecocentric perspective. To be sure, the environmental concerns arising from it hardly represent a sudden outburst of pure altruism. The initial impetus was, for the most part—though with honorable exceptions—a purely utilitarian fear lest our short range greed endanger our long range greed. The initial goal was no more than a more enlightened approach to exploiting the earth for the benefit of one rapacious species.

Yet if our initial analysis has any merit, then this conception is not what is crucial. Rather, perception is, and that perception is shaped by practice as much as it shapes it in turn. Both the shift in the tone of environmentalist writings and in expressions of public perception which I hope to document in detail in a later version of this paper, suggest that those who concern themselves with giving, caring, saving the world also come to perceive it differently, whatever their original motivating conception of it may have been. Even the humblest of creation's orphans, say, the aluminum beer can discarded by the roadside, yet representing an energy saving equivalent to three dcl of gasoline if recycled, can stand out in its worth, as good in itself, in the course of a purely economically motivated practice of recycling. While the posture of greed transforms our perception of the world into one of a store of raw materials, a posture of caring can open human eyes to a perception of its worth, the secular equivalent of the beatific vision: seeing the good.

It would, to be sure, be vastly unrealistic to expect the world of environmental concern—the only world likely to exist a century hence, if any does—will be a world of beatific vision. Yet perhaps a more modest point is justified: yes, even in the world of human artifacts and constructs, not only in the world of God's direct creation, are humans capable of seeing the good, and it is the posture of active concern, giving rather than taking, that makes it possible.

Philosophy, surely, has many tasks. It may not, though, be an overstatement to suggest that in our time one of its most important tasks may well be that of seeking to recover the vision of the good and of teaching the practice that makes that vision possible.

# A Brittle Thesis:
# A Ghost Dance:
# A Flower Opening

## by Michael Peter Cohen

How does one defend a brittle thesis? Not at all: it is the better part of discretion not to defend an old thesis. What I have written about John Muir seemed important to me at the time. Now others can read and judge. When I wrote my story of his spiritual journey through the wilderness, I did not conceive a theory, though the narrative did have a plot. In retrospect, I now recall that many readers of the manuscript were unhappy with the plot; they wanted a more inspirational and victorious ending. More recently, in writing about the Sierra Club, I have had this same problem with plot, which I have discussed elsewhere.[1] In fact, my plots are incomplete because the history of conservation, or of environmental awareness, or whatever we call it, at this time remains *in media res*. If there is a paradigm shift coming, we will all find ourselves in an enlightened world, all deep ecologists, but, as the Oiler says in Stephen Crane's "The Open Boat," we are not there yet. This revolution has not taken place. Nevertheless, one does not want to be caught defending an old thesis.

Caught between the comic and the tragic resolution to my stories, believing, as I do, that Joseph Meeker's theory in *The Comedy of Survival* goes to the heart of the matter, but unhappily finding that my own writing about environmental history tends toward a pattern of incidents most would describe as tragic,[2] exploring the interpretation of patterns of incidents, plots and history, I am not alone in my concerns, and I certainly may not be expert in this craft. I note that Hayden White has been engaged in this problem for a while, with regard to the writing of history, and also Clifford Geertz.[3]

Perhaps my situation is a result of reading and rereading the common theme of certain environmental historians; to name three, Alfred Crosby, Patricia Limerick, and William Cronon. Their message is historical continuity in the midst of cultural diversity: Many of the patterns of events we call history have happened before. Being of a literary background, I have learned to think about these patterns contextually, in terms like those developed by T. S. Eliot for "Tradition and the Individual Talent." Eliot says, "the historical sense involves a perception, not only of the pastness of the past, but of its presence.... This historical sense, which is a sense of the timeless as well as of the temporal, and of the timeless and of the temporal together, is what makes a writer traditional." In Eliot's sense, modern environmental historians mark themselves as traditional writers by their thematic aims. Lately, I have come to see that my work shares something with these historians. When, over the past ten years, exploring the presence of the past, I have asked what John Muir means to my generation, asked what the Sierra Club in the post World War II West has done to change our lives and the way we perceive our lives, like other environmental historians of my generation I turned frequently to Thoreau's method in the writing of *Walden*, where the narrator sees his life and his reading as part of one continuous fabric which links various times and places.[4] But I now see a weakness in the fabric of my thought, which forces me to reconsider the accuracy of my depiction of the presence of the past. Must I begin again, with John Muir newly arrived in California, camping in the great Central Valley?

The rent in my fabric was worse than I thought, worse in more ways than I probably understood, or understand even now. I had been reading in 1979 Muir's "Bee Pastures of California," and Raymond Dasmann's work.[5] I wrote, "It was too late [in 1870] to save the Central Valley. More than even Muir knew, its wild bloom was gone."[6] This was about John Muir's desire to preserve the "plant gold" of California. I believed at the time that Muir's admiration of the flora was limited to his knowledge of the indigenous annual plants, because the perennials were long ago destroyed by Spanish cattle.

These are the things I did not know or acknowledge: I did not know the aboriginal condition of California's grasslands, any more than Muir did. I did not acknowledge the extent to which the man who punned his name with moor unavoidably thought of the California grasslands in European terms. I did not fully realize that one of Muir's major literary symbols, the immigrant bee, was a perfect symbol of the Europeanization of the American continent. I thought of the Spaniards as denuding the grasslands of the Central Valley, where in fact they introduced Eurasian species. Like Muir, or from Muir, I saw incompletely the changes around me. When he counted the flowers around his campsite one morning, Muir was looking at something like a hundred alien weed species. By my time, 63 percent of the grassland species of the Central Valley were introduced plants.[7]

These facts have radical spiritual significance; ecological issues flow into literary issues. If, as some argue, Muir was himself a Neo European weed who did not expose himself to a new ecological world, but sought a Muirland, an imported European world, if he was programmed by American civilization, as historians like William Goetz-mann might argue, if his escape from his father was an expected dialectic, if he can be seen as grist for a psychobiographical mill, then I must reconsider. Perhaps I also downplayed the degree to which writing and books shaped Muir's experience. And with new books between us, I have changed over the last ten years. Has Muir changed?

The implied thesis of my own thinking about Muir may well be untenable, that the thirty year old Muir—not very young, really—was open to the influences of the wilderness he explored, and that wilderness transformed him into an exceptional human being. That thesis entered into my writing strategy when I attempted to judge the older Muir by the philosophical standards of the relatively young man. (Well, of course my method sounds like the 1960s, doesn't it? What else should it sound like?) You know I would not want to have written about Muir if I didn't believe what he wrote about the transformational possibilities of the wild.

So now, the antithesis my recent reading has provided: We humans are all weeds.[8] We can define ourselves as *INVADERS*.[9] As Europeans, we have been amazingly pervasive, if not successful, in the new world.[10] This antithesis applies not simply to European man in the post Columbian world, but to all humans on this continent. From the time of Paleolithic man to ours, humans have been killing off species. The myth of the ecosensitive tribe is just that, a myth. We are aliens, and we are alienated. Thinking of man as a weed has its depressing aspect, but it also has its inspiring side. On the up side, weeds are most successful in an ecosystem which is in the process of being trashed.[11] Weeds save! as Alfred Crosby asserts. Those who hate them are or were wretched ingrates. In a different and intriguing way, Dolores LaChapelle has looked in the same direction, I believe, with regard to gourds.[12]

There is a paradox here: we bring the weeds and we are the weeds. We need the weeds if we are to save ourselves from ourselves. We must love ourselves if we are to save ourselves from our hateful deeds. Well, that is badly put; it is not even something that we choose. If our history has a biological or ecological component, it is by definition a matter of relations—not our intentions—and these relations can only be partially controlled at best. Our history is ineluctable because it is a net which we do not control. Perhaps.

This weeds stuff, this invader stuff is, of course, a metaphor, but a very suggestive one. Following it leads to these kinds of things: first of all, quarantine does not work. One pays for an edenic existence, eventually. The longer a wild place is isolated from the ecological change Euroman brings, the worse will be the disaster when it comes in contact with those forces of Eurobiology and Euroculture. Further, there are going to be new invaders. They will come. We will never be able to "Dream Back the Bison, Sing Back the Swan."[13] According to the historians, we ought to suspect that the ideas we call Deep Ecology may be a sort of philosophical Ghost Dance.[14] In any case, we have been deceiving ourselves. Those animals which we think of as symbolic of "western wilderness" are in fact the most European of the wild species on this continent, which is why they have lasted as long as they have.[15] As weeds, we will always be unconscious of our effects; not us, but what inhabits us, makes the changes in our world.[16] There never was a virgin land, and the indigenous humans we replaced on this continent were not passive. They themselves made the ecological changes on this continent which cleared the way for us.[17]

As this chain of reasoning indicates, clearly enough, the views of environmental historians can provide cold comfort. Colder than Crosby and Limerick, and closer to serving as a direct critique of my own method of study, is the work of William Cronon.[18] Because he begins in New England, Cronon introduces his historical method with "The View From Walden"—Thoreau's view, obtained by leafing through the *Journal*. Cronon finds in Thoreau a central modern concern, the extinction of species, and speaks of Thoreau's "romantic lament for the pristine world of an earlier and now lost time." Digging in, Cronon approaches Thoreau's worst personal fear, the antithesis which threatened to negate the whole direction of his thought. In a certain mood, Thoreau wrote of the diminished ecosystem of his Concord as "an emasculated country," and he wondered, "Is it not a maimed and imperfect nature that I am conversant with?" The point is not lost on Cronon, who sees that "...a changed landscape meant the loss of wildness and virility that was ultimately spiritual in import, a sign of declension in both nature and humanity."[19]

I hope it is clear that this situation is a close analogue to the one I face with John Muir and the Central Valley, for, as Cronon points out, "The view from Walden in reality contained far more than Thoreau saw on that January morning in 1855."[20] When Thoreau sought to know nature, he imagined his "project"—to use a contemporary term—as the reading of an entire poem, but he discovered that he possessed an imperfect copy: "...my ancestors have torn out many of the first leaves and grandest passages, and mutilated it in many places." Here is the central problem which belongs not simply to Thoreau, but to us, and our progeny. And here is Cronon's solution:

> Human and natural worlds are too entangled for us, and our historical landscape does not allow us to guess what the entire poem of which he spoke might look like. *To search for that poem would in fact be a mistake.* Our project must be to locate a nature which is within rather than without history, for only in so doing can we find

human communities which are inside, rather than outside nature. (italics mine)[21]

I cannot help but note that the historian must set up and knock down some pretty contrived straw men in order to arrive at this anthropocentric, yet poetry extinguishing, perspective. (Just how sterile the perspective is becomes clear when one remembers Aldo Leopold asserting, "I am glad I shall never be young without wild country to be young in. Of what avail are forty freedoms without a blank spot on the map." And then one realizes that a twenty-nine-year-old historian seems to be revelling in the declension of a spiritual world.) But I was about to say, the main straw man is contained in the statement: "There has been no timeless wilderness in a state of perfect changelessness, no climax forest in permanent stasis."[22]

Really, I must be incredibly naive; such reading material never fails to shock me, and keep me awake nights. I say to myself, "If this is what historians do, I'm not a historian. If this is where nature resides, it has little interest to me. If this is the way we shall think, I prefer not to." I am tempted to go back to the mountains and dance to these ghostly ideas (no doubt illusions), and live those experiences (no doubt escapist) which seemed to me in the past, and still seem to me, a way to live a life of harmony and joy. Do we not all contain the poem, and is it not time to let it out? Even my poor poem?

But I cannot forget this antithesis, which comes from reading environmental history (and dare I admit also literary theory?) in the 1980s. Not being able to forget is perhaps the major problem in the transition from modernism and postmodernism; it may also be the solution. Can I, under the circumstances, trust what I have thought about Muir, or trust Muir's thoughts as if they are not fictions which I create?

Before I trust myself, I like to consider the naivety of my ideas about wilderness in the 1950s and 1960s, which were often little more than wilderness sloganized. I have had to consider these in writing the Sierra Club History, a process which uncovered many of the sources of my own ideas of wilderness, while I simultaneously uncovered the Club's version of John Muir, a version very different from my own, and one that changed with time. The wilderness movement is moving beyond slogans; seeing how far it has come is pleasing, but I still recognize that combination of naive and sophisticated thought so characteristic of the wilderness movement of the 1960s. On the naive side, we still entertain the Brower definition of wilderness as the place where "the hand of man has not set foot," this slogan linked to a kind of crisis mentality where "What we save in the next few years is all that will ever be saved."[23] On the other hand we have a more sophisticated and recursive notion of Nancy Newhall, that "The wilderness holds answers to more questions than we yet know how to ask."[24] Later in the 1960s, the movement acquired the idea of wilderness as a "scientific" baseline, in hopes of gaining credibility. This one is patently absurd when one considers the extent to which there is no "pristine" wilderness, and we men have always been invaders, carrying our portmanteau biota. Further, our Eurolanguage, as Dolores will point out, often betrays us as much as it allows us to think.

When I think of my own naivety, I see myself in a description of Melville written by Hawthorne after the completion of *Moby Dick*: "It is strange how he persists ... in wandering to-and-fro over these deserts, as dismal and monotonous as the sand hills amid which we were sitting. He can neither believe nor be comfortable in his unbelief...." I recently read, and copied for myself, a darker version of this plight, written by Jorge Luis Borges: "It is as if Melville had written [in Bartleby]: 'It is enough that one man is irrational for others to be irrational and for the universe to be irrational.' The history of the universe teems with confirmations of this fear."

It could be worse. I recall a story told by Gershom Scholem of a certain Hermann Cohen, head of the Marburg school of Neo-Kantianism, and a famous hater of pantheism. Scholem counts Cohen as the author of the most profound short critique of Zionism: "*Die Kerle wollen glücklich sein!* [Those fellows want to be happy!]"[25] No doubt I lack wisdom, and have not achieved enlightenment. It is, of course, Muir I turn to when this perspective gets too pervasive or oppressive. In particular, I recall the Muir who read Emerson's statement, "the impressions of nature fall too feebly on men and are inadequately reported," and who responded in the margin of his edition of Emerson's *Prose Works, I*, "Report what befalls you."²⁶ Which leaves me to go back to the mountains for new inspiration, where I can imagine thinking like a mountain—better than thinking like a weed? Certainly more inspirational than thinking about writing about writing!—especially when, as Muir said, "all the world seems a church and the mountains altars."²⁷ I went back to the mountains last summer, as I do every summer. I report what befell me, giving "some characteristic pictures...strung together on a strip of narrative."²⁸

My report is a little like the story about the elephant: three blind men, each with their own part: trunk, body, leg; an exercise in partial understanding. (Lately, my mate has been reading about elephants, and my nearest colleague at school is Hindu, so I hope you will excuse the metaphor about the elephant. I note that even Clifford Geertz is using it these days.) An elephant, like a mountain. How? Well, as it happens, I summer, these years, in Jackson Hole, a region dominated by what the French referred to as the *Trois Tetons*. Actually, that is the way they look from Idaho, from Pierre's Hole; from Jackson Hole, they are more anglicized, and from one view, the Tetons are called the Cathedral Group. Muir himself made his first enlightening climb of a mountain in 1869, and on a granite horn or tusk called Cathedral Peak. (I climbed it a hundred years later, deliberately on the centennial; at that time, I considered my experience not different from Muir's, but it was, of course. I met Axe Nelson near the summit, but that is another story. The climb of Cathedral Peak also serves as the final capitulation of *The Pathless Way*.²⁹)

The highest Teton is, of course, the Grand Teton. Everybody knows the Grand. Big One. Many people have climbed it many times. I have climbed it several times. The last time I climbed it, when I remembered the old story of the elephant, I also remembered someone saying that elephants were an endangered species.

Six of us met on the Lower Saddle one afternoon. The next morning we would be broken into three groups. There was the intellectual historian—now historian of environmental ethics—Roderick Nash, who would be led by my friend the climbing ranger, avalanche expert, ski patrolman, and occasional guide, Tom Kimbrough. Kimbrough, I had known for many years, since the 1960s, when we climbed in Yosemite, read *Zen Flesh; Zen Bones*, and recited koans over the campfire at night. I think of Kimbrough as my Marlow, "a most discrete, understanding man," as Conrad put it, who "was the only one of us who still followed the mountains," to alter Conrad's words somewhat. There was also Jim Brady, a new Assistant Superintendent of the park, climbing with the supervisor of the climbing rangers, Jenny Lake Subdistrict Ranger Pete Armington. There was the writer of the *Climbers Guide to the Grand Tetons*, Leigh Ortenburger, climbing with— well, he was really guiding me.

We had all done our homework, or were doing it on the mountain. Kimbrough read Nash, because he likes to read, and because he likes to know who he is climbing with. (He borrowed my copy.) I had read Ortenburger, in fact learned to climb in the Tetons two decades ago with Leigh's book literally in hand. Armington had prepared perhaps to impress the new assistant superintendent, who was, in turn, inspecting the Park.

We were not initially there together by choice, or pleasure. Kimbrough was there because he was ordered to be there, as the climbing ranger most likely to make a congenial guide. Ortenburger came to argue with the assistant superintendent about the placement of roads in the valley, all of which can be seen from the summit as if they were on a map, and I was there to see how *Wilderness and the American Mind* fared on the mountain. Well, maybe I wanted to impress Rod. I also wanted to climb with Leigh Ortenburger, at least once in my life. Also, I sometimes feel sort of stupid, living in Jackson Hole every summer and almost never going climbing. Nash wanted to go someplace with Brady, who was an old friend, though why he would choose the Grand was beyond me. Nash had climbed it once, back in the times when he had worked in the Park and learned to row on the Snake River. In fact, three of us, Brady, Nash, and myself, were being guided, though in different ways.

If I was there to learn something from Leigh, and Rod was to be guided by Tom, Leigh expected some conversation on history from me, and so did Tom from Rod. Tom even had his question prepared. He decided to take issue with an assertion Nash makes in the Preface to the

Third Edition: "A few cultures in today's world are still precivilized in the sense of having a nomadic hunting and gathering economy. It is significant that they have no word in their vocabularies for 'wilderness'. . . . Lacking a concept of controlled and uncontrolled nature, [they have] no conception of wilderness." (p. xiv) What then, said Tom, was the Enclosure?

This was a more sophisticated question than Nash would willingly acknowledge. Tom, an inveterate reader of Thoreau, remembered well the lines from "Ktaadn," in *The Maine Woods*:

> The tops of mountains are among the unfinished parts of the globe.... Only daring and insolent men, perchance, go there. Simple races, as savages, do not climb mountains,—their tops are sacred and mysterious tracts never visited by them. (pp. 271-272)

For Tom, Nash's ideas were simply an extension of this Thoreau. As Ortenburger describes the Enclosure, in his *A Climber's Guide to the Teton Range* (1956), it is a ring of rocks sitting on a high rib west of the Grand Teton, at an elevation of over 13,200 feet:

> A worthwhile 15-minute side trip from the Upper Saddle to the Enclosure (this is easier than it appears) will yield an excellent view of the entire west face of the Grand Teton. The curious rock structure on the summit which gave this point its name is of unknown origin, but was probably constructed by Indians long ago.[30]

Ortenburger is fully aware of another possibility, and indicated it in a later version of the *Climber's Guide*: the Enclosure may not have been constructed by Indians who never visited such a place. It may have been built by a French trapper.

One watched Nash respond to Tom. Rod acknowledged the question, seemed to find a category where it could be safely filed, and said, "Ah, the question of primitive, pre-articulate experience," or some such thing. Click, whirr. (If you don't have a word for it, like our word, then you don't have the idea. *WATAM* is filled with cases in point of "the tendency of wilderness enthusiasts to arise from refined, urban situations.")[31]

That August, Rod had already finished *The Rights of Nature*, in which he sides with Euroman and advocates "finding solutions to ecological problems within Western philosophical, religious, and scientific tradition." (p. 219, 220) He spoke, as out of his book, which I had not yet seen, while Tom and I remembered perhaps the texts we read in Camp 4, Yosemite Valley. I had quoted Dōgen, in my attempt to illuminate Muir's journals: "To be in the mountains is a flower opening 'within the world.'"[32] What did that mean, I wondered, and could the Enclosure's meaning to us be opened by this kind of thinking?

Because the Lower Saddle itself is, these days, a refined urban situation, other social affairs intervened. There was dinner to eat in the setting of the sun, and hardware to be racked, and after preparations were done, a great many congenial people—private climbing parties, Exum guides, climbing rangers who descended from the mountains, and the clients too. (Like Muir with the Forestry Commission, I toured, along with Leigh, as an informal part of a government party.) Indeed, the Lower Saddle on the Grand Teton seemed to me, as I sipped some of the Irish whiskey that Kimbrough had carried, a kind of higher Climbers Camp which had replaced the one I first visited at Jenny Lake in 1962, but more cosmopolitan.

But we went to bed early, because there was after all, a mountain to climb, and Ortenburger required the traditional pre-dawn awakening the next day.

Didn't we make up a perfectly bioregional group—mountain dwellers, writers, students of wilderness and wilderness ethics, climbers, and stewards of this rare treasure, the center of a National Park—as we drank coffee and choked down our granola in the dark? Probably not. In the cold morning, the mountain seemed bigger, and intimidated us, as any good mountain should. So we started up the long slopes. The route we would do was the standard, the most popular route on the peak, recommended because of its excellent rock, and its exposure to the sunshine. The Exum Route, pioneered by Glen Exum in 1931, is also the "trade route," used by the guide service founded by Exum, whose guides had departed several hours before us in the dark, with their clients.

One is aware of those who have gone before, and it is one of the charms of the Grand. Every feature of the mountain and the route is named: ascending from the Lower Saddle, passing over the Black Dike, and circumnavigating "The Needle," through "The Eye of the Needle," one finds even individual climbing moves named.[33] Two steps in an exposed place are called "The Belly Roll Almost." There the original route on the peak, the Owen-Spaulding of 1898 which almost everyone now uses for descent, diverges from the Exum Ridge, our route.

Exum had traveled the long, beautiful slanting ledge called Wall Street. And at the end of this ledge, the nature of the climbing changes. Suddenly high up, midway on a steep and long rib of rock, we change to our climbing shoes, and we "rope up." The climbing shoes we choose these days, especially in a dry year, are tight, highly sensitive ballet slipper-like, and wonderfully sticky. They are a far cry from the old mountain boots I wore when I first climbed in the Tetons. And now I must admit something. I have climbed in this range since 1962 and I have never climbed the Exum Ridge; so my introduction to it will be somewhat strange, climbing with the guide, or the author of the *Guide*, and climbing in shoes which, when I first began climbing, were more appropriate to Yosemite, in my home range, than to the Tetons.

But I was about to say about the roping up. The Exum guides, who were far above, had roped all the clients together, into a long line of everybody, which measured the ridge like a huge, though carefully modulated, caterpillar. We, on the other hand, divided ourselves into three groups of two.

As we fumbled with our shoes and ropes and equipment, in the dawn, Tom pointed out the exposure, and the kind of experience the clients had faced, an hour or so earlier, as they stepped out over a sunless void which must have seemed bottomless. On the first ascent, Glen Exum had leaped across. On this day, Leigh and I would go first, because we were expected to go fastest. Tom and Rod were in the middle, and then Jim and Pete. Clouds drifted across the peak, but they were not threatening. In the cold west wind we were wearing all of our clothes.

Leigh spoke. "Use the narrow ledge, which is a continuation of Wall

Street, as a handhold, and climb out on the face below, using footholds in the black rock. With a big step across to the far side...."[34] And I was off, stepping onto the ridge, into the sunlight and wind. Then Leigh was with me again, and then I was on the "60 foot face with knobs facilitating the climbing." (Leigh talks like that!) The rhythm of the climb began, and I gave myself up to it, reading the braille of the rock. "The mountain just kept opening up for us," as Dolores would say, page by page.[35] One passed beyond concern for the weather, and the rock became real — "every rift and flaw"—and it was as it always is when the climbing begins to go well: "...my limbs moved with a positiveness and precision with which I seemed to have nothing at all to do."[36] It was the old story, the same story, high on the mountain on the pink granite, studded with holds, the mica gleaming—so like the granite of Yosemite, so exquisite. I had never climbed this ridge of the Grand Teton before, but I had always climbed it before. Only this time there was Leigh's voice, which was the voice from the book, but was also the voice which said, "I have always known this mountain, and you shall know it, and love it, and love climbing it, as I do, and have always done."

We were alone together, and would be so for the rest of the ascent, climbing "the Friction Pitch," the "V," the "Puff-n-Grunt Chimney," stopping on occasion while Leigh pointed to landmarks, sometimes climbing onto the Eastern face out of the wind, to sit and look at Wyoming. Heraclitean that he is, Leigh would point out, "There are so many variations available on this ridge that it is possible to make two or three ascents and scarcely touch the same rock twice."[37]

Finally, and it almost seemed too soon, we sat on the summit, eating, as one always does on a summit. And I felt sentimental. Of course I felt sentimental about this climb, which has served so many so well as rite of passage, an initiation to the range. Was that permissible, since I had never done this climb before and was now well past that rite? I knew something of its history now, though I had needed no book, having the historian and guide Leigh Ortenburger readily at hand speaking the sentences I vaguely remembered reading. The whole climb was present and history. What, I wondered, were Rod and Tom, Pete and Jim experiencing? I had worked as a guide in Tuolumne once, and Tom had guided, was guiding now, and so was Leigh, though he hardly seemed to realize it. What had this to do with Cathedral Peak, the center of Tuolumne Meadows, its place in Muir's life, or mine, or anyone's? Horns are like hoofs, I thought, and tusks are like teeth, mountains like breasts or cathedrals. This rock was not brittle and was all of one piece; it had grown out of our planet.

Never mind about the summit, or about the view from the summit. I will not show how the various groups arrived, or even introduce the two Canadian women who passed the lost stewards of the park who had grown confused about the route, and arrived just after Tom and Rod. Some day I will sort out the arguments Leigh used when he pled his case about no new roads in the National Park to Jim Brady, who listened and responded thoughtfully. The sun was wheeling to the west, as Muir would say; the important part is the descent to the Enclosure, wherein I took a new guide, my friend Tom.

But first a digression about guides, cults, and leadership. In 1969, Thomas Jukes asserted, "To its leaders, *the Sierra Club is a cult....* John Muir...was the High Priest.... [Now] the older leaders of the Club regard Norman Clyde as the Dalai Lama."[38] Though we might want to call the Sierra Club a cult, Jukes' explanation for the phenomenon is more disturbing:

> Note that feelings of relief are experienced by the business
> executive who submits to the orders of a stockroom clerk
> (from another company of course!) who leads him through
> a mountain survival situation. The business executive
> relinquishes his responsibilities without losing face.

Ergo, the pleasure of being guided consists precisely in relinquishing our responsibility to the guide, without losing face. In a larger sense, being guided means relinquishing responsibility to our history.

In any case, I relinquished my responsibility to Tom Kimbrough, who guided me down the climbing route called the Owen Spaulding, while Leigh took on the responsibility of shepherding the others down the 120 foot free rappell (roping down) which avoids most of the difficult downclimbing on this route. So, moving back in history, Tom and I climbed the route in reverse, moving over the landmarks, traversing the "Cat Walk," where I lost my dignity and took a rope from Tom to descend, down the Owen-Chimney, where I again required a belay, then, without a rope, across the "Crawl," or "Cooning Place," and around "The Belly-Roll." Tom took over Leigh's role, pointing out the landmarks, while I led. But of course, leading while downclimbing roped is really like following when climbing up.

We reached a great sunny block on a ledge in time to see Leigh rope down from the overhang. Like any good guide, he had taken care of the other three first, and was the last to descend. Unlike his charges, Leigh eschewed any mechanical device for rappelling, and used a body-rappell as he descended. (Tom had already set up Rod with a mechanical descending device and advised the others to do the same.) As we sat in the afternoon sun on the Upper Saddle, we were passed by several ascending parties of climbers, some of them visibly apprehensive about the climb above. One party was delighted to discover that Leigh Ortenburger was Leigh Ortenburger. They had not heard of Nash. I thought of Patty Limerick's definition of the "closing" of the frontier, which includes "the popularization of tourism and the quaintness of the folk."[39] Some time ago, to use her terms, "life-threatening deserts became charming patterns of color and light"; but not for everyone. Next we would explore the quaint pile of rocks called the Enclosure.


Although one must climb back up to the Enclosure, it is a short scramble. One may look out over Idaho, or look at the western aspect of the Grand, and the Owen-Spaulding Route, which seems to traverse precariously above the more serious routes on the West Face proper. The Enclosure itself is not much to look at, a small circle of upended dark brittle metamorphic looking rocks, barely enough to break the wind; like a great black manmade flower, its petals opening outward (like old igloos on Denali, which, when their surfaces sublime in the dry cold air, seem to explode open in strange blooms). Not a fortification; hardly adequate as a windbreak, its position is something to think about. One thinks of Neolithic man, or at least Tom did, and as he told Rod, "They did not come here to *hunt*. Hunt what? Big horned sheep?" Why would they be here? Nothing grows here at 13,200. No practical reason for them to be up here, whoever they are or were. Why, then?

Was this a position of power where Shamans sought something? Perhaps. Jackson Hole/Pierre's Hole look flat from this place. There versus here. Difference, perspective, an awe-inspiring viewpoint? I thought so. While the Jenny Lake Subdistrict Ranger wandered down the west ridge, looking for a place where he could relieve himself in private, and Tom discussed the various possible explanations for this phenomenon (the Enclosure), while Lee pointed out the route on the West Face, and we peeked down over the edge at the Black Ice Couloir, I considered that Dolores would surely find this a perfect spot for Tai Chi. Not knowing Tai Chi, not knowing much of anything, except pleasure and the sun on my back, I looked up only to what we had done, the route we had travelled. I knew why the Exum Ridge was the perfect guided climb; it creates the illusion of difficulty with the reality of safety. And then we descended.

Among other things which I have retained from this journey is a copy of the Third edition of *Wilderness and the American Mind*, the copy which Tom had borrowed from me, and which he had carried up to the Lower Saddle. (I was using the book, along with Limerick's, and an anthology called *The Wilderness Reader*, to prepare a course on wilderness which I taught last fall to freshmen.) Nash had seen his book, and inscribed it: "To Tom, in memory of a splendid summit day on the Grand." Tom returned it later that week, commenting that Nash was a "good client." Probably, Nash is the ideal client, cheerful, intelligent, full of information, and aware of the need to keep his dignity. His book sits on my desk, a sort of *memento mori*, reminding me of the pleasure of that day, of good companionship, of our part in the history of the mountains, and the fact that our part in that history will pass away, no matter how hard we try to maintain our prominence in the world.

"We are in need of a flowering of ingenuity equal to that of the Neolithic or, lacking that, of wisdom,"[40] says the student of environment and history. No doubt our relinquishing responsibility is connected with acquiring responsibility; it is an American illusion that we are masterless.

Twenty-five years ago, the historian of the Sierra Nevada noted that Muir's narrative account of the first ascent of Cathedral Peak contained an "equal interest in the abundance of flowers and the structure of the peak."[41] Muir's was a sort of relinquishing of responsibility, a lack of singlemindedness, at least. (Or are Muir's flowers—Cassiope, Scottish heather—only reminders of his European heritage?) Nevertheless, what interests Muir interests us; in that sense, his experience still constitutes a guide and might flower in us; even though we know that the poem we read has many pages missing, even though we know the implications of the metaphor of nature as book, even if we sometimes get the facts or even the structure of our world wrong, Muir's point remains: EVERYTHING IS SO INTERESTING.

# The Disembodied Parasite and Other Tragedies; or: Modern Western Philosophy and How to Get Out of It

by Pete A. Y. Gunter

## 1

How much can a philosopher or historian of ideas do to help us orient ourselves to the contemporary and worsening environmental situation? It is difficult to say. The question underlying any such historical/philosophical quest is simple enough: "How did we get in this mess?" Any answer or answers one might give will hopefully provide further answers—at least some answers—to the inevitable next question: "How can one move towards giving straightforward answers to this question?" The most one can say is that one will try. Even a minimal success, given the seriousness of our problems, would make the game worth the candle.

The period which the present study will primarily explore is the seventeenth century. Still worse, it will center primarily on one figure, René Descartes. The focus will thus (admittedly) be simple—too exclusive. Sir Francis Bacon could have been included, as could Galileo, or any of a number of lesser figures. But in this case I believe the vices of exclusivity are made up for by the virtues of simplicity and clarity. What Descartes wished philosophy to be is clear. (Compare him to Locke, for example, who had many answers to almost any question, most of which answers were mutually inconsistent.) And, in his thought the major outlines of post-Medieval philosophy are conjoined.

Probably the only thing most of us recall about Descartes is his epithet *"Cogito, ergo sum."* (I think, therefore I am.) It seems a trivial saying: a piece of intellectual bric-a-brac to be dragged out and dropped at cocktail parties. In fact, it was to be a fateful utterance, channelling Western philosophy into a systematic subjectivism from which it has never recovered, leaving the so-called "outer world" uninhabited by feeling or value: a situation from which we have never recovered either.

In his *An Introduction to Metaphysics* Bergson notes that "the founders of modern philosophy *were* the founders of modern science." Bergson was clearly not thinking here only of Descartes: but his remark fits Descartes perfectly. I am unable to find a single textbook or history book which does not label Descartes "The Founder of Modern Philosophy." It was he who made the sharpest break with the Medievals; it was he who bequeathed to his successors a series of crucial problems with which we today demonstrably still struggle. If not Descartes, then who?

But Descartes was also an important founder of modern science. The vision which, he tells us, brought him his famous philosophical method brought him his scientific discoveries as well. Descartes was the creator of analytical geometry. He also, and not coincidentally, created a mechanistic physics which was clearly the forerunner of Newton's. And he was a chief collaborator in the creation of a new school of physiology: iatromechanics. Which lead to a new theory of neurophysiology.

A text by text examination of Descartes' writings would shed much additional light on what follows. For our purposes, however, a general survey of his thought will suffice. What did Descartes do, to make himself such a watershed figure? The answer is twofold. Descartes removed man from nature (man the *cogito*, man the thinking, feeling creature). And he "desacralized" nature by conceiving it as simply a geometrical mechanism: a machine, no ifs, ands or buts about it. He even denied that animals—dogs, cats, cattle—have feelings. *They* are only complicated machines. If one cuts off a dog's tail, the dog may seem to be taking a dim view of it, but one is only shortening a machine.

Twentieth-century commentators have pointed out several fundamental features of Descartes' system. For one, it is clearly a compromise, which allows two radically different cosmologies to coexist: the Medieval-religious and the modern-scientific. (If Descartes separated the human mind so absolutely from the human body, it was, as he said, to ensure the survival of the soul after death.) This aspect of his thought (how to save Christianity while allowing science) is featured in most textbooks. Another feature of the Cartesian system has been too often neglected, even though it is clearly the inverse side of the first. *Descartes removed "value" from nature—including the human body—to justify the study, and subsequent exploitation, of nature.*

Let's start with the study of nature. This is far more respectable. We recall Descartes' contribution to physiology, as partially expressed in his *The Passions of the Soul.* The Medieval church forbade carving on cadavers as a desecration of the body. As I recall, it was an offense punishable by death. Descartes and his accomplices, however, braved the ecclesiastical interdict and carved anyhow. We also recall that Descartes defended Copernicus. It was this more than anything else— this and Galileo's condemnation by the Inquisition for the same of-fense—which lead to the retraction of Descartes' never-published *Traité du Monde*, with its defense of the heliocentric solar system. Similarly, the church was Aristotelian, while Descartes' physics was squarely in conflict with Aristotle's. For Aristotle not only did motion need to be helped along, lest it die out: most motion in the universe was supposed to be circular, following the orbits of the great crystalline spheres. For Descartes (who had already stated Newton's first law of motion), once started, motion went in a straight line and was self-sustaining. Not only was there no need of help: only friction, or less subtle counter-effects, could bring movement to a halt. Worst of all, for Aristotle all motion was purposeful, from the rising of steam to the falling of a clod. For Descartes *no* motion was purposeful, since it was simply the effect of a prior mechanical "push."

It would be banal enough to point out that here was a profound conflict. The study of nature was blocked on every hand by ecclesiastical authorities brandishing what could only be described as outmoded, dusty, scientifically indefensible dogmas. What better way to remove the obstacles than by declaring in one majestic gesture that nothing— nothing at all—in nature is sacred, that everything in nature—every-thing whatsoever—is simply geometrical relationships. At one stroke all obstacles to peering, probing, dissecting, measuring, reconstructing, predicting—all obstacles vanish.

Particularly, the obstacles to scientific medicine and agriculture could be overcome. One can, I think, scarcely overestimate Descartes' enthusiasm for these two as yet unrealized disciplines. Speaking of the potential boon for mankind that a scientific medicine would constitute, he states:

> It is true that medicine at present contains little of such great value; but without intending to belittle it, I am sure that everyone, even among those who follow the profession, will admit that everything we know is almost nothing compared with what remains to be discovered, and that we might rid ourselves of an infinity of maladies of body as well of mind, and perhaps also of the enfeeblement of old age if we had sufficient understanding of the causes from which these ills arise and of all the remedies which nature has provided. It was my intention to devote my whole life to this much-needed service....[1]

Clearly, given the context of these assertions, it is not medicine alone which is to be pursued by the new philosophical/scientific methods. A page above he states:

> ...instead of the speculative philosophy now taught in the schools we can now find a practical one by which, knowing the nature and behavior of fire, water, air, stars, the heavens, and all the other bodies which surround us...we can employ these entities for all the purposes for which they are suited, *and so to make ourselves masters and possessors of nature.* This would...be desirable in bringing about the invention of an infinity of devices to enable us to enjoy the fruits of agriculture and the wealth of the earth without labor....[2] (italics mine)

Sir Francis Bacon could not have said it better—or more bluntly. If we are going to profit from the new knowledge, we are going to have to clean house and start over. Should auld Aquinas be forgot? Absolutely.

One final note on the nature which Descartes desacralized in order to study, and studied in order to master and possess. We must not forget the dates of Descartes' life (1596-1650), well into the period of Europe's colonial expansion (roughly, 1500-1900). When it came to mastery and possession, seventeenth century Europeans could scarcely fail to notice what the historian Walter Prescott Webb has termed "the great frontier": an area more than five times the size of Europe, still largely unpopulated, and rich with every manner of resources.[3] Was Descartes influenced, in his vision of a new metaphysical/scientific method, by the vision of, literally, a new world to be conquered? He could not have been ignorant of it. Nor, I think, could he have failed to be influenced by the ferment which the new frontier provoked. It is an interesting coincidence: that the march of modern science began in the West at almost the same time as the march of modern Western exploration/colonization...as Webb notes.[4] I state this point in order to make it clear that if Descartes had wished to provide a rationale for the exploitation of the "great frontier" by Europeans, his philosophy of nature could scarcely have come at a better time. By which I mean: at a worse time, from our point of view.

So much (for the time being) for the Cartesian "other": physical nature, devoid of intrinsic value but rich in resources and eminently conquerable. What of the Cartesian "Cogito, ergo sum"? By separating the cogito from the body, Descartes hoped to accomplish two things. He wanted to be able to argue that the soul, being separate from the body, could survive death. And he wanted to safeguard pure metaphysical concepts, conceived with clarity, distinctness, and certainty.

This is well known. It is in every textbook. But, again, the inverse side of the Cartesian ego is for our purposes all the more interesting, and deserves to be better known. By taking the mind entirely out of nature and elevating it into the center of his method, Descartes not only lowered the value of physical reality still further (by contrast); he assured that the correct approach to nature—the only correct approach—was through the calculating, geometrizing, predicting intellect. Approaches through "feeling," "sensation," the pleasures of the body—any felt kinship with *res extensa* whatsoever, including the non-human creatures—were degraded to the status of at best appearance, at worst illusion. Henceforth the dispassionate viewpoint of mathematical physics (i.e., of mechanics, celestial or terrestrial) was to provide the only valid approach to nature: either to the movements of the celestial orbs or to the passions of the human body. (Surely Voltaire should have vented his satire not on Leibniz, but on Descartes, who bade us approach sex like astronomers.)

But Descartes was no Pythagorean. He was at heart an engineer, interested in the control and design of machinery. (I have always heard, though I've not been able to run it to ground, that his analytical geometry was developed, or at any rate used, to aid in the calculation of weapons trajectories. Think of the equation for a parabola.) And, with Descartes, understanding leads quickly to use. The geometrical and calculating mind becomes for him the manipulating, fabricating (and also the potentially destructive) mind.

Clearly, then, the Cartesian mind/body dualism has as a major purpose to make the mind the controlling agent and possessor (the conqueror) of nature. Once again, the thinker of the mechanisms is clearly superior to the unthinking machinery. Conquest is, then, in the nature of things, justified by the radically contrasting value of the manipulator and the manipulated. The Mind is King—or Emperor.

This is a rather strange mind, the Cartesian "cogito." It appears only in (isolated) human individuals. Dogs don't have it. Neither do chimps or gorillas. It is also *outside the body*, in human beings. If one has a stomach ache, the pain is not in the abdomen, it is not even in the brain; it is outside the body. If one cuts one's finger, the pain is not in the finger. It is "elsewhere," in a non-physical entity, the "soul," somehow related to the brain. (Never mind how the brain is related to the body. Descartes could not explain the interaction of the two, though he presupposed it.)

An in-depth study of Descartes' strangely isolated, all-seeing "cogito" would reveal much about his debts to the Greeks and the Scholastics, as well as about the way he positioned the "cogito" so as to project it into the modern conquest. I wish here, however, to make only one more point. Crucial in Descartes' standpoint is a transformation of what we mean by Reason. Here as in all things Descartes hesitated, trying to harmonize two contradictory views. Here, as always, the Modern view was to win out.

We remember, perhaps, Plato's *Republic*. There a distinction is made between two kinds of thinking: *Noesis* and *Dianoia*. *Noesis* is overwhelmingly general; it seeks overall harmony, balance, coherence. *Dianoia*, by contrast, is more piecemeal; it calculates, geometrizes, predicts. *Noesis* concerns itself with the "justice" of the universe, *Dianoia* with quantifiable natural phenomena, with quantity in general. For Plato *Noesis* seeks the greatest possible degree of peace; but *Dianoia*—strangely, it seems to us—above all sharpens the instruments of war.

Throughout the Greek and Medieval periods this distinction remained in force. However it might be reformulated or reshaped to meet historical contingencies, the distinction between a higher "reason," restraining the otherwise passionate short-sightedness of a lesser mode of understanding, remained central to the philosophical tradition. In the modern world one finds it again in Immanuel Kant's pervasive distinction between Reason and Understanding. Kant continues the tradition, with Understanding doing the computing and Reason (now bereft of the capacity for metaphysical utterance) shaping human moral conduct, aiming at a rationally coherent world. Kant's Reason is Plato's *Noesis*, rather whittled-down; Kant's Understanding is Plato's *Dianoia*, suitably limited to Newtonian categories.

The fact that in Kant Plato's distinction is pared down to epistemological and practical limits, however, ought to warn us that something has happened between the end of Medieval philosophy and the eighteenth century. In terms of the history of philosophy what has happened is already visible, in its basic outlines, in Descartes. For Descartes, Reason (*Noesis*) in its old time-honored sense, is the highest of faculties, ideally guiding the rest. Reason leaps tall epistemologies at a single bound, proving the existence of God (Descartes provides two such proofs), the status of the self as an immaterial substance (the cogito), even supplying us with the most universal laws of nature. But if this Reason is for Descartes the highest of faculties, the most urgently required is the experimental, mathematical, practical thinking which gives us better medicine, better agriculture, better technology. The *urgency*, the sense of expectancy, is associated with the new method; the old Reason merely assures us of our traditions.

We see in Descartes Western civilization inverting its concept of what it is to be "rational." From the Platonic notion of harmony and balance we move to the pursuit of hard, piecemeal, particular projects. From a closed, if coherent and unified world, to a world open, but contemptuous of coherence. When today Western man speaks of reason, he means calculating, predicting, manipulative reason. Reductive. Mechanistic. That is one reason environmentalists have such a hard time making their case. Anything not approaching the modern model of true thought—anything not practical, manipulative, mechanistic—is viewed, by time-honored *modern* tradition, as imaginary, poetic, merely emotional.

## 2

I want, before continuing, to make two apologies: the first *to* philosophy, the second *for* philosophy. If in what has been said so far Descartes has been portrayed primarily as a villain, I want to apologize for the injustice of the portrait. If any of us, possessing his philosophical and scientific genius, were confronted by a diseased scholasticism brandishing outmoded and untested concepts, we too would have reacted much as he did, demanding that "reason" be used to test all statements, revisioning nature as a geometrical machine instead of an Aristotelian organism, insisting on the establishment of new sciences. This was the attitude of all of the bright minds of his century, with the exception of standpat theological conservatives. One can easily see why.

What I am objecting to in Descartes are the evident dangers in his position, dangers which are evident because they have lead, demonstrably, to many of our current problems, or to factors which sustain our problems, and/or prevent them from being resolved. We find in Descartes, in Bacon, in Galileo, in Kepler, in Locke, a revolt, one not unlike the revolt embodied in the Protestant Reformation a century earlier, but on a scientific and epistemological level. One can easily see the justice in the revolt; but there is no need, after the fact, to sanctify the terms in which a revolt is framed, or to reify, in turn, the concepts which revolutionists used as battering rams against then-entrenched institutions.

The second apology is for using philosophy here as a text. If it seems that I have been giving an introductory lecture in the history of modern philosophy—and with it the burden of academic condescension—then apologies are given. Quite possibly the history of technology, or art, or literature, would have provided a more interesting—certainly a more eventful—tapestry. The point, however, has been to find a figure in whom history registers graphically. Descartes is certainly that figure. I know of no better way to tell you about him—than to tell you about him.

He was certainly a man of his generation. And that generation was weary of tired institutions, and—more to the point—could see the outlines of a brave and prosperous new world ahead. The Church, the patron of ancient institutions, meanwhile had co-opted reason: had turned reason to the service of religion. To attack reason, then, was to attack religion; at least, it was to attack "natural theology," as it was inherited from the Greeks *via* Augustine and Aquinas. Reason, then, was in the context of the seventeenth century a conservative option. The new men, then, would emphasize not Reason (*Noesis*) but Understanding (*Dianoia*). But their revolt was to be passionate nonetheless. Rarely has reason—I mean understanding—been so passionately embraced.

This is the meaning of Alfred North Whitehead's very perceptive statement:

> Galileo insists upon "irreducible and stubborn facts" and Simplicius, his opponent, brings forward reasons, completely satisfactory, at least to himself. It is a great mistake to conceive this historical revolt as an appeal to reason. On the contrary, it was through and through an anti-intellectualist movement. It was the return to the contemplation of brute fact; and it was based on a recoil from the inflexible rationality of medieval thought.[5]

Modern science and modern philosophy, Whitehead states, "inherit this bias." That is, they continue a tradition of—it sounds strange to say it—mathematical/experimental anti-intellectualism.

# 3

It appears for our purposes, then, that Descartes' intellectual heritage is threefold: (1) a de-sacralized, geometrized nature, devoid of intrinsic value; (2) a subjectivized self, withdrawn from the body and manipulative in outlook; and (3) a theory of knowledge consisting of a strange melange of cold engineering and hot passion.

It is this unholy trinity that conservationists find themselves having to combat every time they appear at a Congressional hearing or on the evening news. If life on this planet is to endure, this trinity has to be destroyed—or, rather, subordinated, in the way that Plato wished to subordinate *Dianoia* to *Noesis*. Perhaps the best way to conquer the juggernaut is to divide it, and attack one segment at a time, beginning with our concepts of nature, then of the self, then of knowledge. In what follows, I will assume it is clear that I am deconstructing both philosophical and practical assumptions at once. I also hope it is clear that I wish to deconstruct in order to affirm.

## NATURE

To put the matter bluntly, *if we do not eliminate the popular conception that nature is just so much "stuff out there" valuable only insofar as we can use it,* environmentalism can be at best a superficial affair, and at worst will fail utterly. This conception appears in Descartes and is his most unequivocal bequest to us. It was strengthened by the stunning accomplishment of Isaac Newton, and by the reverberations of Newtonianism throughout the corpus of modern science, including biology, psychology, and neurophysiology. Every romantic movement since Descartes and Newton has been a reaction against it.

The examples of Leibniz and Spinoza make it clear that mechanistic views were not *imposed* on philosophy by the new scientific viewpoint. These two thinkers extended the range of consciousness throughout nature, down to the last grain of dust, and argued in holistic, not atomistic, terms. But the course of least resistance was to simply accept the new mechanics as simply mechanical. Descartes had done this, and it was his example that, for the most part, was followed.

Today, however, physics and chemistry no longer suggest a mechanistic view of nature. Already in the 1920s Alfred North Whitehead was able to construct an organismic view of the world using newly developed concepts of quantum and relativity physics. Today—to take *only one* example—non-equilibrium thermodynamics explicitly portrays a universe in which not only irreversible but also indeterministic and holistic processes play a fundamental role.[6] On the basis of this new scientific achievement one can all the more firmly assert the fundamental truth of Whitehead's philosophy of nature: *Nature is everywhere alive, and is exquisitely, inextricably organized.* It is something man feels inherently capable of valuing, for itself.

To assert such viewpoints (and I am saying that they must be asserted) is to deny Descartes' mechanical vision of nature. We owe much to that vision. But it is now not only not necessary to embrace it: Descartes' and Newton's physics, no matter how useful, are simply misleading scientifically. They stand in the way of new insights, not at the vanguard of research.

But how is the new viewpoint—in some respects a very old one—to be gotten over to the popular mind? The basic answer, I am convinced, is through art (meaning all the arts, including literature). Extreme abstractionist art can scarcely convey the brooding power or the lyric grace of nature. Indeed, the main thrust of post-impressionist art has been an often-proclaimed desire to get away from nature, either (and how well Descartes would have understood it) because the human creative psyche is assumed to be absolutely superior to the natural world, or because the natural world is presumed to be clearly not as interesting as the human creative psyche. What I am saying here will make many artists and aestheticians angry. I nonetheless persist in asserting it. One of the first victims of post-impressionist canons was landscape painting. In some form, we need to bring it back.

In the same way, our existential literature is even less helpful in an environmental context than dogmatically abstract art. There is no nature in Kafka. To the extent that Albert Camus finds consolation in nature, he scarcely seems an existentialist. To the extent that Sartre is consistent, he finds nothing in nature that is not ugly, hideous, *de trop*. If anyone can find better in *Nausea* or *Being and Nothingness*, I will be glad to hear about it.

The "modernist" phase in twentieth-century art has, one hears on all sides, come to an end. Perhaps the way out of the contemporary impasse in the arts lies in a profound rethinking of anti-representationalism (that *Leitmotif* of twentieth-century modernism), and a reappropriation of themes stressing the interdependence of man and nature, the grace of the human body, the value of non-human organisms.

I repeat: The arts can convey the message of non-Cartesian attitudes towards nature far more broadly and with greater impact than philosophy ever can. To do so, however, they will have to become *truly* post-modern.

## THE MIND

If the body, including the brain, is comprised of nonthinking matter, then one has two choices. One can proclaim a mind-body dualism, like Descartes, or, like Thomas Hobbes before or J. O. La Mettrie after him, assert a universal materialism. The difficulties of materialism are many; but the problems of Descartes' mind-body dualism are even worse, precisely because he was so certain (*absolutely* certain) that minds and bodies are absolutely different sorts of things.

Obviously the "mind-body problem" cannot be solved in a paper of this kind and length, assuming that it can be solved at all. As a minimum, however, we can make two steps towards avoiding the Cartesian dilemma. With Whitehead and Maurice Merleau-Ponty we can assert the embodiment of awareness: the fact that our thinking, dreaming, remembering are always done from the vantage-point of *this* body, standing, sitting, lying, walking, in *this* environment. We never just "think in general," independently of our physical state. Even dreaming, which Descartes thought provides a basis for skepticism concerning our embodiment-in-nature, involves awareness and interpretation of our bodily state.[7] (In turn, our body is aware of an environment stretching away from it in all directions.)

The other move we might make is to refrain from viewing "the mind" as being purely speculative, essentially mathematical. That is, Descartes time after time urges his readers, if they are to think well, to withdraw their minds from the senses and to concentrate abstractly. In this he was continuing a diatribe against Aristotle, who looked to the senses for fundamental data and who conceived consciousness as being at home in the human frame. In this we must follow Aristotle, understanding what awareness is while involved in the multifarious events of the active life. But we can and must go beyond Aristotle. Our awareness, as Leibniz put it, always includes the whole world, great and small. We are embodied in the whole world. If there is anything that the wilderness experience conveys, indelibly, it is this universal receptivity of experience.

## KNOWLEDGE

When Western thought passed, in the seventeenth century, from *Noesis* to *Dianoia*, from wisdom to computation, a fascinating phenomenon emerged, one scarcely conceivable in a Medieval context: a committed, even dogmatic anti-intellectualism. A convinced modernist would immediately respond that, prior to the seventeenth century, there was no intellect. Hence it could not be attacked. We are now (presumably being "postmodern") in the process of rethinking our approach to modern thought, however. What emerges more and more clearly in this process is not that there was no intellect prior to 1600 (Who, carefully reading St. Thomas' *Summa*, could believe that?) but rather that mathematical/experimental science had not yet gotten onto its feet.

*Dianoia* begat anti-intellectualism. Almost without exception modern anti-intellectualism, whether in Blake, Rousseau, or the young Nietzsche, sought goals that were rational, whether the goals were political, social, religious or aesthetic. In another historical context they would have proclaimed "reason" not "feeling," "intellect" not "passion." But reason in the time-honored sense had been dethroned and deserted along with scholastic metaphysics. It was no longer available for appeal. The romantics appealed instead to "freedom." More often they appealed *against* Newton's quantitative, mechanistic universe. One might say that they protested not wisely but too well. Our culture is soaked in a tired anti-intellectual romanticism, far from Wordsworth, farther still from Beethoven. Meanwhile "high tech" proceeds.

The point I am leading up to is twofold. The first part concerns anti-intellectualism; the second concerns the masquerade of *Dianoia* as *Noesis*.

Environmentalism is strangely situated at the present time. Forced to appeal to the sciences for data, it often projects itself to the public mind as a kind of "green mysticism." Basing itself on the science of ecology, it often portrays itself as standing against the march, if not of modern science itself, then against the march of modern technology. The division is obvious, both in the public mind and in the conflicting goals of environmentalism or, for that matter, the National Park Service.

What I am urging is that environmentalists avoid the temptation towards anti-intellectualism: the anti-intellectualism that we all feel, when confronted with brand new all-efficient technologies which deplete or destroy our environment. Our appeal is to ecology as providing data, and to a standard of reason which transcends all quantification, and which asserts the insanity of diminishing and enfeebling life in order to pile up consumer goods. Reason, in this context, is on our side. Ultimately anti-green intolerance is irrational. We must never let it wear the mask of reason. We must never accept it.

What then of "green mysticism"? One hardly knows what it means. I once had a philosophy professor who defined mysticism as "anything Bertrand Russell didn't like." The definition was catchy, but hardly precise. What I would mean by mysticism is not necessarily what others might mean. But if someone were to say that the Earth and its living things were sacred to them, would that be "mysticism"? And if someone were to argue that on the basis of a new philosophy of nature, it was easy to see how nature *could* be resacralized, would that be mysticism? At any rate, it would be a mysticism consistent with ecology in particular and biology more generally. That would not have satisfied Bertrand Russell. But, then, Russell, for all his bluster, never ceased to be a Cartesian.

Now, the second part, and briefly.

If environmentalism must effectively defend itself against both the mask and the substance of irrationalism, it needs to unmask the pretense of *Dianoia* to be the whole, or even the major part, of reason. Above all it needs to unmask the greed and ignorance that propels so much applied science. I will give one clear-cut example. In countless public debates with lumber company executives I have had to fight the allegations of ignorance and emotivist irrationalism. The constant corporate complaint has been that conservationists are either ignorant of economics, or ignorant of forestry. To back up their point lumber p.r. men like to point to their new "even-aged forest management" and their decades of experience with scientific silvicultural studies. But the truth is that lumber companies support and control not only special forest research institutes but *all* schools of forestry. Projects to determine how much damage is done by "even-aged forest management" (i.e., clear-cutting) to soil, to soil nutrients, to air and water quality *are simply not done at*

*American forestry schools. They are not allowed.* (I know of only one such study, at Yale, over twenty years ago.)

What needs to be done in this area is not less scientific research but *more*. This is the reality that needs to be "unmasked"; *Dianoia* parading as *Noesis*. *Dianoia* posing as unafflicted by greed.

# 4

And so, to conclude. Journals, books, magazines, even high-toned talk shows are today full of proclamations proclaiming the end of "modernism" and the painful birth of a new "postmodern" age. A little thoughtful reading leads one to doubt, however. The proclamations are dramatic, but the tone of voice of the proclaimers strongly suggests that what is being heralded is just one more installment of modernism: neomodernism, if we want to dignify it.

In fact, to get beyond modernism we would have to dismantle and demote the modernist project whose major outlines are so clear in Descartes. To do this we would have to rescue the world from valuelessness, the ego from sheer subjectivity, and the future from a technological *hubris* that contains much power but, God knows, little wisdom.

# Not Laws of Nature
# but *Li* (Pattern) of Nature

by Dolores LaChapelle

> We did not think of the great open plains, the beautiful
> rolling hills, and the winding streams with tangled growth,
> as "wild." Only to the white man was nature a "wilder-
> ness" and only to him was the land "infested" with "wild"
> animals and "savage" people. To us it was tame. Earth was
> bountiful and we were surrounded with the blessings of
> the Great Mystery. Not until the hairy man from the east
> came upon us and the families we loved was it "wild" for
> us. When the very animals of the forest began fleeing from
> his approach, then it was that for us the "Wild West"
> began.                          —*Luther Standing Bear*[1]

### Introductory Remarks

The on-going historical sense of the superiority of European
thinking over all other forms of thinking is so strong that even before
beginning to discuss Chinese thinking it is necessary to point out the
following discoveries of ancient China—some of which were made
1500 to 2000 years before the West "discovered" them:

> *Modern* agriculture, *modern* shipping, *modern* astronomi-
> cal observatories, *modern* music, decimal mathematics,
> paper money, umbrellas, fishing reels, wheelbarrows,
> multi-stage rockets, parachutes, hot-air balloons, manned
> flight, brandy, whisky, the game of chess, printing, and
> even the essential design of the steam engine, all came
> from China.

Robert Temple[2] is summarizing four decades of research by Joseph
Needham. There are now thirteen volumes of Needham's *Science and*

*Civilization in China* with more to come, all published by Cambridge University in England.

I first discovered Needham back in the 1950s. Since I have been studying him ever since, I was delighted to find that Temple had produced a popular account of his work. I find, however, that most people refuse to believe it, saying, "If all this is true why haven't we heard of it before?"

The reason why no one before Needham, either Chinese or Western, had put this together was first, many Chinese inventions reached Europe by way of the Arabs, and second, many basic philosophical and mathematical ideas came to Europe by way of letters from Jesuit missionaries in the 1600s and most especially from a book titled *Confucius, Sinarum, Philosophus Sive Scientia Sinensis*, published in Paris in 1687. This book was widely read all over Western Europe and influenced most major European thinkers of that time. For example, it was a major influence on Leibniz who for fifty years had been reading about Chinese works, since first corresponding with Jesuit missionaries in China. Rousseau got his concept of the natural goodness of man from Chinese sources and his actual models from the idealized Indians of Baron Lahontan.[3]

The reason little of this is known is that in the two hundred years during which Europeans were colonizing China, they had no idea they could learn anything from these inferior "colonial people." Then in 1911 came the Revolution in China when the Chinese deliberately disowned their own culture. E.R. Hughes writes of this time as "the fury of the revulsion against the past and all its works, which characterized the period following the 1911 Revolution...."[4] Hughes was a missionary in the interior of China from 1911-1929; later he was Professor of Chinese language and literature at Oxford. His book was published just as the second World War began; hence it was lost.

Needham, a biochemist at Cambridge in 1937 accidentally discovered the importance of Chinese science from Chinese graduate students at Cambridge. Needham met Hughes in China in 1943, when Hughes was working with Dr. Wen, a Chinese historian, later killed in the revolution. Essentially, we owe a great deal to these two scholars, who first opened up the real culture of China to the West.

My final point is that, since China was so far ahead of us in science, yet never destroyed their environment at the rate we have done it during the last century, perhaps they have an underlying approach to nature that we lack. This paper is an attempt to show this underlying difference between Western thoughts on nature and Chinese. We do not follow nature's law; that is a human construct imposed onto nature. Instead we learn to fit into the *li*—the pattern of nature. In wilderness we find the clearest expression of this *li*.

## PART I—Law vs. *Li*

### Development of the Concept of Natural Law in the West

One of the oldest ideas of Western civilization was that just as earthly lawgivers enacted codes of law to be obeyed by humans, so God had laid down a series of laws which must be obeyed by minerals, crystals, plants and animals and even the stars in heaven. This concept of Laws of Nature became more widespread over a period of centuries so that by the seventeenth century it was common. For instance:

> "Praise the Lord, for he hath spoken,
>    Worlds his mighty voice obeyed;
> Laws, which never shall be broken
>    For their guidance he hath made."[5]

The basic root of this celestial law-giver making rules for non-human natural phenomenon comes from the Babylonians. In Tablet No. 7 of the Later Babylonian Creation Poem we find the sun-god Marduk (who was raised up to central importance around 2000 B.C.) giving laws to the stars. It is he who "maintains the stars in their paths" by giving "commands" and "decrees."

The pre-Socratic philosophers in Greece wrote of necessity but not of law in nature. Law in the most general sense is first mentioned by Demosthenes, who was not a philosopher but an orator and politician, who lived some two hundred years after the last of the early Greek philosophers, Anaximander. Demosthenes wrote that "Since also the whole world, and things divine, and what we call the seasons, appear, if we may trust what we see, to be regulated by Law and Order."

Aristotle never used the idea of Laws of Nature. Plato used it only once in the *Timaeus*, where he wrote that when a person is sick his blood picks up the parts of food "contrary to the laws of nature." Not until the time of the Stoics do we find the idea of the whole world ruled by Law. "The Stoics maintained that Zeus (immanent in the world) was nothing else but Universal Law." This concept began with Zeno in 320 B.C. and continued on down to Diogenes, who died 150 B.C. Needham claims that this Stoic development resulted from Babylonian influences due to the fact that from about 300 B.C. astrologers and star-clerks from Mesopotamia began to spread through the Mediterranean world. The clearest statement on natural law comes from Ulpian, a leading Roman jurist (d. A.D. 228) whose work makes up so large a part of the Justinian *Corpus Juris Civilis* of A.D. 534. Ulpian wrote:

> Natural law is that which all animals have been taught by Nature; this law is not peculiar to the human species, it is common to all animals which are produced on land or sea, and to fowls of the air as well. From it comes the union of man and woman called by us matrimony, and therewith the procreation and rearing of children....

The other main source of the idea of Natural Law came from the Jews into Christianity, where it became so identified with Christian morality that St. Chrysostom (early fifth century) wrote of the ten commandments as a codification of natural law.

By the 17th century with Boyle and Newton the concept of Laws of Nature, "obeyed" by everything from chemicals to the stars and planets, was fully developed.[6]

## Origins of the Concept of *Li* in China

In China there was never the idea that one God ruled everything; instead they referred to "Heaven and Earth" as the origin. When the Chinese use the word *Tien* (heaven) they are not referring to heaven in the Christian sense but rather "the mysterious government of the blue sky at noon."[7] This concept seems to have come right down from the early Chou peoples who lived on the high steppes of Northern China. Needham explains: "The order in Nature was not an order ordained by a rational personal being (God) and hence there was no conviction that rational personal beings would be able to spell out in their lesser earthly language the divine code of laws which he (God) had decreed aforetime." By the time we come to the era of the Taoists, we find that they "scorned such an idea as being too naive for the subtlety and complexity of the universe as they intuited it."[8] Needham goes on to say: "With their appreciation of relativism and the subtlety and immensity of the universe, they were groping after an Einsteinian world-picture without having laid the foundations for a Newtonian one."[9]

Later in China we have the positive law (*fa*) of the Legalists which referred only to humans. The Confucians "adhered to the body of ancient custom, usage, and ceremonial, which included all those practices, such as filial piety, which unnumbered generations of the Chinese people had instinctively felt to be right—this was *li*(a), and we may equate it with natural law.... Neither of these words, *fa* or *li*(a) was easily applicable to non-human nature."[10]

As Wei-ming Tu explains: "*Li*(a) is a concept pregnant with ethico-religious connotations. The mere fact that it has been rendered as ceremony, rites, propriety, good custom...and a host of other ideas suggests the scope of its implications.... In the ages prior to Confucius *li* had a function typical of ritual. Rites were sacrifices performed on certain solemn...occasions." For example, sacrifices to *Tien*, Heaven [the sky] or the spirit of the soil, *She*, or the ancestor spirit.[11]

Gimello explains: "By Confucius' time the very distinction between sacred and profane had ceased to be a crucial one...Confucius, and later, Mencius and Hsun Tzu, as well as the authors of the *li Chi*, all maintained that...sage rulers observed meet and just *patterns* in the world and sought, in turn, to devise formal rules of conduct that would enable men to make those same patterns explicit in their own lives." These aspects of *li*(a) were later incorporated into the *li* of Chu Hsi.[12]

## *Li* as Pattern

There was no word in Chinese documents for the idea of Natural Law as we in the West conceive it. Actually Natural Law is a human construct imposed onto nature. Instead the Chinese had the concept of *li*(b), which is the only *li* I will be dealing with below. Although many European translators of Chinese documents such as Bruce, Henke, Warren, and Bodde, used the word "law" and sometimes the word "principle" in translating "*li*," the better translation is "pattern," according to Needham.

> The word, *li*, in its most ancient meaning, signified the pattern in things, the markings in jade or the fibres in muscle; as a verb it meant to cut things according to their natural grain or divisions. Thence it acquired the common dictionary meaning, "principle." It undoubtedly always conserved the undertone of pattern, and Chu Hsi himself confirms this.

Needham sums up by saying that "*Li*, then is rather the order and pattern in Nature, not formulated Law. But it is not pattern thought of as something dead, like a mosaic; it is dynamic pattern as embodied in all living things, and in human relationships and in the highest human values." Contrasting *li* and a law at the fundamental level, Needham points out that for the Roman jurist "Ulpian and the Stoics all things were 'citizens' subject to a universal law; for Chu Hsi all things were elements of a universal pattern."[13]

According to Izutsu: "The extraordinary importance of this technical term *li* may be guessed from the fact alone that the entire philosophy of the Sung dynasty Confucianism is known as the 'science of *li*'."[14]

**Historical Development of *Li***

Taoism provides the easiest access to true knowledge of the "old ways": our animal heritage and the heritage from primitive cultures. The Taoists were intellectuals who left the imperial courts in the Warring States period. Already the hydraulic river valley culture of China was well developed and cities thrived; but once they went off into "the wilderness" to learn from nature they also learned from the primitive cultures still existing intact on the higher ridges between the valleys. Fortunately there were no airplanes or cars so such cultures were effectively cut off from the developed valleys. What the early Taoists learned was not a "system," and there was really no name for it until after Buddhism invaded China, forcing Chinese scholars to re-examine their own Chinese roots.[15]

Needham remarks that on the one hand "there was Buddhist metaphysical idealism...interested neither in human society nor in Nature." Confucianism, of course, was mainly concerned with relations between humans and although this was always within a context of nature it was not explicitly stated. Summing up, Needham writes:

> From all this there was only one way out, the way that was taken by the Neo-Confucians culminating in Chu Hsi, namely, to set, by a prodigious effort of philosophical insight and imagination, the highest ethical values of man in their proper place against the background of non-human Nature, or rather within the vast framework (or, to speak like Chu Hsi himself, the vast *pattern*) of Nature as a whole.[16]

A noted authority on China, de Bary of Columbia University, states that this achievement called for "uncommon confidence" after "centuries of Buddhist despair and detachment.... It depended upon a conscious ordering of human priorities—a deliberate discipline and regulated growth which would lead upward and outward, uniting man to Heaven-and-earth and all things."[17] This problem which faced the Neo-Confucians seems remarkably similar to our problem today.

The precursors of Chu Hsi were the early thinkers of the Sung Neo-Confucian movement such as Chou Tun-I, the two Ch'eng brothers and the uncle of these brothers, Chang Tsai, who were contemporaries of Avicenna and Omar Khayyam. (To place Chu Hsi within our European time frame, consider that he died about 1200 A.D. and that Thomas Aquinas was born 1225 A.D. Aquinas put together Christian and Aristotelian thinking and thus further locked us into the mind/nature duality and the trap of narrowly rational thinking.) Then came Chu Hsi, one of the greatest men in the whole development of Chinese thought: according to Needham, his was a philosophy of "organism." Thus these Sung dynasty thinkers had attained a "position analogous to that of Whitehead, without having passed through the stages corresponding to Newton and

Galileo." Needham continues by explaining that they had a dialectical logic "thereby anticipating Hegel, without ever having passed through the logic of Aristotle and the scholastics."[18]

Thus, some seven centuries ago, these Neo-Confucians were addressing questions we are just beginning to ask now with the present efforts to define "eco-philosophy" and "eco-sophy" or deep ecology. All of the writings concerning the Sung dynasty philosophy of *li* were gathered together by the Ming emperor Yung-lo in about 1415— seventy-seven years before the discovery of America! Naturally, this philosophy of *li* has been attacked from all sides. In general, Christians have been outraged because it denies a personal God; theologians don't like the "pantheism" in it; others say it is materialistic. Fundamentally, though, *li* provides us with a workable concept for the main task facing us now—the "need to reinsert humanity into nature."[19]

With the above introduction I return to the specific details of the historical development of *li.*

## The Diagram of the Great Ultimate

The first major Sung scholar to begin the process which eventually became the philosophy of *li* was Chou Tun-I (1017-1073). Although he did not write a great deal he put together a diagram which proved very useful for all later Sung scholars. In this diagram a circle at the top illustrates The Ultimate Tao—also called "The Supreme Pole (*Tai Chi*)." This motionless beginning "moves and produces the *Yang.* When the movement has reached its limit, rest (ensues). Resting, the Supreme Pole produces the *Yin.* When the rest has reached its limit, there is a return to motion.... The *Yang* is transformed (by) reacting with the *Yin,* and so water, fire, wood, metal and earth are produced." These two statements have nothing to do with a time sequence but are to show paradoxical identity. These five elements make up the "myriad things" of the universe.

The Supreme Pole (circle at the top) is called "*Wu chi erh tai chi,*" which means "That which has no pole! and yet (itself) the Supreme Pole!" (The importance of this phrase comes from the fact that it consolidated Taoist and Confucianist ideas.) The *Wu chi* comes from the Taoist *Tao Te Ching* document. The word *chi* here can mean ridge-pole but also is the technical term for the Pole Star around which all the universe revolves.

Commenting on the diagram, a later Sung scholar wrote:

> The term *tai chi* expresses the majesty of the universal Pattern [*li*]. The word *chi* means axis or pivot, root, or basis. The word *tai* means so great that nothing can be added to it and expresses the fact that it is the Great Pivot of the Universe. All things, however, which bear this name, such as the North (celestial) Pole, the ridge of a house, or the four compass-point directions, have visible forms and locations to which we can point, but this *chi* alone is without form, and has no relation to space. Master Chou therefore added the term *wu (wu chi)*, expressing the fact that it is not (confined to) any *form*...but is really "the great Basic Root of the Universe."

This diagram illustrates the conception that the entire universe is a single organism. The Taoist *wu chi* (*wu* meaning not or non) showed that "the entire universe depended on no such cardinal point, for every part of it took the leadership in turn." Master Chou (Chuang Chou) had long before used a little story about the natural processes in the animal or human body, uncontrolled by consciousness to show that the Tao needs no consciousness to bring anything about.

> It might seem as if there were a real Governor, but we find no trace of his being. One might believe that he could act, but we do not see his form.... But now the hundred parts of the human body, with its nine openings and six viscera, are all complete in their places. Which should one prefer? Do you like them all equally? Or do you like some more than others? Are they all servants? Are these servants unable to control each other, but need another as ruler? Or do they become rulers and servants in turn? Is there any true ruler other than themselves?

The Taoist element (*wu chi*) in the "*wu chi erh tai chi*" has been explained above; the other element of this explanation comes from

Confucian thought. Thus *tai chi* shows the "immense power informing the wholeness of the universe and present everywhere within it." Both of these statements together show the world as a single organism, no particular part of which can be identified as permanently "in control." Events neither randomly occur as in mechanical materialism nor are they directed on their paths by divine intervention as in spiritualism. Instead, as Needham explains: "all entities at all levels behave in accordance with their position in the greater patterns (organisms) of which they are parts."

At the next level of the diagram we have the *yin* and *yang*. These terms are now well-known in the West but with our propensity for dualistic thinking they have become forced into dichotomies: male/female, light/dark, etc. Actually, in the old form of these Chinese characters, *yang* is represented by the sun together with the character, *fu*, meaning "mountain." The character for *yin* was a coiled cloud with the character, *fu*. Thus *Yang* describes the "sunny side of the mountain" and *Yin*, "the side in the shadow." Thus we see in these terms the changing relationship of sun and mountain. In the morning, when the sun is behind the mountain, the trees are dark—almost black; while later in the day when the setting sun shines directly on these same trees, they are bright and glowing with light. Hence *Yin* and *Yang* refer to a constantly changing inter-relationship.

The last principle of the "Diagram of the Supreme Pole" states: "The true (principle) of that which has no Pole, and the essences of the two (Forces) and the Five (Elements), unite (react) with one another in marvelous ways, and consolidations ensue...reacting and influencing each other, change and bring the myriad things into being." The universe of things and events comes about by the mutual interaction of all these different forces—the idea being more like the force-field of modern physics than Newtonian mechanics.

Above is a brief description of the basic "Diagram of the Great Ultimate."[20] I cannot take the time to go further into the history of this "Diagram" other than to point out that the Ch'eng brothers gave Neo-Confucianism its enduring structure. They accomplished this, as Koller explains, "by making principle (*li*) the basis for their philosophy. Building upon the work of Chou Tun-I, who had been their teacher, they replaced the concept of the Great Ultimate—which impressed them as being too abstract and excessively Taoist—with principle."[21]

The last, and most difficult concept for Westerners is that of *chi*. In explaining *chi*, Needham says *chi* is matter-energy and *li* is organization. As Needham points out: "Today we know (too surely for our peace of mind) that matter and energy are interconvertible."[22] *Chi* is essentially what *li* operates through or what *li* organizes. Dr. Siu, a modern Chinese biochemical researcher, has written a book titled *Chi: A Neo-Taoist Approach to Life* published by the Massachusetts Institute of Technology Press in which he uses modern modes to explain *chi* and *li*. In explaining Chu Hsi's thought, Dr. Siu writes that the "universe is formed by the *chi* as activating essence" acting on the *li*.[23]

For example no one shapes trees. It is the *li* of the tree that does the shaping along with *chi*—gravity, sunlight, water, wind, etc.—all working together toward greater life for that particular tree. To use Chinese terms, the *chi* (energy/matter) flows through and shapes along with the pattern (*li*) of treeness. But there's no mechanical law involved here. No two trees are identical in nature. At high altitude this is very clear. A seed lodges in a crack in the rock where the water stays a bit longer and so begins to grow. First it grows up toward the sunlight and then, being blown by a continuous southern storm wind, it begins to lean out toward the north. However when it has grown so large that gravity begins to pull it down over the cliff, it turns and begins growing again toward the sun. If the site is very exposed this may occur often during the tree's life, producing the exquisite form which the Japanese developed into the art of bonsai trees. Essentially on this earth *li* is influenced by gravity, sun and the space of the surroundings.

## Concluding Explanations of *Li*

Needham explains that the "two fundamentals in the modern view of the universe, as the natural scientist and the organic philosopher sees it, are Matter-Energy [*chi*] on the one hand, and Organization or the principle of Organization [*li*] on the other." Needham makes it very clear that *li* is in no "strict sense metaphysical, as were Platonic ideas and Aristotelian forms, but rather the invisible organising fields of force existing at all levels within the natural world."[24] No god is needed in the on-going creative world of *li* and *chi*.

The only "law" in *li* is the "law to which parts of wholes have to conform by virtue of their very existence as parts of wholes." Such parts need to fit into place along with the other parts in the whole organism which they compose. Such thinking did not lead to laws such as "the statutes of a celestial lawgiver analogous to an earthly prince, but arose in the thought of the Neo-Confucians, directly out of the nature of the universe."[25]

For further explanation here I must return to Chu Hsi, who told his students: "The term *tao* refers to the vast and great, the term *li* includes the innumerable vein-like patterns included in the Tao."[26] Another way to say this is: "Gathered up, all things are unified in the one supreme Principle of Being. Separated from one another every one of them has each its own supreme Principle of being."[27] To make this very concrete, think of the parts of the body. The foot becomes the foot by following its *li*—to uphold the body. If it followed some other *li* it would no longer be a foot.[28]

In the final paragraph of his work on *li*, Needham points out that there was no concept of obeying laws of nature as in the west but instead, in the Chinese world-view, "the harmonious cooperation of all beings arose, not from the orders of a superior authority external to themselves, but from the fact that they were all parts of a hierarchy of wholes forming a cosmic pattern, and what they obeyed were the internal dictates of their own natures. Modern science and the philosophy of organism, with its integrative levels, have come back to this wisdom, fortified by new understanding of cosmic, biological and social evolution."[29]

## PART II—Aspects of Human *Li*

Universal harmony comes not by the celestial fiat of some God, but by the spontaneous cooperation of all beings in the universe brought about by their following the *internal* necessities of their own nature.... All entities at all levels behave in accordance with their position in the greater patterns (organisms) of which they are parts.[30]

The larger organism which humans can relate to in their daily life is the ecosystem or bioregion in which they live. There is a pattern for each *place*—everything in that place must conform to that pattern. Each human or animal or plant has a *li* of its own but develops into the fullest potential of that *li* by conforming to the pattern of the place as a whole. For example, there is a different *li*—a different pattern—for the rainy northwest than for the dry southwest. If you follow the *li* of your place in your daily life as well as follow the *li* within you then there is "power" or "empowerment" as some call it, which is the *chi* energy freely moving within. Tribal people call this "walking in a sacred manner."

According to the ancient Chinese thinker, Mencius, *li* is a movement or a continuous process of extension, such as when a fire suddenly flares up or a spring wells up. Mencius wrote: "If *li* is fully extended it will be sufficient for a man to serve all within the four seas. If it is not at all extended, it will not even suffice for man to serve his own parents."[31] According to Wei-ming Tu this means that the "dichotomy of human self and nature has to be understood in a new perspective. The self must be extended beyond its physical existence to attain its authenticity."[32] This concept has been called the "Principle of Extended Identity" by Silas Goldean, a modern thinker. "It seems to me that a watershed offers the ideal neural and social scale for the exploration of human identity; it is complete enough to yield a pattern of relationships, and large enough to be inexhaustible. Practiced in place the extension of identity becomes the realization of community, and as place turns into home, the heart opens like a seed."[33]

An ancient Chinese concept, "Heaven, Earth and man have the same *li*," means that since the universe is good, man by nature must be good. According to Mencius, "Salvation lies in fully developing one's original good nature."[34] There's no thought that man came into the world with Original Sin on his soul as in the Christian tradition.

C. G. Jung, who did so much to explore the collective unconscious, developed the concept of archetypes. As Maier explains: "The archetypes are the psychological aspects of the biological facts, the patterns of behavior by which we live and are lived."[35] Each of the archetypes has its own *li* and the fully developed human being is the one in which each of these patterns within is allowed to flourish (follow its own *li*) rather than being supressed and distorted. Talking about how each emotion within the human is good and necessary because we acquired it through the long history of evolution, Konrad Lorenz says: "These behaviors are labelled by such words as love, friendship, hate, jealousy, envy, lust, fear and rage." He continues by explaining that the "question of whether any of these motivations be 'good' or 'bad' is quite as much beside the point as to ask whether the thyroid gland be good or bad." It is a matter of harmonizing each of these behaviors so that the individual flourishes. Lorenz sums up by saying: "There is no human vice which is anything else than the excess of a function which, in itself, is indispensable for the survival of the species."[36] Jung himself said:

> Intellectuals think that we live in a world of ideas which we invent—a domesticated world where we create new hybrids of grains, new breeds of cattle and new ideas—it's all the same—they come out of our ideas. But deep inside us is a wilderness. We call it the unconscious because we can't control it fully so we can't will to create what we want from it. The collective unconscious is a great wild region where we can get in touch with the sources of life.[37]

This is the wilderness within. The Chinese method, which allowed full expression of all these patterns within (*li*) and yet have them contribute to the overall harmony of the *li* of the person, was the concept of *yin* and *yang* together with five elements or physical forces operating in the universe. All of this was symbolized in the *I Ching* where "the lines in the hexagrams and the trigrams were regarded as centers of energy continually acting and reacting on each other according to their relative positions." Hughes says that "these thinkers' minds led them to concentrate more on categories of relationship than on categories of substance." The hexagrams have to do with the phenomenon of heaven and earth, weather, seasons, human relationships etc. Summing up, Hughes writes: "This explanation was on the basis that no entity in the universe could be static or self-contained. One triumph of the system was what can only be called its ecological good sense. The plumage of birds and the pigmentation of animals' skins were guessed at as deriving from the nature of the terrain in which the birds and animals lived."[38]

Each being has its own *li*—its own pattern—thus it's not who is higher and lower on the ladder, it's merely how each animal or plant best fulfills its own *li* or pattern within the ecosystem as a whole and thus contributes to the health of all. When the aboriginals in Australia hold their "increase ceremonies" they not only sing for the plants or animals they use for food but for the insects who sting them because those insects are part of that place and that insect's *li* contributes to the flourishing of the whole place. In *Mind and Nature*, Gregory Bateson has this to say about patterns or *li* when he writes:

> We have been trained to think of patterns, with the exception of those of music, as fixed affairs. It is easier and lazier that way but, of course, all nonsense. In truth, the right way to begin to think about the pattern which connects is to think of it as *primarily* (whatever that means) a dance of interacting parts and only secondarily pegged down by various sorts of physical limits and by those limits which organisms characteristically impose.[39]

## The New "Science of Chaos"

The continuing trends in Western science during the last few decades bring it closer to the ancient Chinese "science of *li*" with its connotation of organism or self-organizing universe. Chaos theory moves away from what is called reductionism in terms of constituent parts, whether they be called atoms, quarks, chromosomes, or whatever and toward relating the obscure and the familiar through the commonality of shared patterns observable on all levels of being. As one physicist puts it: "Relativity eliminated the Newtonian illusion of absolute space and time; quantum theory eliminated the Newtonian dream of a controllable measurement process; and chaos eliminates the Laplacian fantasy of deterministic predictability."[40]

The early chaos theorists, those who "set the discipline in motion, shared certain sensibilities. They had an eye for pattern, especially pattern that appeared on different scales at the same time. They had a taste for randomness and complexity.... They feel that they are turning back a trend in science toward reductionism."[41] Near the end of Gleick's book, *Chaos: Making a New Science*, he states: "Pattern born amid formlessness: that is biology's beauty and its basic mystery." Gleick summarizes by writing: "Nature forms patterns. Some are orderly in space but disorderly in time, others orderly in time but disorderly in space.... Pattern formation has become a branch of physics and of materials science."[42]

Two of the most common patterns in nature are spirals and branching patterns. "Spirals are found where harmonic flow, compact form, efficient array, increased exchange, transport, or anchoring is needed."[43] Thus we see spirals in clouds, water, the pattern that branches make, when growing from the trunk of a tree, etc. There are many different sectional patterns within the tree model itself. Most of these recur in other natural forms as well. These patterns have become the basis of fractal geometry. Mollison continues: "Benoit Mandelbrot assembled his own insights, and the speculations of others, to found a mathematics of fractals (his term, from the Latin *fractus* or shattered), which is evolving to make sense of irregular phenomena, as Euclid did for more regular and measurable forms."[44] Fractals are common in nature, occurring in clouds, forked lightning, neuron nets and their signals, snowflakes, tree branches and roots, earthquake patterns, etc.[45]

In both the book *Chaos* and the TV Nova special on chaos, statements are frequently and arrogantly made about this "new science," forgetting again that the Chinese were there centuries before us with their "science of *li*." Nature, of course, neither plans nor uses mathematics nor does she produce patterns; rather, she lets patterns produce themselves.

## Experiencing *Li* by Living in Place

If one could feel the entire universe then one could see how the *li* of each part fits within the *li* of the whole; but no one can know the entire universe. All we can really learn to know fully is our own place on earth where we live. We can learn to recognize the patterns which underlie the seeming chaos. For example, after living many winters here in my own place in the San Juan Mountains and gathering firewood each autumn, I now know why the trees are smashed and piled up by an avalanche just there—in that particular place. No modelling on a computer could ever tell how that one particular ripped-up living tree was forced between two standing dead trees.

*I carry water and I carry firewood. This is in itself something absolute. This precisely is a miraculous working of the Tao.* (ancient Chinese text)

If one is working only from within the narrow human ego while carrying firewood one remains stuck in the narrow, merely human mode. Only when I carry firewood with "devout respectfulness"—feeling the grain, feeling the dryness or dampness, fingering the cut edges and the bark, then I am once again back in the place where this particular piece of firewood gave itself to me and immediately I am once again in the Tao. I am back where I picked it up in the avalanche path in the fall.

That previous winter it was torn out from the tree, hurled down by the awesome beauty of vortexes forming within the avalanche of snow and smashed between two standing trees, to hang there drying through spring wind and summer sun waiting for me. I am feeling again the sharp, clear air of fall surrounded by the glittering yellow leaves of the aspen.

I am filled with devout respectfulness that the awesome avalanche gave me this gift of already ripped to size firewood, while above me I see the last remnant of the avalanche path melting away on the side of the hill. Tiny marsh marigolds shoot up in the narrow runnels from the melt, barely out of the ground and blossoming already. They know from the quality of light that they haven't much time to finish their cycle and get the blossoms out and seed set before winter sets in. I am conscious, also, that at this very moment, the very tiny point of the reversing of the *yin/ yang* cycle is beginning high above me. Next winter's avalanche is already forming in the few inches of early snow lying there through the warm glorious fall weather and the icy cold nights, already beginning the process of turning these soft snowflakes into the dreaded "depth hoar"— those fragile skeletons of October's snowflakes waiting, waiting while winter snow piles up inexorably—until at a precise moment (which no way of measuring and no amount of human research can ever predict), when there's enough weight, the whole thing collapses, tearing to shreds the thin skeletonized structure, releasing the mass of snow above it. And all at once the whole side of the mountain begins to move, bringing all the accumulated snow on top of it down in one thunderous mass sometime during the winter. Gravity forces it up to an incredible speed, sucking in all the nearby trees and spilling them out below.

I remember the big snow which finally brought the depth hoar avalanche down last February. Seeing ski tracks coming out of the edge of the avalanche, high up on the mountain above town, I went frantically grabbing for the binoculars. I saw, with relief, that they came over the piled up mass of rock hard snow—not coming out from under, which would mean certain death with that much snow above. This same slide smashed several trucks and gravel-making machinery left far out from the avalanche path the previous fall. After all, that slide doesn't ever cross the Animas River, so why move things?

Cradling this avalanche gift of wood in my arms, walking back toward the house, I look up at Sultan Mt. towering above me with its massive cornice still hanging there—the mountain still breathing out winter over our early spring valley and the Evening Star, regal Venus glowing out of the early evening twilight above it.

Filled with joy—again—this very world, this valley, these mountains in which I live is the true reality. This world is filled with life, filled with the creative energy of the Tao, never ceasing to flow, bringing into being an infinite number of things one after another. Again, I am played by the very air moving down from Sultan, the light of Venus, the feel of the wood and again I *know li*.

Living in this high mountain valley year after year, I recognize the concept of *li*. During the years of studying *li* I've found Izutsu very helpful. He writes of the *li* of humans coming into existence only when the human species began; but we humans are inextricably bound to the *li* of all the other beings in our place. We're all in it together—us and all the other mammals, us and fragile eroded tundra, us and the mountain and snow and avalanches. The animals *know* this all the time but we have to keep *remembering* that we know.

When the rational part of the mind thinks on the subject of *li*, automatically there is the subject, the thinking mind, and the object being thought about—*li*, in this case—and it's all disconnected, but when one experiences *li* it's another matter. I first became immersed in the process of *li* when skiing powder snow. When I felt the snow lift my skis for me and the gravity pull them down in the next turn and on and on with no thought or effort on my part, I began to re-cognize the individual *li* (or pattern) in each of the participants—the *li* of me as a human, the *li* of snow, the *li* of gravity—all moving together with no conscious cutting off by the rational cortex. Then, "Mind suddenly realizes its identity with the absolute *li* of all things."[46] When, for any reason one finds oneself being skied by the snow and gravity, being played by the wind, being moved by the very air as in *Tai Chi*, a breakthrough occurs "and the distinction between subject and object becomes completely obliterated, and the mind and *li* are realised to be one."[47] Gregory Bateson says it this way: "The concept of 'self' will no longer function as a nodal argument in the punctuation of experience."[48] This is when the wilderness within meets the wilderness without.

One doesn't get there all at once—to this depth of understanding. As Izutsu explains—the progress of the "investigation of things," which is the Chinese way of explaining this, "is not a horizontal but a vertical process. It is a vertical process in the sense that the mind goes on being deepened step by step." I know this is difficult to understand so I will end by quoting Izutsu again: "In the view of the Neo-Confucian philosophers, *li* is not only in each of the 'external' things, but it also exists in the mind, and—more important still—that the *li* of things and the *li* of the mind are ultimately one and the same *li*."[49]

In our domesticated, normal human world we are always imposing ourselves on nature. So how is it possible for us to once again know the true *li* of the land in our place? It is possible only in a true wilderness and then only if one moves quietly and with respect. Here one can fully experience the *li* of each of the non-human beings around one and so once again recognize the *li*—the pattern—out of which each of us comes. The early Taoists could easily walk up onto the intervening ridges between the settled agricultural valleys and thus experience the *li* of all beings. Today with airplanes there are no easy ways to experience wilderness. Further, there are so many humans that if we all went to the wilderness there would be no other species left anywhere. But ritual is another way to bypass the "merely human" aspect of us and again experience the relationship of our human *li* with all the non-human outside the so-called boundary of our skin. In the next section I will go more into ritual.


## PART III—Concerning Rituals and Nature

In ancient China the Emperor made a ritual tour of inspection to each of the four quarters of the Empire. There he performed the Imperial *Feng* and *Shan* Ceremony (translated as veneration of Heaven and Earth) at each of the Sacred Mountains.[50] The Chinese word for the actual rites performed is *li*(a). It has a different Chinese written character from the *li* discussed above, but is pronounced the same way. According to Wieger, *li*(a) means propriety, ceremony, worship, the external exemplification of internal principles, feeling of reverence, respect.[51] The ancient *Feng* and *Shan* Ceremony exemplified the connection between the mountains and the internal life of the people—both coming from the Supreme Principle—*li* (Pattern). In other words out of what the people and the Emperor felt or knew within from this ceremony, the mountain enabled them to make decisions concerning human life in that place. Nature gives us the "organizing" principle (*li*) for our human society. It comes out of nature in that place, not from the human rational mind. This underlying ritual meaning of the word *li*(a) later became part of the *li* of Chu Hsi—the Supreme Principle, *li*(b).

To begin to "reinsert humanity back into nature" we need a wilderness space in each bioregion where a few people could go ritually each year to perform "*Feng* and *Shan*" ceremonies, to learn from the land itself what it needs from us humans so that we "fit" into the ongoing life of that place.[52] To give an example of the inherent beauty of this concept I will explain *Feng Shui* (translated as Wind and Water).

The concept of nature as the Organizing Principle led to the *Feng Shui* experts in China who sited houses, tombs or temples at a site where nature chose. In other words they followed the patterns (*li*) of the land in all aspects above and below and the wind and water to determine where nature itself provided the best place for that human construction. They did not change the land to fit human needs. *Feng Shui* has been mistranslated as *geomancy* (essentially an Arab word with occult overtones) and modern Europeans think of it as some sort of unimportant magic. Actually it was considered so essential in China that at the time when various European countries had taken over different cities in China for their trade (called "factories") and wanted to build railroads between them, the extremely corrupt Chinese officials of that time refused to allow the railroads—no matter how high the bribe. It was considered that important not to cut the energy lines of the earth. Another illustration of the importance of *Feng Shui* principles is that of the Ch'in general Meng Tien who had directed the construction of the Great Wall at the end of the Third Century B.C. He later swallowed poison and killed himself, because as he said: "I am guilty, and assuredly should die. From Lin-t'ao all the way to Liao-tung, a moated wall.... In the course of this work I cannot have avoided cutting through the earth's veins."[53]

The perfect "fit" of these *Feng Shui* structures into the land became the base of later Chinese landscape painting. These paintings are often assumed to be artist's license but actually it's the way human structures can look when fitted into the *li* of the land and not imposed onto it.

Summing up, the word *li* in this sense means not only ritual but "external exemplification of internal principles" which is related to *li* as principle.

Here in our country, until only two decades ago, it was assumed that although humans had sacred rituals there was absolutely no connection between rituals and the land itself. In the growing concern over the destruction of the last wild lands left in the United States it was felt that only "practical" approaches such as politics or buying up the land as the Nature Conservancy does were the only hope we had. The last thing anyone ever considered when it came to saving the land was rituals. But in 1968 the publication of Roy Rappaport's *Pigs for the Ancestors* led to increasing studies in this relatively new field so that now we can say that for most of humanity's time on earth, ritual has perhaps been *the* regulator in human/nature interactions.[54] Further, the importance of ritual grows as we become more aware of the limitations of language.

Language is a function of the dominant hemisphere of the neocortex. But the human brain consists of two older brains as well—the limbic or animal brain and the old brain (sometimes called reptile brain).

Neither of these brains is subjected to the dualistic distortions of our Western European language because these brains do not communicate in spoken language. But if we can't talk to our unconscious how does it communicate with us? In our culture it communicates only with great difficulty and only occasionally and individually, by means of music, dreams and great poetry. In primitive or traditional cultures the unconscious communicates by means of ritual, continuously and to all present.

Essentially, according to Rappaport, ritual refers to a class of communication events which occur among both men and animals in which one or more participants transmit through conventional signs or symbols information concerning their own physiological, psychological or sociological and ecological status to other participants, human and non-human. Thus ritual enables sharing of the *li* or pattern within the group as well as enabling the humans to learn the *li* (or pattern) of their surrounding ecosystem. For example, the Tukano people, who live in the central part of the Northwest Amazon river basin along the Vaupes river, view their universe as a circuit of energy in which the entire cosmos participates. The Sun-Father, a masculine power, fertilizes a feminine element, the earth. The basic circuit of energy consists of "a limited quantity of procreative energy that flows continually between man and animal, between society and nature" [somewhat similar to the Chinese concept of *chi*]. Their ritual cycle goes on all the time and has to do with this circuit of energy. In terms of ecological theory, the Tukano thus conceive of the world as "a system in which the amount of energy output is directly related to the amount of input the system receives." The Tukano have very little interest in exploiting resources more effectively but are greatly interested in "accumulating more factual knowledge about biological reality and, above all, about knowing what the physical world requires from men."[55]

To return to Roy Rappaport's work among the Tsembaga. In his book, *Pigs for the Ancestors*, he shows how ritual regulates the relationship between people, pigs and gardens. It protects people from the competition of their pigs for basic resources and also indirectly protects the environment by helping to maintain large areas in virgin forest. Rappaport observes that the "latent function" of the rituals has to do with allocation of scarce protein. Rappaport mentions the systems aspect of ritual when he explains that the rituals of the Tsembaga affect the size of the pig population, the amount of land under cultivation, the amount of labor expended, the frequency of warfare and other components of the system. All of this is done without the help of bureaucratic planners, computers or boring meetings. It happens of itself through the ritualized on-going system and largely by means of celebrations at festivals.[56]

After his work with the Tsembaga, Rappaport continued his research on ritual and in 1979 his *Ecology, Meaning and Religion* was published. Although his book has been generally overlooked by both ecologists and environmentalists it actually shows us the way we can begin to save the wilderness. As there is no way I can show the depth of this book in such a short paper, I will merely give a few important quotes. "Because knowledge can never replace respect as a guiding principle in our ecosystemic relations, it is adaptive for cognized models to engender respect for that which is unknown, unpredictable, and uncontrollable [excellent definition of wilderness]." He continues:

> A model dominated by, let us say, the postulates of economic rationality would propose that an ecosystem is composed of elements of three general sorts: those that qualify as "resources," those that are neutrally useless, and those that may be regarded as pests, antagonists, or competitors. In contrast the Ituri Pygmies take the forest encompassing them to be the body of God. These two views of the world obviously suggest radically different ways of living in it.[57]

Another quotation provides a very concise explanation of the importance of ritual.

> I am asserting that to view ritual as simply a way to fulfill certain functions that may as well or better be fulfilled by other means, or as an alternative symbolic medium of expressing what may just as well—or perhaps better—be expressed in other ways is, obviously, to ignore that which is distinctive in ritual itself. Moreover, it becomes apparent through consideration of ritual's form that ritual is not simply an alternative way to express certain things, but *that certain things can be expressed only in ritual.*

And of particular interest for this study of *li*, he says that ritual "does not always hide the world from conscious reason behind a veil of supernatural illusions [as many believe]. Rather, it may pierce the veil of illusions behind which unaided reason hides the world from comprehensive human understanding."[58] Here we have another insight into the Chinese idea that human *li* can understand the *li* of nature because we are both the same *li*.

It is important to realize that understanding the *li* of place depends not on how long one stays in a certain place but on how well the rituals are done. The Apaches and Navajos have only been in the Southwest area since about 1380—in other words only two hundred years before the Spanish came into the area.[59]

Tribal people in general, as well as the pre-Socratics and the early Chinese, all make use of patterns as their basic thinking. As far as tribal people are concerned, we tend to call them illiterate but that is because we do not realize that their patterns, songs and dances are valid literature and accurate recording systems. Mollison notes that,

> having evolved number and alphabetical symbols, we have abandoned pattern learning and recording in our education. I believe this to be a gross error, because simple patterns link so many phenomena that the learning of even one significant pattern, is very like learning an underlying principle, which is always applicable to specific data and situations.

For example, he continues,

> Polynesians used pattern maps, which lacked scale, cartographic details, and trigonometric measures, but nevertheless sufficed to find hundreds of island specks in the vastness of the Pacific! Such maps are linked to star sets and ocean currents and indicate wave interference patterns.... The Pitjantjatjara people of Australia sing over sand patterns and are able to "sing" strangers to a single stone in an apparently featureless desert.[60]

Carpenter wrote of the Eskimos on the Aleutian Islands who had never seen an airplane until the battles in the Second World War. These

Eskimos would look over the shoulder of the airplane mechanics and finally say something like "why don't you try to connect that with this" and the engine was fixed. Their knowledge of pattern allowed them to instantly understand the pattern of the engine.

I will conclude this study of *li* by recounting one of the famous Taoist "knack" stories of Chuang Tzu. These stories show how human *li* can so understand the *li* of the natural world that the result seems like magic.

The Duke of Lu, seeing an exquisite carved wooden bell stand asked the carver, Ch'ing, how he had learned this art. Ch'ing replied that he knew no special art at all and went on to say: When I am about to work on a bell stand, I try not to waste my *chi*. I fast in order to preserve serenity of mind. After three days I cease to cherish any desire for prize, money, or official glory. After five days the ideas of praise or no praise and the question of workmanship just leave me. After seven days I attain to a state of absolute serenity, forgetting that I have a body with arms and legs. At that moment I forget that I am working for the court. My sole concern is about my work, and nothing of external interest disturbs me. I now go into the forest and select the most suitable tree, whose natural frame harmonizes with my inner nature. I know that all I have to do is begin helping by carving the bell stand out of the confinement of the tree. I then apply my hands to the work. When all these conditions are not fulfilled I do not work for I perceive that only when the *li* (in Nature) unites with *li* (in Man) can my bell stand occur.

According to Kuang Ming-Wu, the Taoist, "Chuang Tzu...defines, or rather focuses, the limits of humanity. It is perhaps the most universal problem that he is immersed in: how to replace in us the will to form with the will to accept natural form."[61]

I will close with the well-known quotation from Hsun Tzu: "Through rites (*li*) heaven and earth join in harmony, the sun and moon shine, the four seasons proceed in order, the stars and constellations march, the rivers flow and all things flourish."

We now know, because we, today, are destroying nature, that without an understanding of *li* the seasons falter, the sun darkens, and rivers die.

# The Blessing of Otherness: Wilderness and the Human Condition

by Michael Zimmerman

When I was a boy growing up on the outskirts of a small Ohio town, I would spend countless hours climbing on the glacial cliffs in the woods nearby. The woods were mysterious and enchanting. I could almost detect the presence of the Native Americans who were so much at home there not so long ago. I imagined that they lived easily in the woods; they could follow its paths for miles; they knew what bushes had berries and what streams provided fish; they understood what weather the oncoming clouds portended and they also knew what caves afforded quick shelter. I envied the Native Americans for their ability to live upon the land, and I envied the fact that it remained unspoiled by industry, urban sprawl, and pollution. During those boyhood years, I experienced a faint measure of the natural glory in which they had moved about on moccasined feet. Much of those woods had been cut for timber long before I arrived, but many respectable trees had returned to them and there were few visible marks of human presence. The tall trees, the creeks running down the hillsides, the cries of squirrels and birds, the pungent scent of decaying leaves, the wind in the autumn leaves, the sun shining on a snowy cliff face, the bright blue wildflowers promising an early spring—these were all astonishing and sacred to me. Raised in the Roman Catholic Church, I had no vocabulary that would allow me to worship in the woods, but worship I did. I knew that God was there, though I sensed that adults had long forgotten this overwhelming fact.

My attempts to explore the "wildness" of those woods was mediated to some extent by my encounter with an old man, "Chief" Cochran, so named because of his love for and familiarity with Native American rituals and narratives. From him and from others involved in Scouting, I gained a limited vocabulary that helped to express my awe and respect for the woods and all things wild. Years later, as a college student, I read Wordsworth's poetry, especially *The Preludes*. I instantly believed that I shared his intuition of the "brooding presence of Nature," though I have always found it difficult to articulate it. The Romantic vision of nature as a mysterious and sacred power was confirmed by my own early experiences in the Ohio woods. Despite my appreciation of the "presence" of wild nature, its spontaneity, repose, and infinite depth, I also realized that nature was alien to me, in its lack of personality, language, and self-awareness. Somehow, I knew that I shared in infinite depth of wild nature, but my self-awareness seemed to be lacking in the wild things. When depressed by some defeat at school or by some family quarrel, I would often head for the woods in hopes of finding solace. I was never able to find the reassurance that another human being could provide for me; moreover, I often experienced what I took to be a sense of indifference on the part of the sky and water, trees and ants toward me. I now believe, however, that the woods were not so much indifferent to me as they were enthralled to their own rhythms and movements that were far vaster and older than those of my personality. At some level, my body identified with those ancient movements. To feel at home in the woods, I would have to expand beyond the limits of my personality-structure, which had been created and reinforced by a family and a social order that had grown wary of and hostile toward the woods and toward everything merely "natural." Yet in this very process of expanding beyond myself, I also came to understand something of what I myself was: a civilized being.

Many Native American tribes seem to have developed a profound relationship to wild nature. This relationship, mediated by language, symbol, custom, and ritual, helped to define the very identity of the members of those tribes: civilized humans living as brothers and sisters of the animals who shared the world with them. Members of Western culture also gain their sense of identity in part by virtue of how they define their relationship to nature. Because of the influence of the Greek and Hebraic traditions, however, Westerners have tended to make a radical distinction between humanity and nature, the former involving a rational and historical dimension not to be found in the latter. Western identity was shaped by the view that nature is not sacred and thus is a resource to be used for human ends. In its dominant trends, Western civilization had little interest in developing a "relationship" with wild nature, apart from the economic benefits that exploiting it might bring.

The ways of our civilization became all too clear to me when, right after my sophomore year in high school, my family moved to a mid-sized Ohio city. There, removed from easy access to the woods, I learned the political and economic justifications offered for the increasingly rapid pace at which the woods were being destroyed. I could entertain the science fiction fantasies of human omnipotence that would supposedly be made possible by the miracles of modern science and technology. But I never felt convinced by them, for these fantasies wrongly presupposed that humans could ultimately separate themselves from nature, or at least completely control it. I believed that in exchange for the material benefits associated with an ever-expanding economy, we were losing something very precious, something that no amount of material goods could replace. The loss of wilderness and woods was and is a source of great sadness to me. Today, while the Amazon rainforest is disappearing, while acid rain and other forms of pollution increase, and while arable land decreases while human population increases exponentially, I am more than ever convinced that my original intuition was correct: that humans learn something essential from the wild nature, something which they lose only at their own peril. Wilderness is a direct reminder that not everything can be reduced to the status of a human product, project, or construct; wilderness is the "other" which reminds humanity of its own dependency on the powers at work not only in wilderness, but also in humanity itself. Not wishing to be so reminded of its relation to and dependence on nature, we modern humans set out to "conquer" the wilderness—and in the process risk not only the loss of our own identity as humans, but also the destruction of the ecosphere as well.

One reason I studied philosophy was in hopes of finding the wisdom needed to cope with the loss I experienced in the face of the industrial transformation of the earth and of the human world. I have found that much of Western philosophy seems to have contributed to the very anthropocentrism and dualism responsible for the technological domination of nature. Moreover, most contemporary philosophers seemed uninterested in exploring the possible relation between Western philosophy and the environmental crisis. In the past fifteen years, however, an increasing number of philosophers have begun exploring that relation. In the present essay, I examine how three different thinkers address the question of the relation of wild nature to humanity: Martin Heidegger, Susan Griffin, and Hans Peter Duerr. I choose these particular thinkers in part for autobiographical reasons: each of them represents a position from which I have learned a great deal. All three authors seek to discover an alternative to the domineering way in which Western humanity treats the natural world. They all agree that wild nature represents "otherness," but they differ regarding the character of this otherness and its relation to humanity. For Heidegger, nature is radically "other" than human existence, for only humans are supposed to be able to understand what it means for things "to be." For Griffin, wild nature is "other" than the patriarchal ego, which yearns to escape from mortality and dependency. That nature, however, is also allegedly the source enabling women to form an alternative to the identities prescribed for them under patriarchy. For Duerr, wild nature is "other" than the civilized ego, but that ego can understand the meaning of being "civilized" only by virtue of passing over into the state of wildness. All three authors, then, conclude that nature, and in particular wild nature, constitutes an otherness that may play a role in human identity. Let us examine each of their views briefly in turn.

## Heidegger's View of the Nature-Humanity Relationship

Heidegger's thought has been influential among environmentalists, especially deep ecologists, who regard him as an example of a thinker who broke out of the anthropocentric and dualistic attitudes that have led to the age of the technological domination of nature.[1] Anthropocentric Western humanity, rejecting the notion that wild nature is intrinsically worthy of respect, maintains instead that nature is valuable only as a resource for human ends. Many environmentalists, by way of contrast, maintain that we must protect wild nature not only for the prudential reason that it might be necessary for preserving the ecosphere (and thus human life itself), but also for the reason that all things have the "right" to flourish and to live out their evolutionary destiny without undue interference by humans. Preserving wilderness is vital, so environmentalists tell us, not only for prudential reasons, and not only because it allows at least some beings to follow out their destinies without human interference, but also because it stands as a radical "otherness" to technological civilization, an otherness that serves as a reminder that humanity is itself inextricably bound up with a natural environment that cannot ever be fully "controlled."

Heidegger's thought seems in many respects consistent with the environmental attitude of respect for nature. Especially in his later writings, he spoke of the necessity of learning to "let beings be," instead of dominating beings by treating them as raw material for the sake of increasing power for its own sake.[2] According to Heidegger, the technological way of disclosing things is the final stage in the history of metaphysics, i.e., the history of the search for ultimate foundation, ground, or "being" of entities. Conceiving of being as a foundation, however, overlooks the radical difference between being as such and all entities whatsoever. To depict being as a "foundation" is to conceive of it as a superior entity, such as "God." For Heidegger, however, "being" names the "presencing" or "manifesting" of an entity, not the entity which founds, grounds, or causes other entities. For an entity "to be," i.e., for it "to be manifest," a site or clearing is required. Human existence constitutes the temporal-historical clearing ("absencing") in which being ("presencing") can occur. Without human existence, entities would not "be" in the sense that they would be concealed, unmanifest. Heidegger insisted that human existence is unique, in that only humans can understand the fact *that* entities *are*. Despite radically distinguishing humans from all other entities, he also criticized the anthropocentrism of the Western tradition.

Heidegger argued, for example, that Descartes's dualism greatly contributed to this anthropocentrism by making human rationality the ground or foundation for all things. For Descartes, in other words, "to be" meant to be re-presented to and controllable by the self-certain subject. Descartes initiated the era in which humanity encountered everything merely as an object useful for enhancing human security and power. Whereas the pre-Socratic Greeks had been in awe of the sheer presencing of entities, technological humanity is in awe of nothing and seeks to control everything. Time and again, Heidegger railed against the destructiveness of the anthropocentric dualism at work in the technological understanding of things. He spoke of the darkening of the world, the destruction of the earth, and the flight of the gods; and he also urged humanity to prepare itself for the possibility of a "turning" in history, a turning that would free humanity from its lust for power and would usher in a new, non-technological way of disclosing things. Heidegger claimed that humanity's highest possibility is not to dominate entities, but to care for them, to let them be what they are, to join with things in the dance of the "fourfold," of earth and sky, gods and mortals.

Understandably, deep ecologists and other environmentalists have found much with which they can agree in Heidegger's critique of Western culture's anthropocentrism and dualism, as well as in his view that humanity must learn to dwell appropriately on earth. Unfortunately, despite these important points of contact between Heidegger's thought and contemporary environmental attitudes, there are basic differences which call into question the plausibility of making his thought a "source" for deep ecology. First of all, environmentalists appeal to the science of ecology for support for their contention that humanity is only one dimension of the complex organic fabric which composes life on earth. Heidegger, however, believed that science could say nothing essential about the humanity-nature relationship, since it misconceived humanity and nature. Moreover, he argued that modern science is intrinsically bound up with the anthropocentrism and Will to Power present in the technological domination of the earth. Heidegger, then, radically opposed any attempt to define humanity in terms of the categories of modern natural science. Proclaiming that humans are not animals, he maintained that the West went astray with Aristotle's definition of humanity as the "rational animal," a definition which supposedly culminated in Nietzsche's "blond beast," the embodiment of the power-hungry technological human "animal." The technological domination of nature, in other words, resulted from the fact that Western humanity conceived of itself naturalistically, as somehow akin to animals! So opposed is he to that conception that Heidegger speaks of an "abyss" lying between humans and animals.[3] Because this "otherness" of animals (and of the rest of nature) is absolute, Heidegger would hardly be willing to speak of humanity's place in the "biosphere" or "ecosphere."

Karl Löwith, one of Heidegger's best students, argues that his mentor's fierce anti-naturalism is proof that he did not overcome the anthropocentrism and dualism of modern humanity, but in fact had simply found a new vocabulary for them.[4] Claiming to be renewing the insights of the Greeks, Heidegger nevertheless rejects the Greek vision of humanity as a constituent of the cosmos. Sounding remarkably like certain Gnostic theologians, Heidegger sometimes talks as if humanity were essentially other than or outside the cosmos, a stranger "thrown" into the world of nature, hence, hardly a "filament" in the "web of life."[5] According to Löwith, Heidegger's thought remains anthropocentric and dualistic in radically distinguishing humanity from and privileging it above everything else. Such anthropocentric dualism is, in Löwith's view, a residue of the Christian and Cartesian traditions in Heidegger's thought.

Heidegger's anti-naturalistic attitude, which is completely foreign to contemporary environmentalism, led him to support what he believed were the true aims of National Socialism, while simultaneously criticizing the Nazi reliance on biologistic and racist doctrines.[6] Such racism revealed that primitive Nazism was unwittingly in league with the very naturalism responsible for the rise of modern technology, against which Nazism supposedly stood. Heidegger believed that Western history, which culminated in modern technology (common to capitalism and communism), was a history of decline, not progress. The technological quest for power for its own sake, for infinite production and consumption, was the inevitable expression of a humanity which had been reduced to the status of a clever animal (Nietzsche) striving for nothing more than security and control. Rejecting Nazi racism and biologism, Heidegger nevertheless maintained that the "inner truth and greatness" of National Socialism lay in its attempt to initiate a new beginning for the West that would save it from the one-dimensional, animalistic future threatened by the spread of the technological world. Heidegger maintained that the Nazi yearning for "blood and soil," for "instinctual renewal," for "wild passion," were manifestations of a misguided naturalistic and atavistic attitude which was foreign to the "essence" of the movement. National Socialism portrayed itself, at least until its commitment to industrialization and rearmament in 1936, as being consistent with the romantic longing for a pre-industrial world in which the German *Volk* had a proper relationship to nature. Capitalists and communists raped the land and thus destroyed the *Volk*, but National Socialism would heal the land and restore the proper balance between the natural-instinctual and the cultural-technical. The fact that much of National Socialist propaganda favored what today might be called an "environmentalist" or "green" attitude helps explain the fact that environmentalists today are sometimes accused of being proto-fascist in their talk of "healing" the humanity-nature relationship.[7] Critics remind environmentalists that the natural "wildness" admired by many Nazis turned out to be a lust for blood and violence, a lust which translated into a murderous quest for the very technological power which Hitler had at one time condemned.

In his lectures on Hölderlin and Nietzsche during the 1930s and 1940s, Heidegger sought to develop an alternative, non-racist, non-biologistic conception of the German *Volk* and of the "soil" in which it was "rooted." He was particularly intrigued by Hölderlin's attempt to develop a non-metaphysical, non-naturalistic (i.e., non-reductionistic, non-scientistic, non-materialistic) conception of "nature." Hölderlin, in Heidegger's account, rejected Christianity's disenchantment of nature.[8] Hölderlin spoke of primal nature as the "All-One" and as "holy wildness." His destiny was to encounter and to sing the primal presencing of things, their "holy wildness," in a way that would initiate a new, non-metaphysical, non-technological age for the German *Volk*.

Heidegger interpreted this holy wildness to mean something like the Greek *physis*, ordinarily translated as "nature." For Heidegger, however, the ordinary conception of nature, as the totality of material particles interacting within space and time according to nomic necessity, failed to express what *physis* really meant. *Physis* names the overpowering presencing, appearing, or "being" of entities. This presencing cannot be understood in terms of causal-material processes, however, for those very processes can only "be"—in the sense of being-manifest—within the disclosive event of "being." While denying that he was in any way a subjective idealist, Heidegger nevertheless moved toward what amounts to a linguistic cosmology: being and language are internally related. He never satisfactorily explained the relationship between the "presencing" or "self-manifesting" of an entity, on the one hand, and the biological processes involved in the self-growing of a living entity, on the other. Which is more primal: the *growth* of a plant or the *disclosure* of that growth within a historical-linguistic world?[9] Heidegger attempted to address such issues in his complex doctrine of the "strife" between earth and world. He conceded that earth represents the domain which can never be made fully present within any world; hence, he concluded that no matter how much modern technology attempted to make everything completely present for exploitation, it could never violate the self-concealing mystery of the earth.

While Heidegger raised profound questions regarding the humanity-nature relationship, he also distinguished the essence of humanity so radically from what we normally regard as the organic matrix of life on earth that his thinking may not be easily reconciled with much of contemporary environmentalism. For Heidegger, what is most worthy of devotion is not natural beings as such, but rather the self-concealing presencing or the "holy wildness" by virtue of which they can be manifest. Wild nature, for example, in the form of wild animals, is radically "other" than the essence of humanity. Animals are said to be "world-poor," because they lack the language necessary for there to be a historical-linguistic world in which entities can reveal themselves.[10] Western humanity's mistake, in Heidegger's view, was to have failed to see the extent of the difference between itself and animals. While animals are tantalizingly close to us in certain respects, we can learn nothing about ourselves by studying them, or by studying the "animality" within us, for what is essential to human "existence"—its capacity for understanding the being of entities—bears no relation to the categories of "life."

By contrast, the "otherness" of being reveals something essential about human existence. Being does not name an entity of any sort, but rather the presencing of entities. Humans are the clearing required for that self-concealing presencing. Ultimately, the "essence" of human existence is no-thing at all, but is rather the state of being the linguistic "between"—the cleft between being and entities. Having forgotten that it is no-thing, Western humanity conceived of itself as a peculiar kind of thing, a rational animal, or a self-certain ego-subject. Having conceived of itself in such a way, humanity set out to protect itself against the threats posed to its survival by other things. Paradoxically, then, Western humanity's attempt to dominate nature is the direct result of humanity's attempt to conceive of itself as a natural thing. The end to such domination, then, requires that humanity conceive of itself as non-natural, while also conceiving of nature in an entirely different way than that proposed by naturalism and materialism.

## Susan Griffin's View of the Humanity-Nature Relationship

Susan Griffin has written several books, but here I shall focus on her most noted and influential work, *Woman and Nature: The Roaring Inside Her*, originally published in 1978. In this remarkable book, Griffin argues that patriarchy has been responsible for the subjugation of nature within and without.[11] Since women have traditionally represented both nature-within (feelings, intuition, all that cannot be reduced to linear rationality) and nature-without (child-bearing, lactation, caring for the animal needs of other humans), they have borne the brunt of the patriarchal denial of and attack upon everything merely organic, fleshly, imperfect, material, and mortal. The first part of her book, "Matter," deals with the ways in which man "regards and makes use of woman and nature." In a series of remarkable vignettes and observations, Griffin shows the extent to which man's domestication and subjugation of nature (land, timber, animals, wind, matter in general) through the processes of science and technology runs parallel to his domestication and subjugation of woman. Griffin shows that much of Western philosophy is guilty of supporting and justifying the various dualisms—mind vs. body, man vs. woman, man vs. nature—which have led to the domination of woman and nature in patriarchy. So totally have women been dominated by men, we are told, that women have had their voices as well as their bodies taken away by men: women have been dispossessed of the language necessary to experience and to define their own bodily existence. Women, in other words, have been violently separated from themselves; hence, they live in a state of constant alienation and diremption.

According to Griffin, this alienation is but one instance of the divisions cleaved by patriarchal man. Terrified of his own mortality, patriarchal man establishes radical binary oppositions between his immortal, eternal essence—his soul, his mind, his culture—and the mortal dimension that is not essential to him—flesh, blood, bone, feelings, sex, children, nature: "Separation. The clean from the unclean. The decaying, the putrid, the polluted, the fetid, the eroded, waste, defecation, from the unchanging. The changing from the sacred.... Death from the city. Wilderness from the city. Wildness from the city."[12] From Plato's soul-body dualism, to the Christian and Cartesian versions of such dualism, to the industrial world of modernity, patriarchal man has invented categorical schemes devised to assure him that he is more than mere flesh—that he will transcend the organic dimension from which he in fact arose. Rendered paranoid by his fear of death and limitation, patriarchal man must control everything which reminds him of the lie that he leads. Just as he conquers the wild animal, and subdues the wildness in his woman, he must also conquer the wildness within himself: his passion, his desire, his unfathomable yearnings that resist all attempts at being "clarified" by rationality.

Wildness threatens the integrity and boundaries of the patriarchal ego. In the face of the wild sexual desires aroused in him by woman, for example, he is threatened with dissolution, with the loss of the rigid ego-boundaries he has erected to protect himself from being "absorbed" by the all-devouring female. "He faces annihilation in her, he says. He is losing himself to her, he says. Now, he must conquer her wildness, he says, he must tame her before she drives him wild, he says."[13] He puts woman in a domestic cage that resembles the zoo into which wild animals are placed—to be observed and appreciated, now that their wildness has been minimized. In a striking passage, following closely upon a lengthy account of how abstract scientific rationality reduces "nature" to a complex set of material events acting in accordance with nomic necessity ("nature" confined within the intellectual cage of science), Griffin depicts the dreams which may accost coolly rational men on "wild nights":

> On these wild nights, he saw himself in her body and then he moaned at the injustice of finding his humanity concealed and trapped in this way until he would wake up screaming in terror. And into those first moments of waking crept this doubt that seemed to edge into him and stick, creating an unnatural space between his soul and his flesh, the doubt of the justice of things after all. Suppose that gesture of hers meant her soul was like his.... Suppose there is no difference between them except the power he wields over her. And suppose that in an instant of feeling himself like her, he let this power go, then would he not become her, in his own body even. And some part of him seemed to know what it would be to be her in his body, and how he came to know this he does not choose to remember.[14]

Recoiling against the possibility of discovering the wildness of female/nature within himself, the man violently pushes away from this intimacy by pushing back space as far as he possibly could, to the very ends of the material universe. In that infinite space, he could find some breathing room: "She had too much eagerness to attach meaning to things. He reminded himself that all this life is determined, that what meaning there is in the movements of matter is indecipherable. That he himself is made up only of particles in space. And beyond this, he tells himself, only the stars burn, burn in a dreadful void."[15] Griffin starkly contrasts the abstract, linear rationality of the patriarchal ego, which reduces everything to the either/or status of impediment or contributor to the project of overcoming mortality and corporeality, with the long-repressed female mode of awareness, which is particularizing, narrational, meaning-giving, and celebratory of passion, feeling, embodiment—life itself.

In the final part of her book, having addressed the separation imposed by patriarchal man upon nature and humanity, Griffin deals with the phenomenon of healing: rejoining what has been separated. Here, she attempts to recover an essential female voice, to free it from the perversions and concealments thrust upon it by patriarchy, and thus to liberate women and nature from centuries of repression. The work of healing requires the transformation of patriarchal categories regarding nature, space, time, the body, feeling, sexuality, knowledge, the sacred, death, and life. For example, as opposed to the patriarchal vision of "reality" as a perfectly clear and ordered realm, a totality of material particles running in accordance with eternal "natural laws," Griffin depicts a "reality" constituted by chaos, by constant movement, unfolding, generation, development, interpenetration, none of which is subject to external "laws," but rather is expressive of an inner generative-creative power. Such a reality involves none of the rigid boundaries, borders, limits, divisions so fanatically erected and defended by men terrified to recognize the female within them. The "roaring inside her" names the creative turbulence, the dynamic origin of the web of life, a wild source which threatens all merely logical structures and all technological achievements. The masculinist drive to name, to know, betrays what cannot be named: "Behind naming, beneath words, is something else. An existence named unnamed and unnameable."[16] Instead of controlling, the emerging woman learns to let things be; she surrenders to the hidden movements that can never be brought to the surface, but without which there would be no life, no meaning, no joy. What seems to patriarchal consciousness to be merely turbulent, dark, confused, indeterminate, irrational, insane may be in fact a deeper clarity repressed for centuries, as in the terrifying witch trials which corresponded with the dawn of modern science. The witches gave voice to and embodied the wildness, the wilderness, so horrifying to the patriarchal ego bent upon excluding all disorder from view. Griffin remarks, however, that

> We are disorderly. We have often disturbed the peace.
> Indeed, we study chaos—it points to the future. The oldest
> and wisest among us can read disorder....

> Many of us who practiced these arts were put on trial.
> We stood at the gates of change, but those who judged us
> were afraid. They claimed the right to order the future.
> They would have had all of us perish, and most of us did.
> But some kept on. Because this is the power of such things
> as we know—we kept flying through the night, we kept up
> our deviling, our dancing, we were still familiar with
> animals.... And even if over our bodies they have trans-
> formed this earth, we say, the truth is, to this day, women
> still dream.[17]

The witches represented the female capacity for opening to the
wisdom of the body, to the wisdom of disorder, chaos, wildness. Healing
Mother Earth can only occur if women move once more into the dreamworld, surrender themselves to the unconscious movements and apparently inchoate unfoldings essential to all life, growth, and creativity.
Affirming their own bodies, their sexuality, their disorderly desire, their
capacity for pleasure, women must learn to affirm their ties with the
Earth and with all generative wildness/wilderness. In wonderfully
poetic passages, Griffin describes the interpenetration of the processes/
domains of the Earth with the processes/domains of the female body. For
example, we hear that

> when I let this bird fly to her own purpose, when this bird
> flies in the path of his own will, the light from this bird
> enters my body, and when I see the beautiful arc of her
> flight, I love this bird, when I see, the arc of her flight, I fly
> with her, enter her with my mind, leave myself, die for an
> instant, live in the body of this bird whom I cannot live
> without, as part of the body of the bird will enter my
> daughter's body, because I know I am made from this
> earth, as my mother's hands were made from this earth, as
> her dreams came from this earth and all that I know, I know
> in this earth, the body of the bird, this pen, this paper, these
> hands, this tongue speaking, all that I know speaks to me
> through this earth and I long to tell you, you who are earth
> too, and listen *as we speak to each other of what we know:*
> *the light is in us.*[18]

There are evident similarities between Heidegger's and Griffin's thought. Both depict Western history as the story of the domination of nature, the destruction of the earth, the darkening of the human world; both are suspicious that science and technology are inherently geared toward such domination; both call for a new mode of behavior that lets things be. Despite such points of contact, however, for a feminist like Griffin Heidegger's ontology represents a more subtle version of the ancient masculinist fear and hatred of the body. The radical distinction Heidegger drew between humans and animals, for example, and his identification of "holy wildness" not with the wildness of the organic body but rather with a non-organic event of "presencing," are symptoms of his mistrust of the organic realm. His concern about the earth, his talk of "letting beings be," his critique of modern technology are admirable so far as they go, but they don't go far enough, for they remain rooted in a humanity-nature dualism: Heidegger could not accept the fact that humans are *essentially* organic. Moreover, Heidegger discounted the sexual difference; he maintained that human *Dasein*, his term for the essence of human existence, is sexually neuter. By emphasizing the ontological difference instead of the sexual difference, he betrayed his ignorance of the real source of the technological domination of the earth: patriarchy. It is man, according to Griffin, not "humanity" or "human *Dasein*," who is responsible for subjugating women, nature, and the woman within man himself. Not a neutral *anthropocentrism* or human-centeredness, as Heidegger had suggested, but instead *androcentrism* or male-centeredness is the source of our environmental crisis and our related social crisis as well.

Griffin's critique of Heidegger (at least as I have imagined it) has some validity. He never reflected upon the character and origins of patriarchy, because in the end he concluded that Western history was not the product of anything human, including human social-sexual arrangements or categories, but rather of the self-concealing movements of being itself. He retained a nature-humanity dualism which prevented him from taking seriously the organic dimension of human existence, and which also prevented him from discovering "self" in the "otherness" of wild nature. Without such recognition of humanity's profound kinship with other beings, how could Heidegger hope that it would ever be possible to "let beings be"?

Yet, despite its insight and style, Griffin's own book has disturbing limitations. First of all, Griffin reveals herself to belong to the camp of "essentialist" feminists, those who believe that women are endowed with certain capacities which men do not share, or else share only in a secondary, derivative sense, by virtue of their being born from and raised by women. These essential traits involve a proximity to the wildness of nature, to the unconscious forces at work beneath the veneer of scientific rationality, to the chthonic powers that can never be controlled by life-denying men. Feminists who are critical of such "essentialism" have argued that it reinforces the stereotypical view of woman as irrational, emotional, particularizing, body-bound, incapable of rational reflection, unable to control her voracious sexual desires. Moreover, essentialism also divides the human species into a "good but oppressed" and a "bad and oppressing" duality, one that cannot be sustained in the face of the fact that women in Western industrialized countries are deeply implicated in the consumerist economic system which sustains the industrial technological domination not only of "nature," but also of "development," which subjugates not only women of color in our own country and women in Third World countries, but non-ruling class males as well. While speaking about the need to heal by rejoining what has been separated, essentialist feminists (including Mary Daly as well as Griffin) in fact proclaim a new version of the old female-male binary opposition, with the difference being that females are now the favored pole while males occupy the marginalized pole. A true healing would require that women and men alike learn to recognize their own strengths and weaknesses, and seek to establish new, non-sexist social norms and institutions, rather than continuing in the old practice of regarding one gender as being "better" than the other. Today, more than a decade after the publication of *Woman and Nature*, many feminists have moved away from essentialist feminism and have conceded that factors other than gender—such as race, class, and ethnicity—shape human identity in different times and places.

Griffin's essentialist views influenced some "eco-feminists," who have argued that only women have the relationship to nature that is requisite for healing the ills of patriarchy.[19] If only women were freed from the constraints imposed by patriarchal men, some eco-feminists claim, those nature- and body-loving women would eventually develop the kinds of social institutions needed to heal the Earth. Other eco-feminists have questioned this essentialist stance, emphasizing that the origins of the environmental crisis and the subjugation of women are too complex to be explained in terms of one causal-historical principle: patriarchy.[20] Further, the notion that men are somehow incapable of "relating" to nature in an "intuitive" and "direct" way is belied not only by the practices of tribal men, but also by the significant number of male artists and writers, from East and West, who have expressed in moving ways their experience of nature and who have condemned the human exploitation of it.[21]

Finally, we may wonder about the extent to which Griffin is willing to place her trust in primal wildness, in the indeterminate turbulence of desire and passion. Some critics would say that they have heard this kind of thing before, for example, in National Socialist rhetoric about the necessity of overcoming the artificial categories of civilization which undermine the relation between blood and soil, between the passions of the *Volk* and the uncontrollable urges of wild nature. Such critics would not suggest, of course, that Susan Griffin's book is a proto-Nazi tract! Nevertheless, they would caution about the dangers of a feminism that fails to appreciate the extent to which National Socialism was an example on the national scale of the psychological phenomenon of recollectivization.

Recollectivization occurs when the ego-strength of individuals is not sufficient to withstand the anxiety and tensions associated with modern industrial civilization; such individuals seek to "escape from freedom" (E. Fromm) by identifying themselves with "natural" or "instinctual" forces that are supposedly "deeper" and "more primal" than those of "rationality" and "atomistic individualism."[22] This is the way in which National Socialists appealed to Germans who were terrified by economic collapse and political disintegration. Those who accuse ego-consciousness of being merely a patriarchal formation and who call for the overthrow of the political framework which guarantees individual people relative freedom from interference run the risk of inviting a new instance of recollectivization. There is no doubt that ego-consciousness, unfortunately, tends to deny, repress, and dissociate itself from the primitive, including the body, feelings, woman, nature.[23] The task facing us, however, is not to return to a more primitive level of awareness, one for which many of us continue to yearn, but rather to integrate the more primitive with a more developed mode of awareness. Only by evolving beyond ego-consciousness toward a more inclusive, less dissociative mode of awareness will humankind be able to integrate what has been previously denied by ego-consciousness. It may well be, however, that this more integrated awareness, one that promises to harmonize in a non-regressive way the individual with community, male with female, humanity with nature, cannot be achieved without passing through the stage of dualistic and dissociative ego-consciousness. If so, then efforts should be made to consolidate the stage of ego-consciousness, although simultaneously to rid it wherever possible of its patriarchal dimension. There may be no chance for humanity to move to a more inclusive level of awareness if much of the human population remains trapped in social-political formations which subordinate individual rights and needs to the good of some collective agency, usually the "nation-state." There is no guarantee that the movement beyond the semi-collectivistic consciousness of nationalism toward ego-consciousness and beyond will succeed. Indeed, the stage of ego-consciousness itself, especially in the patriarchal mode it has assumed in human history, is highly threatening to the well-being of the entire ecosphere. For the ego tends to conceive of nature either only as a source of power or as a threat.

Griffin seems to regard rational ego-consciousness as intrinsically patriarchal and as irredeemably offensive to wild nature, to "the roaring inside her," which represents the "other" to the patriarchal mode of awareness. She elevates the once-marginalized and repressed "other" to a position of superiority, representing it as the essential dimension of the female. Griffin claims that witches, attuned to the instinctual and wild modes of awareness, show the way to the future, but critics may wonder whether they are calling for a reversion to a more primitive mode of awareness which is not consistent with a level of differentiated awareness that many women, not to mention men, would be unwilling to give up. It is one thing to call for an end to the subjugation of feelings, of woman, of the body, of nature; it is another thing to call for what at times seems to be an unmediated surrender to "instinctual" powers.[24] If Heidegger saw too little of "self" in the "other" of wild nature, perhaps Griffin sees too much "self" in wild nature and not enough "other."

## Hans Peter Duerr's Vision of the Humanity-Nature Relation

Duerr is a German scholar who published an extraordinary book in the same year (1978) that Griffin published *Woman and Nature*. Duerr's book, translated into English by Felicitas Goodman, bears the title of *Dreamtime: Concerning the Boundary between Wilderness and Civilization*. Like Griffin, Duerr has much to say about witches. Indeed, he begins by addressing the question of whether witches really possessed the magic salve which enabled them to fly about on broomsticks, as reported by many apparent eyewitnesses in documents produced at witch-trials and other such tribunals. Duerr maintains that the witches did "fly," but the difficulty comes in understanding the nature of such flight. Like Griffin, Duerr affirms that humanity excludes and dominates wilderness only at a great price. Unlike Griffin, however, Duerr maintains that the capacity for entering into wilderness is not restricted to women or witches, although they have often been associated with such practices, but is possible for men as well. Second, Duerr argues that surrendering to wildness, entering the wilderness, is not an end in itself, but serves the purpose of revealing to us what it means to be civilized humans. Hence, while we should be willing to cross over the boundaries erected by civilization against wilderness, we must recall that what distinguishes humans is precisely their civilized, i.e., non-wild, condition.

> In contrast to our own [modern] culture, the societies possessing what we called "archaic" cultures have a much clearer idea about the fact that we can *be* only what we are if at the same time, we are also what we are *not*, and that we can only know who we are if we experience our boundaries and, as Hegel would put it, if we thus cross over them.
>
> What this does not mean, however, is that we should endlessly move our fenceposts further and further into the wilderness and ceaselessly clear, work, categorize what is

"out there". It means instead that we ourselves should turn wild so as not to *surrender* to our own wildness, but rather to acquire in that way a consciousness of ourselves as tamed, as cultural beings.[25]

The desire to categorize what lies beyond the boundary of civilization must be resisted because only insofar as we can encounter untamed wildness, that which is other than civilization, can we understand what it means to be civilized. Witches were feared and persecuted because they reminded people of the wildness that could not be controlled by reason or by force, but these persecutors failed to understand the extent to which "civilization" is only possible because it is differentiated from "wilderness." Regarding wilderness as evil and horrible in comparison with the goodness and order of civilization is a typical instance of a binary opposition, one pole of which is made subordinate to or marginalized in relation to the other, privileged pole. Other instances of such binary oppositions include man vs. woman, mind vs. body, humanity vs. nature, black vs. white, capitalist vs. communist. What is overlooked by those who identify themselves with the privileged pole in such oppositions is the fact that such "identity" is dependent on the "difference" embodied by the marginalized pole. Male identity, for example, is bound up with the fact that it is other than female; civilization is defined by its difference from wilderness. Instead of trying to repress the "other" that is so central to our own identity, and instead of trying to gain control over it by rational processes, we must learn to acknowledge and to appreciate this otherness or difference *as such*, especially if we hope to understand the characteristics of our own identity. Yet if attempts to dominate wilderness are destructive to identity, so is the alternative of surrendering to it as an end in itself. According to Duerr, we should surrender to wilderness in order to discover what it means to be civilized humans. Being "civilized" would mean having integrated and accepted the wild dimension of oneself. Such integration allows a person to *differentiate* the human from the wild. Ordinarily, what passes for a "civilized" person is an ego that radically *dissociates* itself from everything merely "wild" and "natural." This dissociative attitude owes to the fact that the ego emerges only by means of a struggle against the more primal and wild forces that are present in humans at birth. Dissociating itself from wild nature is one way in which the ego seeks to protect itself from "regressing" into a more primitive level of awareness. The dangers of such regression are real enough, which is why Duerr recommends caution in any expedition into the wilds of the "dreamworld."

The wisdom of the witches lay in the fact that they knew that exploring their wildness was a precondition for their being civilized women, a fact that was simply incomprehensible to men (and to many women as well) who had attempted to live solely behind the fences of civilization. As Duerr argues, to the archaic mind the fence between wilderness and civilization was not insurmountable:

> At certain times this fence was, in fact, torn down. Those who wanted to live *consciously* within the fence, had to leave the enclosure at least once in their lives. They had to roam the forests as wolves, as "savages". To put it in more modern terms, they had to experience the wilderness, their *animal nature*, within themselves. For their "cultural nature" was only one side of their being which by destiny was inextricably bound to their animal *fylgia*, visible only to him who stepped across the dividing line, entrusting himself to his "second sight".[26]

According to Duerr, we are mistaken to think that whatever lies "beyond" ourselves is merely a projection of our own psyches or culture; such cannot in fact be the case, because without the otherness of wilderness, culture itself would not be possible. Culture and wilderness are mutually definitive; they exist in a dynamic unity. We must learn to see them not in terms of a binary opposition, in terms of which wilderness is marginalized and subjugated, but rather in terms of a vital partnership, one that invites the narrow civilized ego to shed its boundaries to discover what lies beyond. In such boundary-shedding, according to Duerr, consists "dreamtime," the time that lies between civilization and wilderness. In dreamtime, one can "fly" in the sense of being immediately at other places, because the world is always already interrelated in a way that transcends the physical separateness which seems so inescapable to those encapsulated within the confines of rigid ego-consciousness.

Duerr's understanding of the extraordinary transformation of space-time involved in "dreamtime" brings him into a certain proximity with Heidegger, despite the fact that Duerr maintains—unlike Heidegger— that humans are internally related to wild nature. Heidegger insisted that humans are not things, neither atomistic ego-subjects, nor clever animals. Rather, he argued that human existence amounts to a temporal-historical clearing within which things can appear. For the most part, the temporality of this clearing is constricted by virtue of the fact that it has been identified with the ego-subject, which itself is but another phenomenon appearing within the clearing. When human existence is authentic, however, temporality is transformed in such a way that an experience of eternity-within-finitude occurs. Past, present, and future become a spiraling cycle that enable things to manifest themselves in remarkable new ways.[27] One no longer feels compelled to explain things in terms of "cause and effect" or "before and after," but rather experiences things without the need for rendering a rational account of them.

Similarly, according to Duerr, in "dreamtime," space-time is transformed; one experiences the eternal in the finite. One is no longer constrained to a particular physical or temporal location, but rather one is both everywhere and nowhere: "The drop of water which after taking a consciousness-expanding drug is no longer seen *as* something, but as something *in and of itself*, which is *sunder warumbe*, 'without why,' as Master Eckhart says, changes into the universe, into *everything* and thus into *nothing*."[28] In "dreamtime," then, these radical alterations of space-time correspond to extraordinary transformations of the "self." No longer an atomistic ego, no longer "located" in a physical body, the "dreamer" flies beyond the constraints of everyday physical categories.

For Duerr, then, "nature" clearly cannot be understood merely in terms of the physical processes described by the contemporary science of ecology, for in "dreamtime" one discovers that the very categories presupposed by science are suspended or irrelevant. Science has validity within the appropriate domain, in Duerr's view, but not in all domains. Like Heidegger, Duerr argues that science assumes that for something to be "real," it must be explicable in terms of categories that we already accept. What cannot be quantified, for example, cannot be said "to be." What the "dreamer" discovers, however, is that wilderness and civilization are both fibers in the complex "thread" of reality.[29]

Duerr argues that shamen and witches have played such important roles in "primitive" societies because they have learned to "fly," that is, to explore the boundary "between" the curious "order" of wilderness and the "weirdness" of civilization. In the process of flying, one discovers that wilderness is not radically other, but is profoundly related to civilization itself. It is important to emphasize that wilderness or "wild nature" is not chaotic, confused, irrational, and disordered. Instead, the natural world displays an extraordinarily complex, recursive order which is clearly visible in complex animal societies as well as in the behavior of organisms of all kinds. In my view, at least, human society and human awareness are more complex variations of what has already emerged within the animal world. We are part of the same continuum of nature unfolding the possibilities inscribed within it.[30]

To some extent, writers like Griffin depict the "wild nature" within us as a disordered, boundless, shapeless, inchoate flow of energy which is in some ways reminiscent of what Freud meant by the id: the quantum of erotic and aggressive psychic energy which had to be repressed and sublimated in order to build civilized society, but only at the cost of the profound gratification associated with unmediated expenditure of that energy. Deploring the constraints which society imposes on human sexual desire and emotional spontaneity, social critics sometimes hope (and fear) that eliminating such constraints would grant us the boundless "wildness" and "freedom" which is supposedly characteristic of animals. Such a vision of boundless freedom is, however, a projection onto animal life. Animal behavior is not "wild" in the sense of being "boundless," for animals behave in accordance with their own instinctual and social limits. What overly repressed humans require is not an end to socialization, but rather the advent of appropriate socialization, the kind which would provide the rituals and social forms necessary for expressing emotions and fulfilling desires.

What distinguishes human awareness from animal awareness is humanity's heightened capacity for self-reflection and for reflecting upon the objects which confront it. Such capacities, of course, may belong to other creatures as well, including whales, porpoises, and extraterrestrials. The capacity of such self-conscious beings to think "rationally" is, arguably, an extension of the self-organizing, self-delimiting power at work in the rest of the natural order. What Duerr argues is that we cannot understand the self-awareness with which we are endowed unless we are able to step back from it, while at the same time not becoming completely submerged in the animal state. As don Juan told Carlos Castaneda, one can only learn to "see" the "lines of the world" if one is able to exist between worlds.[31] Hence, Duerr would caution against Griffin's yearning to become "one" again with "wild nature," especially if by "wild" she means boundless, indeterminate, frenzied. For Duerr, "crossing over" into the wild must be a temporary journey undertaken for the sake of an insight not available in *either* reality—wilderness or civilization—on its own. Hence, Duerr maintains that " 'not being able to be the other person' [or to become one with wild nature] is not a *hindrance* to, but rather the *prerequisite* for understanding the strange."[32]

Duerr warns of the terror and danger involved in crossing over to the wilderness dimension. The terror is connected with loss of one's civilized "identity." The danger concerns "running wild" after one has crossed over. Running wild would include committing acts of "bestiality," although as I suggested above, beasts generally do not behave in a wild, boundless, "bestial" way. Duerr offers many descriptions from folklore and historical documents of the apparently mad behavior of people who have allegedly passed over to the side of wilderness. How are we to assess such behavior? If a person abandons his or her civilized dimension, does he or she cross over into "wildness," or simply into madness? Does the "crossing over" which Duerr describes as "flying" and "dreamtime" really get us beyond the boundaries of the human, or merely beyond the boundaries of sanity? Is madness, in effect, a dimension of civilization? If the shaman and witch are not mad, what then do they encounter in their moments of rapture? Do they gain insight into the deeper kinship of humans and the rest of nature? Is this kinship a manifestation of the Divine?

While I do not feel qualified to answer these questions, I do feel qualified to say something about how my wanderings in the woods of the

Ohio countryside shaped my personal identity. My sense of "selfhood" was molded in part by my attempts to "go over" to the "otherness" of the squirrel-self; my human "voice" was shaped by my identification with the wind in the summer leaves; my sense of civilized "time" gained definition by contrast with the peculiarly timeless-yet-cyclical seasons of the woods; my capacity for "love" was strengthened by the all-embracing repose which I at first experienced as cosmic indifference. In my limited attempts to "cross over," I experienced fear that I might lose my identity and place, for there were no animal societies to which I could attach myself, no group to which I could gain admission in the Ohio woods. If I left behind my civilized identity, I would simply perish in the woods. To survive in the woods, I would have to remain human: self-conscious, given over to prediction and planning, and so on. And yet, I also knew that I was deeply attached to the ways of the woods and to the creatures who composed the woods—and that I was losing something by leaving them behind. The woods were a blessing to me because they were "other" to me even while they were also kin to me. In my attempts to "go over" to the woods, I was also restored to my own humanity in a way that would not have been possible without those journeys.

Looking back, I see that my encounter with wilderness would have been enormously enriched by a *social ritual* that would have enabled me to solidify my sense of identification with the forest families—trees and squirrels, deer and birds—and that would have simultaneously initiated me into a human family which had appropriate respect for and relationship with the other families of the forest. Lacking such an initiation, I have never really felt "at home" in the world: not truly a member of any family, human or non-human. In one sense, I am "between" worlds, but have not yet learned to fly. Dolores LaChapelle has argued that my condition is typical of Western people who have lost the rituals needed to gain human identity and to situate themselves within the larger cosmos.[33] Paul Shepard has claimed that human identity has evolved in relationship to the landscape in which humans dwell. As that natural landscape is replaced by human constructions, the future of humanity is threatened, for the young require contact with wild nature in order to become sane human beings.[34]

Today, one thing seems sure: if we humans are to avoid the self-destruction stemming from ecospheric catastrophe, we must relinquish our hostility toward and fear of wilderness. We can give up such fear, however, only so long as some measure of wilderness is preserved. Yet wilderness preservation also requires that the requisite human rituals be re-established, for the proper identification with wild nature necessary for civilizing people cannot be achieved by reading books or by legislation. Of course, the journey into wild nature can be understood to include our own bodies and feelings and intuitions; hence, a journey into wilderness can involve an exploration of these domains of ourselves. Even such a journey, however, needs a social, ritual context if it is to have lasting significance.

All this helps me to understand why I experience such profound sadness and terror in the face of the disappearance of wilderness around the world. If the wild world vanishes, then so too vanish all those plant and animal families with whom I yearn to be in relationship. If the wild animals and plants disappear, this planet will be profoundly impoverished. And our capacity for being human—fully, appropriately human—will be greatly diminished. So-called primitive people understand the importance of the continued presence of the other forms of life on Earth; they know about the blessing of otherness, the blessing which enabled them both to be human and to be members of the larger cosmic community.

# Wilderness, Civilization, and
# Language

## by Max Oelschlaeger

*Where is the literature which gives expression to Nature?
He would be a poet who could impress the winds and
streams into his service, to speak for him; who nailed
words to their primitive senses, as farmers drive down
stakes in the spring, which the frost has heaved; who
derived his words as often as he used them,—transplanted
them to his page with earth adhering to their roots; whose
words were so true and fresh and natural that they would
appear to expand like the buds at the approach of spring,
though they lay half smothered between two musty leaves
in a library,—aye, to bloom and bear fruit there, after their
kind, annually, for the faithful reader, in sympathy with
surrounding Nature.*          —Henry David Thoreau

In a seeming paradox, this essay concentrates on language, that
most human and civil of all things, and thereby appears to obliterate that
apparently most inhuman of all things: the wilderness. More specifically, it studies the language of some of the great wilderness poets and
philosophers. But there is no paradox in focusing on language, for
therein lies the route back to the *source*, to the wildness in which lies the
preservation of the world. Marjorie Grene argues that "we still have the
image of a human world shorn of any roots in nature and a natural world
devoid of places for humanity to show itself." Accordingly, we need to
show from within the hermeneutic circle how "historicity, as necessary
condition for, and defining principle of, human being, can be within, not
over against, nature...."[1] In other words, if it is through language that we

271

have been alienated from nature, then reconciliation might also be effected through language.

The literature of the wilderness is almost invariably misunderstood by the modern mind as a romantic longing for forested mountainsides, unpolluted blue skies, and free-flowing streams. A vivid and most recent example is Anna Bramwell's *Ecology in the 20th Century: A History,* where she argues that the European environmental movements (roughly equivalent to our own deep or foundational ecology movement) collectively represent a *socio-political program* to return to primitivism, and therefore are a threat to the continuing advance and progress of Western civilization.[2] For Bramwell, as for all modernists, the wilderness is one thing and civilization another, and the lines of distinction are drawn with an assumed Cartesian distinctness and clarity. The future, she believes, lies ahead, that is in the perfecting or fine tuning of that cultural process which has brought us to the point of holes in the ozone and the Exxon Valdez. Accordingly, she clearly represents the modern mind at work: her worldview is that of Parmenides and Plato, of Descartes and Newton, of Locke and Adam Smith, of that human project which is Euro-culture. But language speaks, and those who equate the literature and rhetoric of the wilderness with primitivism thereby reveal the outlines of their preconceptions of the world. What Bramwell and her ilk fail to recognize, as Paul Shepard so clearly explains, is that cultures are mosaics or composites built up over the longueurs of space and time.

While at present no definitive natural history of culture has been written, there is a significant body of evidence that confirms the reality of gene-culture coevolution. Alternatively stated, human culture is genetically framed, and culture itself has had an influence on the evolution of the human gene pool. Crucially, there is a Lamarckian element to the evolution of human culture, and language itself is the primary instrument or mechanism of Lamarckian evolution. As Paul Ricoeur points out, it is through language that humankind has a world, "a world, not merely an environment," a created space in which humankind dwells.[3] Here we cannot argue to substantiate the assertion that language speaks, that language is an ontogenetic, world-making activity that is deeply and inextricably entwined with the human project.[4] However, insofar as that premise is legitimate, then the claim of the modern mind that the wilderness poets and philosophers advocate a return to primitivism, that is, to go back in time, cannot be sustained. As Shepard points out, the modernist claim that we cannot go back "is a disguise for several assumptions, which in turn may hide ways of perceiving or preconstructing experience. One is the paradigm of uni-direction, the idea that time and circumstance are linear."[5] Another presupposition of the modern mind, as Pete Gunter explains, is that of "the postulate of the Environmental Infinite," an unquestioned belief that "suits our historical penchant for manipulation" and ignores the ecological reality that the exploitation of nature occurs not in a void but in a world "of which the exploiter is a relational part."[6] Here, then, we see the essential role of language studies, for they are fundamental to exposing and then overcoming the presuppositions which entrench the distinction between nature and culture.

Let us proceed slowly, for the hold of Euro-language on the modern mind is stronger than it appears. Indeed, is it not the case that we are all culture-dwelling animals and consequently enframed by language? Neil Evernden argues that the common core of the entire panoply of contemporary ideas of the relation between wilderness and civilization is *environmentalism itself.* "We call people environmentalists because what they are finally moved to defend is what we call environment. But, at bottom, their action is a defence of cosmos, not scenery. Ironically, the very entity they defend—environment—is itself an offspring of the nihilistic behemoth they challenge. It is a manifestation of the way we view [and speak of] the world."[7] Environmentalism, in whatever guise, is a way of speaking about the world which enframes our way of seeing, valuing, and thinking about the wilderness and its relation to us as human beings. Evernden invites us to abandon that "nihilistic behemoth," that is, the modern worldview which underlies contemporary "environmentalism," and to think like postmodern men and women. And yet paradox of paradoxes, we are people who presumably must think of the world in terms of the learned categorical scheme of Modernism. There appears to be no possibility of understanding any alternative view, for to do so appears to entail abandoning the cultural project upon which we have been so long embarked. From this funda-mental human predicament there is seemingly no escape. And yet there is the possibility of understanding something beyond "environmentalism" for those who are willing to stand within the hermeneutic circle, and to listen to the words of our thinking poets and poetic thinkers.

Heidegger argues that in order to see philosophically we first must take a step back. Reason is ordinarily understood to eventuate in thoughts: a product of human cognition in social context. Yet, as Heidegger observes, "We never come to thoughts; they come to us."[8] To allow thought to be bounded by social context is not to think philosophically at all, but to be human is to be linguistically and historically enframed. As the postmodern hermeneuts have come to understand, the only way forward is the step back: the rediscovery of our roots, our grounding in a wild and free source from which *Homo sapiens* has come and to which any true human beingness remains attached. Again the modern mind stumbles on paradox; yet, as was written thousands of years ago, "Reversion is the action of Tao." As was written recently by Stanley Diamond, the search for the primitive, for the wild, is a quest for knowledge of our primary human potentials, and only insofar as we rediscover these is there any hope for a reconciliation of civilization and wilderness.[9] Even more basically, as Dolores LaChapelle argues, the recovery and celebration of the behaviors and practices of archaic cultures are essential to some future wholeness, for through these ceremonies we can re-create "the experience, unique in our culture, of neither *opposing* nature or *trying* to be in communion with nature…but of *finding* ourselves within nature…."[10]

## Thoreau's Evolutionary Fables and Quest for the Source

J. B. S. Haldane remarked of nature that things are not only strange, they are always stranger than they seem. This is true too of the world of ideas and criticism, since several generations of scholars have dumped Thoreau into the basket of transcendentalism. Elsewhere I have deconstructed that thesis in detail.[11] But no one who has read Thoreau's texts sans literary theory can maintain that opinion. Thoreau has been almost universally misunderstood by the modern mind, so eager to reduce this most critical and penetrating of American philosophers to comfortable Emersonian cubbyholes.

Let me come to the point: Thoreau rejected categorically the cultural mainstream in which Ralph Waldo Emerson swam, and a brief comparison of Emerson's "Nature" with any of Thoreau's works is enough to establish that fact. For Emerson went into the wilds seeking that "original relation to the universe" with preestablished ideas or conclusions as to what he would find. What he found was Genesis 1, proof positive that God existed and had given to his most cherished creation, Lord Man, the natural world to rule over. So viewed, Emerson merely went over again the ground covered by the physico-theologists, such as John Ray and William Paley, discovering purpose and final cause in nature.[12] The organic world blooming with life, full of sights and sounds and smells, was through his philosophical spectacles mere appearance, a visible promontory obscuring something more fundamental, more real than the phenomenal face of nature. Namely, mind itself, and ultimately God, who unified all seeming diversity into the One. Emerson believed that through personal encounter with nature human beings could recapture a prelapsarian condition and transcend the dichotomies that separated them from nature and God.

> *Know then, that the world exists for you.* For you is the phenomenon perfect…. As fast as you conform your life to the pure idea in your mind, that will unfold its great proportions. A correspondent revolution in things will attend the influx of the spirit. So fast will disagreeable appearances, swine, spiders, snakes, pests, mad-houses, prisons, enemies, vanish; they are temporary and shall be no more seen…. *The kingdom of man over nature*, which cometh not with observation,—a dominion such as now is beyond his dream of God,—he shall enter without more wonder than the blind man feels who is gradually restored to perfect sight [emphasis added].[13]

This passage, the conclusion of "Nature," clearly and unmistakably reveals Emerson's orientation toward the natural world. The position is conventionally anthropocentric and androcentric, enframed totally by a Judeo-Christian, Baconian-Cartesian perspective: nature was but mere putty in human hands, bestowed by God upon his most favored of all creation, *man*.[14] *Man* was to be master and possessor of all that laid before him, the phenomenon perfect for whom the world was made. How comforting such an "argument" must have been to the Boston Brahmans and to Emerson, with a supreme being restored to the cosmological throne through "Transcendental Reason," rescued from the clutches of blind faith resting on Scripture and the revealed word— a God reached not through the Bible but through recitation of a transcendental litany. While "Nature" does not commit the *petitio principii*, a fallacy inherent in all biblically based arguments for God's existence, the argument is tortured, to say the least, and inconsistent with Kant and the German Idealism from which New England Transcendentalism sprang.[15]

In his deconstruction of Emersonian transcendentalism, Thoreau wrote that we are so early weaned from Mother Nature's breast that we forget that primordial world from which we've come. Through a lifetime of living in close contact with nature Thoreau actually achieved that original relation to the universe about which Emerson only wrote. And he discovered the essential superficiality of the tired Emersonian argument from design. While Charles Darwin and George Marsh were finally to lay Emersonian transcendentalism and the like in its grave, Thoreau beat them to the punch by more than a decade.[16] Thoreau knew that "Nature" was not a philosophical inquiry but a literary exercise designed to rest a preestablished belief in God on rational, rather than scriptural, footing: the conceptual focal point was the human soul and God, and not nature itself. For Emerson a wilderness odyssey was nothing more than an occasion for mind to discover a reflection, first, of itself (nature as a system of laws, concepts, and commodities), and then, finally, to confirm God's existence.

We must here rely on only a few examples to illustrate the thesis that Thoreau broke out of the mold of Emersonian transcendentalism, and thus Euro-language, into a new way of speaking and minding nature. Thoreau's critical insights into language and its effect on thought are evident even in his earliest work. He saw that language enframed human consciousness within a realm of *conventional* meaning that had divorced itself from or lost touch with the *elemental* grounds of language. Consider the first paragraph of *A Week on the Concord and Merrimack*

*Rivers*, a book written while Thoreau was actually staying at Walden and had not yet entirely escaped Emerson's influence.

> The Musketaquid, or Grass-ground River, though prob-
> ably as old as the Nile or Euphrates, did not begin to have
> a place in civilized history, until the fame of its grassy
> meadows and its fish attracted settlers out of England in
> 1635, when it received the other but kindred name of
> CONCORD from the first plantation on its banks, which
> appears to have been commenced in a spirit of peace and
> harmony. It will be Grass-ground River as long as grass
> grows and water runs here; it will be Concord River only
> while men lead peaceable lives on its banks. To an extinct
> race it was grass-ground, where they hunted and fished,
> and it is still perennial grass-ground to Concord farmers,
> who own the Great Meadows and get the hay from year to
> year.[17]

Thoreau obviously gave priority to the word "Musketaquid" as the true name of the river, as one rooted in a natural and therefore authentic and fundamental rather than merely cultural history. Something in the Indian name captured his imagination as it flowed from the tongue with a fullness not characteristic of English. He wrote that "If we could listen but for an instant to the chaunt of the Indian muse, we should understand why he will not exchange his savageness for civilization."[18] The name "Concord," although kindred, was merely conventional, an imposition of a recent social order upon natural fact, as the last two sentences of the passage make clear. For Thoreau the Indian designation predated the more recent appellation, and, therefore, had an organic relation to nature that the European term lacked. By emphasizing the name of the grass-ground river as Musketaquid, Thoreau attempted to disclose a presence concealed by the conventional Anglo-Saxon designation, a pre-social (primitive, aboriginal) and therefore genuine meaning. Further, as grass-ground river, the river endured beyond the merely contingent reign of civilization. From our point of view Thoreau had achieved a hermeneutical insight into the linguisticality and historicity of the human predicament. Aborigines, he believed, possessed a living rela-tion with the river that civilized people had lost. Of course, even archaic languages are at one level conventional, although here we cannot explore the many complications involved in this thesis. The point is that Thoreau understood language from a proto-hermeneutic standpoint: his imaginative quest was for knowledge of the *bios* underlying the merely conventional designations of Euro-language. Crucially, Thoreau was encouraging an evolutionary spiral from within the hermeneutic circle, finding in aboriginal words a culturally renewing source of primal meaningfulness and significance. By emphasizing an aboriginal name as over against its conventional name, as Fred Turner suggests, "The vicious circle of expectations governing perceptions which in turn confirm expectations, reproducing each other without novelty, is broken by the idea of evolution; the hermeneutic circle becomes an evolutionary spiral; and experience, in Thoreau's sense, is the locus both of mutation and selective testing."[19]

Throughout *A Week* Thoreau wove archaic languages—as in selections from Virgil—through quotation into the text, attempting to disclose or reveal layers of meaning obscured by Euro-language. Unlike his earlier writing, such as "A Walk to Wachusett," where citations from the classics seemed mere ornamentations, in *A Week* the material was functionally integrated into the philosophical exploration, essential to the discovery of meaning, to the realization of nature's priority to the conventional and cultural. Ancient and dead languages fascinated Thoreau, particularly when they recognized living nature, for that element bonded him with those people lost in the mists of the past, and thus with a time before the enfolding of meaning within the merely present. "These are such sentences as were written while grass grew and water ran." Thoreau sought to write sentences like those written when the grass first grew and the water ran fresh and clean, unsullied by humankind. These were what he called "perfectly healthy sentences," though he recognized they were extremely rare.[20]

There can be no question that Thoreau knew what he was attempting to do with language even in *A Week*, and by the time he finished *Walden* and "Walking" he had become a master of his medium. His use of ancient languages that referred to nature and Indian words in *A Week* gave way in *Walden* to his own imaginative re-creation of the elemental or primordial dimension of human experience that grounded language.

Thoreau explored such lines of thought in his journal as he was writing *Walden*, and the entry for May 10, 1853, brought together his hermeneutical insights with his quest to become a person of Indian wisdom.

> He is the richest who has most use for nature as raw material of tropes and symbols with which to describe his life. If these gates of golden willows affect me, they correspond to the beauty and promise of some experience on which I am entering. If I am overflowing with life, am rich in experience for which I lack expression, then nature will be my language full of poetry,—all nature will *fable*, and every natural phenomenon will be a myth. The man of science, who is not seeking for expression but for a fact to be expressed merely, studies nature as a dead language. I pray for such inward experience as will make nature significant.[21]

These statements are remarkable. *I pray*, Thoreau wrote, *for such inward experience as will make nature significant.* Such visions would animate a new natural mythology, and help to recapture a sense of meaning that scientific generalization and philosophical abstraction had stripped from the world. Thoreau was a seeker of religious experience who, by going into the wilderness, emptied the mind of conventional wisdom, and prepared to receive life through primary experience: an epiphany that would reveal an eternal mythical present, a revelation of dimensions of being hidden from ordinary consciousness. *And all nature will fable, and every natural phenomenon will be a myth.* If we view *Walden* through this lens, then it represents a new wilderness mythology, an alternative to all that "Nature" and Nature were, a world beyond nineteenth-century conventionality. Wild nature will fable (from *fabulari*, meaning to talk), that is, speak through a person if that person will but let natural phenomena have voice, and such a speaking will be as if literally true, alive, and organic. The facts of science, imprisoned within a conventional language (including mathematics), were dead and inert human inventions, at best useful fictions. So too was

Emersonian transcendentalism, and the idea that nature had been created by a transcendent God for *man*.

Here we cannot dwell on the many nuances and implications of Thoreau's use of language in *Walden*, but must examine only a single metaphor: the melting sandbank. In all likelihood Paleolithic people first associated spring with nature's coming again to life after winter's austerity. In any case, Thoreau's choice of spring as the season for summing up his wilderness philosophy was clearly deliberate. He used a melting sandbank as a natural metaphor to illustrate "the principle of all the operations of nature," that is, to exemplify the principles of evolution itself.[22] As the sun warmed the air, the frozen sand came to life, and once freed from winter's grip, began to shift, slowly but inexorably into a virtual cosmological scale of forms that seemed to reenact the entire history of the universe. Thoreau's observations were richly imaginative, and the melting sandbank a trope of gargantuan proportions: for it was an evolutionary fable or natural mythology that recounted that original insight of the Paleolithic mind into nature-in-its-order-of-operation. Charles Darwin almost simultaneously was making his epic cruise to South America and, ultimately, to Galapagos. However, Darwin's imaginative leap, given the abundance of organic forms with which he worked, was perhaps not so great as Thoreau's, for the latter derived his principle of the mutability of natural form from inorganic material alone.

Thoreau's concern and fascination with evolutionary process had intensified during his stay at Walden, and it became a deep and abiding theme, the central theme of the book. He interrupted his stay at Walden to go on the first of his three journeys to the Maine woods, and this first trek was to prove of singular significance in breaking free of Emersonian transcendentalism. While astride Ktaadn's ridge Thoreau realized that the Emersonian argument from design was an empty shell, and that humankind was not the plan perfect. "Talk of mysteries!" "Think of our life in nature." "Contact! Contact! Who are we? Where are we?"[23] Emerson's kingdom of "man over nature" was an illusion engendered by the very language of transcendentalism, and there could be no empirically or logically consistent supernatural account of either the human species' origin or its relation to the world itself. In the Maine woods, face-to-face with the vast titanic chemistry of the universe, Thoreau had intuitively realized the broad outlines of a profoundly evolutionary perspective on the world.

If we believe that his constant quest was to elucidate the relations between human consciousness and nature, then "Ktaadn" clearly defined one endpoint: brute facticity, the material world, even his own material body within which his consciousness existed, could be alien. Thoreau's journey to Ktaadn rekindled that Paleolithic coming to consciousness of humankind's naked rootedness in and absolute dependence upon nature. If *Walden* and "Walking" later articulated and refined his explorations and definitions of the positive side of the wilderness, then "Ktaadn" deals with its negative and foreboding promontory. Yet this is not an entirely accurate account of what "Ktaadn" represents, since the encounter was crucial to the evolution of Thoreau's idea of wilderness. Positively viewed, (i) the position developed in "Ktaadn" was antithetical to Emerson's philosophy, the final step in Thoreau's development from orthodox transcendentalism to his own genuine relation to the universe. His writings, hereafter, carried the mark of his singular experiences, of his unique vantage point on the wilderness, and of his genius. But, more importantly, (ii) the encounter with Ktaadn helped Thoreau sharpen the nature of his understanding of the interrelations between humankind and nature. By the time he returned to Walden Pond he was enroute to developing—independently of Darwin and Wallace specifically, and the paleontological and geological advances in Europe more generally—an evolutionary perspective on nature; by the time *Walden* was published Thoreau had worked this problem through.

The fable of the melting sandbank was a bold step toward resolving the paradox of Ktaadn, for it metaphorically extended the sweep of evolutionary process from elemental matter itself through life to culture and ultimately cosmos. All of nature, indeed, cosmos, was alive and in flux, virtually a living continuum in which the higher was merely an elaborated or articulated arrangement of the lower. "The earth I tread on is not a dead, inert mass. It is a body, has a spirit, is organic, and fluid to the influence of its spirit, and to whatever particle of that spirit is in me. She is not dead, but sleepeth."[24] He recognized that the natural world, though already made was still in the making, and that the evolution of natural forms was pervasive, and not only within the inorganic realm. The organic had evolved from the inorganic, and evolution continued within the organic realm, a process clearly including the human animal.

> What Champollion will decipher this hieroglyphic for us [Darwin's *Origin of the Species* was not published until 1859, some five years after *Walden* went to press], that we may turn over a new leaf at last? This phenomenon is more exhilarating to me than the luxuriance and fertility of vineyards. True, it is somewhat excrementious in its character, and there is no end to heaps of liver lights and bowels, as if the globe were turned wrong side outward; this suggests at least that Nature has some bowels [a Thoreauvian arrow aimed at the very heart of Emerson's "Nature"], and there again is mother of humanity.... It [the melting sandbank] convinces me that Earth is still in her swaddling clothes, and stretches forth baby fingers on every side. Fresh curls spring from the baldest brow. There is nothing inorganic. These foliaceous heaps lie along the bank like the slag of a furnace, showing that Nature is "in full blast" within. The earth is not a mere fragment of dead history,...but living poetry like the leaves of a tree, which precede flowers and fruit,—not a fossil earth, but a living earth; compared with whose great central life all animal and vegetable life is merely parasitic.... You may melt your metals and cast them into the most beautiful moulds you can; they will never excite me like the forms which this molten earth flows out into. *And not only it, but the institutions upon it, are plastic like clay in the hands of the potter* [emphasis added].[25]

Thoreau clearly realized the fanciful nature—viewed from a strict scientific perspective—of his fable. Indeed, he was having fun with *word play*, as the many puns reveal (turn over a new leaf, excrementious in character). But through that word play Thoreau drove home the point that humankind was only an element within rather than the reason to be of that spectacle unfolding before his eyes. "What is a man but a mass of thawing clay? The ball of the human finger is but a drop congealed.

The fingers and toes flow to their extent from the thawing mass of the body.... Is not the hand a spreading *palm* leaf with its lobes and veins?... The nose is a manifest congealed drop or stalactite. The chin is still a larger drop, the confluent dripping of the face."[26] A Thoreauvian dissertation on foetal development would be sure to fail, since he sought meaningful comparisons grounded in immediate experience rather than scientific explanations based on theory. Thoreau was seeking only to become that *true person of science*, a person of Indian wisdom, to which he had alluded in "A Natural History of Massachusetts." Science per se studied nature as a dead language; Thoreau sought to make nature his language—to speak the fables obscured by Euro-language. He wrote in his journal that "Nature is reported not by him who goes forth consciously as an observer, but in the fullness of life. To such a one she rushes to make her report. To the full of heart she is all but a figure of speech."[27]

## Muir's God-talk

In his 1967 essay entitled "The Historic Roots of Our Ecologic Crisis," the historian Lynn White, Jr. laid much of the blame for Euroculture's unrestrained exploitation of nature on the door step of Judeo-Christianity. Elsewhere I have criticized that viewpoint as intrinsically superficial. In present context "Historic Roots" serves only as an introduction to understanding John Muir's god-talk. Had White understood Muir's work he would have realized that (i) by ignoring or overlooking not only a relevant but more penetrating source of insight into the grounds of environmental malaise than his own he committed the cardinal sin of historical research, and (ii) that he egregiously simplified the question of religion's role in ecological crisis. While we cannot here discuss at length the methodological inadequacies in White's essay, we would be remiss to not point out perhaps the major underlying difficulty. As more than one scholar has pointed out, the Genesis stories did not lead to an agricultural revolution, but rather provided a legitimating narrative for a socio-cultural process already some millennia old. White holds Judaism culpable in that "Christianity inherited from Judaism not only a concept of time as nonrepetitive and linear but also a striking story of creation."[28] The biblical notions that humankind is the special creation of God, who gave *man* dominion over the earth, along with further injunctions to be fruitful and multiply do indeed appear to create an aggressive, exploitative orientation toward nature. Yet such support does not warrant the conclusion that Judeo-Christianity is the root of current environmental malaise.

Historical explanations such as White's are modelled on a paradigm naïvely derived from classical science, i.e., cause (Judeo-Christian attitudes) and effect (ecological crisis). Such a model is patently inadequate if for no other reason than the *post hoc ergo propter hoc* fallacy is not methodologically controlled. Judeo-Christianity is no one thing that serves as a primal explanatory substance, but is bound up with the mutable human world—*both influencing and influenced*. As contemporary students of Judeo-Christianity we are involved in a hermeneutic circle. As Norman Gottwald explains, "when we exegete the Bible we are also exegeting ourselves in our own socio-literary worlds."[29] In short, there is no definitive or Archimedean point from which any scholar can pass judgment on Judeo-Christianity. Indeed, we are ourselves bound into a Judeo-Christian cultural tradition from which there is no escape, but only the possibility of reinterpretation. Any scholarship which ignores these salient facts is therefore questing for a kind of Olympian objectivity and absolute truth which simply does not exist. We too are caught up in the human predicament—history and philosophy books begin and end, but the ongoing socio-cultural process does not. Judeo-Christianity is best understood as a further elaboration of an historical process already thousands of years old rather than as a new point of departure. Such a perspective does not absolve it from any responsibility for ecological crisis, but rather emphasizes that any attempt to seize upon a single historical variable to explain the Western world's relentless humanizing of the wilderness is intrinsically narrow. Indeed, only in the context of the agricultural revolution, and the later Protestant Reformation, and industrial, scientific, and democratic revolutions, can any sense be made of White's claim that environmental malaise reflects certain Judeo-Christian presuppositions. Paradoxically, such an exegesis undercuts the cogency of that very hypothesis.

A weakness of White's model is that alternative hypotheses, such as Paul Shepard's thesis that the agricultural revolution itself is the crucial variable underlying ecological crisis, are thereby ignored. Indeed, as Muir's contemporary biographers have made abundantly clear, one cannot claim to understand John Muir's idea of wilderness without attending to the intertwining of *agri*-culture and religion in his childhood.[30] For through the very circumstances of his life young John Muir re-enacted the Fall of Humankind. Pressed into a life of virtual servitude by his father, he had been forced to lay waste to the wilderness which was from a very early age a source of solace and renewal. And this was justified by his father in the name of God (see Genesis), for set in context, the ideology of Judeo-Christianity has essentially served as a justifying and legitimating rationale for *agri*-culture. Muir's god-talk, we suggest, becomes most meaningful when seen against the enframing backdrop of religious convention and *agri*-culture. Muir was no apostate, but a deeply religious human being who abandoned the tired anthropocentric religious truths of the modern mind, and put in their place a scintillating religious vision: a profound evolutionary pantheism. His journals and published works from 1867 on confirm his rejection of a Judeo-Christian based anthropocentrism and an affirmation of a biocentric perspective on wild nature. Clearly, if this be the case, then Muir's wilderness philosophy is radically disconcordant with Euro-culture. Regrettably, casual readers easily overlook the paradigmatically revolutionary aspect of Muir's work, particularly since his language seems so often to be conventionally religious. Muir's wilderness theology is, in fact, often confused with the orthodox theism that undergirded Emersonian transcendentalism.[31] Critically read, Muir's words stand the religion of Daniel Muir, indeed, the theistic mainstream of Western culture, on its head, for he entirely rejects the cosmogony of Genesis 1 and the idea of a supernatural God standing apart from creation. Further, he entirely rejects the anthropocentrism of Euro-culture, that is the attitude of *Lord Man*, as Muir called it, that blinded humans to their true role in the flux of life and the continuing creation.

Historically viewed, Muir realized much of what Benedict Spinoza, the great seventeenth-century pantheist, had realized: a God apart from nature was no God at all, but a mere abstraction from the *lived experience* of the natural world.[32] Like Thoreau, Muir saw through the frailty of the argument from design: this was a mere figment of the human imagination. His *evolutionary pantheism* revivifies the outlook of *Homo religiosus*: nature again becomes a sacred and living world. The natural world, the beasts and plants, had not been created by a God apart from the world for humankind to rule over; rather God was an all-inclusive and continuing process of natural creation. The world itself was divine, and everything in it. Muir had realized that humankind had no special dispensation, and consequently he abandoned the doctrine of special creation and any supernaturalistic account of the human soul.[33] Thereby he overturned that dramatic inversion of Western culture by which the palpable reality of nature had been denied as mere appearance and the phantasmagorical, that is, the supernatural, God and Heaven, claimed as reality. In other words, Muir bracketed *History* as the locus and meaning of value and grounded life in the *divine source*—an eternal mythical present in everlasting process.[34]

While the idea that everything that exists is unified in a single process pervades his journals as early as 1867-1868, he had not at that time clearly reached the conception that the all-inclusive unity was itself divine. In *A Thousand-Mile Walk* he prayed to God for a revelation, and in *My First Summer* he spoke of angels, implying that he yet believed in a divine and celestial realm independent of nature. By 1873 Muir's belief in a supernatural realm entirely distinct and existing apart from the natural world appears to have been transcended; he had reached a middle ground, that is a *panentheism* that allowed for both the divinity of creation and a separate existence for a divine cosmic presence. (A pantheist believes that "God is merely the cosmos, in all aspects inseparable from the sum or system of dependent things or effects...." A panentheist believes that God "is both this system and something independent of it...." A traditional theist believes that God "is not the system, but is in all aspects independent.")[35] Muir had studied the wonder of the ongoing creation in Yosemite valley, the myriad species, and the interrelations among animate entities and the land, and he now saw *nature itself as divine*, for "all of the individual 'things' or 'beings' into which the world is wrought are sparks of the Divine Soul variously clothed upon with flesh, leaves, or that harder tissue called rock, water, etc." A few months later Muir observed that the "rocks and sublime canyons, and waters and winds, and all life structures—animals and ouzels, meadows and groves, and all the silver stars—are words of God, and they flow smooth and ripe from his lips." While the journals for 1873 allow no definitive interpretation (Muir is not given to philosophical argument and sharp distinctions), they unquestionably reveal a pantheistic perspective in that nature itself is conceived as a temporal manifestation of a Divine Soul. Read in context (that is, his conversion experiences of 1868 and 1869, and the evolutionary framework underlying his research), there is some reason to think that he was verging on denial of any absolute distinction between Divine Soul or God—as eternal and infinite—and creation itself—as temporal and finite. "What is 'higher,' what is 'lower' in Nature?... All of these varied forms, high and low, are simply portions of God radiated from Him as a sun, and made terrestrial by the clothes they wear, and by the modifications of a corresponding kind in the God essence itself."[36] Such a position—that the *essence of God is changed*—cannot be reconciled with any variant of pantheism which conceptualizes nature as a divine *but merely temporal* manifestation of a divine *and eternal* cosmic spirit.

By 1879 Muir had reached a conception of God as incarnate; ten years in the mountains, free of religious convention while living in the very immediacy of evolutionary process, had revealed that *God was nature*, a sacred living *temporal* presence in *everlasting* process. This was a speculative insight of sweeping proportion, for it restored wholeness and meaning to a world rent asunder by Modernism. Loyal Rue has recently and cogently argued that there is little hope for transcending our cultural crisis unless we generate new conventions of meaning that effectively reintegrate cosmology and morality, and that in order to do this two essential criteria must be satisfied: religious distinctiveness and scientific plausibility. Muir would agree, and his work can be plausibly interpreted as so doing. For once we accept the reality of natural evolution (scientific plausibility) then the idea of God as supernatural creator is untenable. So viewed, Muir abandoned the stage of *History* and became once again *Homo religiosus*. To use Rue's compelling phrase, Muir formed a *new covenant* with God, one that was both scientifically plausible and religiously distinctive: that is, fully consistent with the evolutionary paradigm of Darwin, Lyell, Agassiz and the

other great nineteenth-century evolutionists *and* with the covenant
tradition of Judeo-Christianity. "The New Covenant might be called an
agnostic mode of piety, a mode in which there are no metaphors for
discourse about God apart from metaphors about creation and history."
Muir's agnostic piety or wilderness theology was compatible with the
wilderness sage tradition of the Old Testament, for "the divine name
YHWH was not regarded by early Hebrews as expressive of God's nature;
his nature was not at their disposal. Yahweh was known only by his
works [activity]."[37] As John Muir first realized through a religious
conversion in the wilderness, and then confirmed through his observations, God is a natural phenomenon in everlasting process.

Such a realization testifies to Muir's genius, for to reach that insight
required the most critical of insights. Joe Barnhart explains that historically "pantheism has not known what to do with the idea of God as
experiencing succession. Pantheism [typically] cannot recognize time
as a part of God's own being."[38] But Muir appeared to have accepted the
reality of time and its implications, for he no longer believed in a
transcendent God remaining apart from nature. Genesis was but a
religious metaphor, for through actual encounter with wild nature Muir
knew that "the world, though made, is yet being made; that this is still
the morning of creation." While in the mountains of Alaska, observing
its glaciers and life processes, he knew that he was witnessing an
ongoing process of Creation itself. He felt that "in very foundational
truth we had been in one of God's own temples and had seen Him and
heard Him working and preaching like a man."[39] This analogy must not
be taken lightly, for it is radically incongruent with conventional Judeo-
Christian theology (theism). Rather than affirming the existence of a
supernatural and eternal God, Muir's God was like a human being—
temporally bound with a natural world. The heresy here was overwhelming, for the human being was not made in God's image, but God
was fashioned by analogy in precisely the same fashion as humankind,
and thus grounded in process: birth, life, and death. Muir had committed
the same apostasy as Giordano Bruno: by making divinity incarnate,
providence was necessarily denied, since there could be no transcendental spectator apart from the world with a divine plan in mind. And
humankind itself could no longer be the chosen species awaiting eternal

salvation while bearing witness to an essentially meaningless because preordained passage of time. Time for Muir had become real and irreversible, and the human being—both body and soul—bound totally with time.

The rejection of supernaturalism was fundamental to the evolution of Muir's biocentric outlook on the natural world and his bracketing of Lord Man. In a passage (written in 1873) that parallels to a remarkable extent Thoreau's passage in *Walden* on the melting sandbank, he explicitly posited the terrestrial origin of *Homo sapiens*. The human species, Muir argued, "has flowed down through other forms of being and absorbed and assimilated portions of them into...[itself], thus becoming a microcosm most richly Divine because most richly terrestrial, just as a river becomes rich by flowing on and on through varied climes and rocks, through many mountains and vales, constantly appropriating portions to itself, rising higher in the scale of rivers as it grows rich in the absorption of the soils and smaller streams.... [ellipsis in original]."[40] Unquestionably, Muir believed in God throughout his life, but his God was neither the Cosmic Hitler of Daniel Muir nor the Transcendental Oversoul of Emerson, but a God incarnate and in process. Muir's completed wilderness paradigm thus represents a veritable theological antithesis of orthodox Judeo-Christianity. Clearly, Muir's outlook is more a wilderness theology than metaphysics. A thoroughgoing evolutionary metaphysics was not to appear until the process-relational philosophers of the twentieth-century (Alexander, Whitehead, and Hartshorne).[41] Indeed, Muir was not versed in the technical language of metaphysics; alternatively stated, he was more interested in working through theological questions, about creation and the origin of the human species, than scholarly metaphysical issues. But, as Alasdair McIntrye so clearly sees, "pantheism as a theology has a source, independent of its metaphysics, in a widespread capacity for awe and wonder in the face both of natural phenomena and of the apparent totality of things." As Michael P. Cohen observes in a deconstructive reading of his own interpretation of Muir in *The Pathless Way*, "everything is so interesting."[42] Precisely! The incredible beauty of things captivated Muir totally and completely, and in that evolutionary reality he saw God. "It is at least in part because pantheist metaphysics provides a vocabulary which appears more adequate than any other for the expression of these emotions that pantheism has shown such historical capacity for survival."[43]

There are, of course, those who would point to Muir's wilderness theology—his pantheism—as generally muddled thinking, an unlikely combination of faith and reason. Environmental malaise is one thing, so this critique might argue, but there is nothing to be gained in our endeavors to clean up the earth and conserve the wilderness by bringing in god-talk.[44] Yet Muir's god-talk, we have argued, is perfectly comprehensible, providing a scientifically plausible yet religiously distinctive alternative to Modernism, which employed the orthodox God of Judeo-Christianity to underwrite the truth and certainty of classical science and *Homo sapiens'* right to exploit nature. Understood in the context of his life, Muir's pantheism was a reflective outgrowth from his religious indoctrination in childhood, his religious experiences in the mountains as an adult, and his evolutionary perspective on nature. Pantheism allowed him to see the world steadily, and see it whole, and thus was entirely complementary with both his psychic needs and intellectual commitments. Indeed, these all were woven together into a virtually seamless fabric. As Robinson Jeffers wrote, and surely Muir would agree, "There is but one God, and the earth is his prophet. / The beauty of things is the face of God: worship it; / Give your hearts to it; labor to be like it."[45]

## Leopold's Thinking Like a Mountain

One way to develop a perspective on Aldo Leopold's language, his new way of minding nature and seeing the world, is to survey briefly the variety of opinion that has been offered on his work. Some, such as the deep ecologist George Sessions and the wilderness philosopher Holmes Rolston, III, find in the land ethic the leading edge of a new paradigm for thought and action. Others, such as L. W. Sumner, characterize the land ethic as dangerous nonsense, and Paul Taylor uses the land ethic to exemplify the naturalistic fallacy. David Ehrenfeld, surely sympathetic to Leopold's philosophy, nonetheless finds reason to believe it remains at base anthropocentric. In a reading opposite to Ehrenfeld's, John

Passmore finds the land ethic to be so biocentrically centered as to be without meaning. And these are only a few of many judgments made of Leopold, a melange of inconsistent and often contradictory interpretations. Can it be that Leopold's ideas are so hopelessly ambiguous and vague as to defy consistent interpretation or so poetic as to permit virtually any interpretation?[46] Or is it, as Whitehead remarks, that "almost all really new [or paradigmatically revolutionary] ideas have a certain aspect of foolishness when they are first produced"?[47]

Although inconsistent with our dominating cultural tradition which insists upon truth as single and permanent, diversity of opinion is not necessarily bad, and may have positive benefits by promoting discussion of the land ethic and Leopold's ecological perspective. Indeed, part of his genius is that through his creative use of fresh terms like "the land ethic" and "thinking like a mountain" he invites us to reconsider our relations to the land—to a larger, older, and enframing community of life on earth. And therein arises the possibility of the many alternative *interpretations* made of Leopold's writings, since his work meshes with the tides and forces of Western culture: the land ethic is a twentieth-century manifestation of perennial questions about our relation to wild nature, questions to which—given the inescapable reality of evolutionary process—no final answers may be given. Indeed, as Leopold realized, even attempting to answer such questions requires an exacting synthesis of theoretical knowledge, practical experience, and creative vision. His novel, indeed, poetic way of speaking about these perennial questions, his argument that we should mind the world by "thinking like a mountain," is cogently and perhaps best understood as part of a continuing intellectual and cultural revolution that began long, long ago in dim and obscure mists of the Paleolithic. For in that Paleolithic coming to self-consciousness, as Joseph Campbell so beautifully argues, was "the *first* revelation of Nature-in-its-manner-of-operation to the inner eye of the mind of *Homo sapiens*, c. 40,000 B.C.[E.]."[48]

Leopold was cast into the world just as the nineteenth-century scientific revolution engendered by Darwin and Clausius was beginning to influence the way that humankind conceptualized nature and its order-of-operation. If we view Leopold as theoretically and existentially immersed within a cultural matrix transcending merely disciplinary issues, then the land ethic can be understood as part of an ongoing twentieth-century struggle to discover passage between the descriptive world of science and the prescriptive world of ethics.[49] In learning how "to think like a mountain," Leopold necessarily reconceptualized science in general, and ecology in particular, since he clearly recognized that science is inextricably entangled with the culture that sustains it. Only a few intellects of our century have grasped this crucially important fact: Einstein was one and Leopold another. However, there is more to Leopoldian ecology than this alone, for his perspective was also deeply informed by an ancient understanding of things wild and free. Leopold's revivification of these primordial sensibilities is most apparent in writings such as "Marshland Elegy," but also show themselves even in such ostensibly pragmatic works as *Game Management*. In such writings we find Leopold entertaining crucial questions of natural history and ethology, of psychology and biology, as to just who and what we are. To borrow a phrase from Paul Shepard, Aldo Leopold represents a wildlife ecologist turned philosophical anthropologist, and thus he epitomizes that as yet unfinished project to reconcile the deep seated linkages and affinities between humankind and nature with the reality that we are also culture-dwelling creatures.

Consider, for example, Leopold's lifelong avocation of hunting and his professional involvement in game management. To some, if not the majority of erstwhile conservationists and preservationists, this involvement in "blood sport" taints if not totally undercuts other aspects of his work, such as the land ethic.[50] Clearly, the concept of game management entails associated activities of hunting and killing game animals for human recreation. Consequently animal rights advocates and preservationists who believe that wildlife and lands should be absolutely free from human encroachment are not likely to find *Game Management* philosophically appealing, since the book appears to be a practical scientific treatise that serves only the values of efficiency and utility. A close reading, however, reveals that the case is not so simple as critics like to think. In the first place, the text reveals some of the breadth inherent in Leopold's idea of wilderness, such as his concern with the preservation and management of non-game species, his recognition that all species of wildlife, not just game-species, were adversely affected by unrestrained humanization and exploitation of the land, and his realization that the quality of human life was enriched by preserving wildlife and land.

> The objective of a conservation program for non-game wild life should be...to retain for the average citizen the opportunity to see, admire and enjoy, and the challenge to understand, the varied forms of birds and mammals indigenous to his state. It implies not only that these forms be kept in existence, *but that the greatest possible variety of them exist in each community*.
>
> In times past both of these categories of opportunity [to hunt game and to observe non-game] existed automatically, and hence were lightly valued. Both are now, by reason of their growing scarcity, perceived to be immensely valuable. Conservation is nothing more or less than a purposeful effort to perpetuate and extend them as one of our standards of living.[51]

*Game Management* made explicit that the rationale for game management was at least in part predicated on human use of the game produced, but implicit in Leopold's argument was a *deeper rationale*, neglected by his antihunting critics, lying in human nature itself. He asserted at the beginning of the text that "the practice of some degree of game management dates back to the beginnings of human history." Always cautious when advancing from facts to conclusion, he may have been somewhat ahead of his data in making this claim. He cited a dubious source suggesting that "tribal taboos" regulated hunting even among primitive people. "The tribes observing taboos which were biologically effective in preserving the game supply were more likely to survive and prosper...than the tribes which did not. In short, hunting customs, like plant and animal species, were evolved by a process of selection, in which survival was determined by successful competition. Game laws grew out of these hunting customs."[52]

The veracity of the specific anthropological details is not crucial here, but rather the ecological insight per se: humankind has interacted with natural species since prehistory, and ideology has always been fundamental in determining the character and quality of that association.

By predicating discussion of the origin of tribal taboos and game laws on ecology (the study of dynamic interrelations between the human species and game), Leopold was opening the door to the study of the biology of human behavior—a persistent theme that, once loosed, he pursued tenaciously. Only a few other twentieth-century thinkers have probed so thoroughly in such directions (for example, Konrad Lorenz, Arthur Koestler, Paul Shepard, José Ortega y Gasset, and Loren Eiseley). Leopold was keenly aware of antihunting sentiment, but effectively met such critiques by raising difficult and complicated issues that go to the very nub of human nature. Indeed, so complex are these considerations—involving questions of human genetic structure, psyche, and brain—that they have yet to be answered. But the evidence is unmistakable: as early as 1933 Leopold was probing into the biological bases of human behavior.

> Hunting for sport is an improvement over hunting for food, in that there has been added to the test of skill an ethical code, which the hunter formulates for himself, and must live up to without the moral support of bystanders. That the code of one hunter is more advanced than that of another is merely proof that the process of sublimation, in this as in other atavisms, is still advancing.
>
> The hope is sometimes expressed that all these instincts will be "outgrown." This attitude seems to overlook the fact that the resulting vacuum will fill up with something, and not necessarily with something better. *It somehow overlooks the biological basis of human nature,—the difference between historical and evolutionary time-scales* [emphasis added]. We can redefine our manner of exercising the hunting instinct, but we shall do well to persist as a species at the end of the time it would take to outgrow it.[53]

Clearly Leopold saw civilization, or *History*, for what it was and is: a thin veneer overlaid on a human nature fashioned in the geological and biological longueurs of time. While we may quarrel with Leopold's hypothesis of a "hunting instinct," there can be no quarrel with the underlying rationale: humankind has a nature, a genetic structure that is essentially fixed when viewed against a cultural time line, and this nature has—regardless of any and all illusions that deny our grounding in nature—

profound implications for human behavior. In other words, as Paul Shepard so succinctly puts the point, "Human culture, being genetically framed and environmentally adapted, is also an integrated conglomerate," in effect a composite of "linked and harmonious but separable parts."[54]

Leopold's insights both parallel and extend Thoreau's earlier understanding of the biological bases of human behavior and culture. Thoreau wrote that "I love the wild not less the good" and that he "felt a strange thrill of savage delight" as he contemplated seizing a woodchuck and devouring it raw. Such expressions are much like the primordial sensibilities Leopold articulated in "Marshland Elegy" or "Thinking Like a Mountain." *Game Management* also rang with reverberations of Thoreau's insights into lives of quiet desperation, for modern people were—by implication—living lives out of synchronization with nature. Throughout the Paleolithic era the human species had existed in an ecologically balanced relation with the land. But Leopold realized that agri-culture, with its "Abrahamic concept of land," was a cultural overlay on a long natural history that upset the integrity and stability of natural ecosystems.[55] *Game Management* simultaneously advocated the preservation of wild species while recognizing that hunting—within even the context of advanced industrial society—was not only a natural human activity, but often essential to the well-being of game species and even entire ecosystems.[56]

The Darwinian revolution dramatically undercut, as Leopold understood, both an Abrahamic concept of land and the Cartesian dream that *Homo sapiens* could be the master and possessor of nature. Genesis provided no adequate answer to the question of humankind's relation to the land, but only rationalized what already existed and postponed further inquiry. And so too did physics and the Cartesian project fail, for humankind, despite its illusions, could not stand above but remained always a part of nature. Yet Western culture continued on its historic course, and refused to reconsider the absolute presupposition that nature was anything more than a stage upon which the human drama was performed. The land ethic flew in the face of the conventional wisdom, and explicitly recognized humankind as a self-aware participant whose choices had profound impact on the course of evolution. Leopold did not deny that the human species was in many ways qualitatively distinct from the rest of nature; indeed, the land ethic presupposes sentience. He also recognized that the scientific revolution gave the Western world a new ability to intervene in natural process, and that burgeoning human populations were just one manifestation of that change. More ominously, he saw that the unbridled socioeconomic exploitation of nature was leading toward an ecological apocalypse. Thus he came to the only feasible conclusion: since *Homo sapiens* was organically and therefore irreversibly bound with nature, any meaningful change in the course of events would require an unprecedented—because deliberate—evolution in human consciousness.

Consequently, Leopold argued that advanced industrial culture must refashion its traditional decision-making matrix—one dominated entirely by the unquestioned aims of the Modern Age—and adopt a biocentrically informed decision-making matrix, that is, an ecology based land ethic. In a fashion sure to engender its rejection by the modern mind, the land ethic combined science, history, and philosophy. The land ethic was grounded in ecological science because it factually presupposed a systems-level understanding of biotic communities. Nature's way serves as an alternative to the ideology of Modernism, for the human animal is compelled by the verdict of science to consider the reality of its membership in both global and regional biotic communities. Furthermore, the land ethic presupposed historical judgment, the possibility of meaningful changes in human consciousness, and the potential for an expansion of ethics beyond an Abrahamic (anthropocentric) concept of the land to include wild nature. Finally, the land ethic presupposed the philosophical judgment that human beings were obliged to adopt as an ideal (the ought) a standard of behavior that transcended the manifest deficiencies (the is) of contemporary culture. The philosophical rationale for the land ethic might be summarized as follows: Evolutionary potential exists (open future consistent with the stream of influence); ethical choice is an informed choice (an idea fundamental to and rooted in the classical foundations of Western culture); and reflective thought underlies choice (direction of evolutionary potential), thereby inextricably fusing (teleonomically, not teleologically) the is and the ought.[57] So viewed, the land ethic is a remarkable anticipation of the judgment rendered by Ilya Prigogine some thirty years later. "We can no longer accept the old a priori distinction between scientific and ethical values. This was possible at a time when the external world and our internal world appeared to conflict, to be nearly orthogonal. Today we know that time is a construction and therefore carries an ethical responsibility."[58] We are the children of Creation, and we now know that our actions irreversibly ripple throughout the entire life-world. In other words, the metaphor of evolution does truly integrate cosmology and morality.

Leopold's premises—that science and ethics intertwine, and that perhaps at base they are divergent streams from the same source—coincides with the advance of postmodern reason. Indeed, to view Leopold from the standpoint of modernistic interpretative schema is to miss the underlying tension, the period of revolutionary science, that animated his life's work. While often thought of as definitive, the land ethic is perhaps best understood as the product of a *transitional ecology*, part of an unfinished intellectual process that began with the Darwinian revolution and has yet to work its way through. As our century begins to give way to the next, an unfinished socio-cultural agenda awaits. Leopold wrote that "Ecology is an infant just learning to talk, and, like other infants, is engrossed with its own coinage of big words. Its working days lie in the future."[59] Wallace Stegner echoes Leopold's assertion, observing that the "land ethic is not a fact but a task. Like old age, it is nothing to be overly optimistic about. But consider the alternative."[60] However procedurally perilous the path upon which he embarked, his philosophical judgment has been vindicated, for the issues central to Leopold's thought are fundamentally entwined with the contemporary course of events: questions about the limits of economic growth, water and air quality standards, designation of wilderness areas within the national forests, and protection of endangered species are a reflection of Leopold's land ethic.

The point is simply this: Leopoldian ecology is cognitively revolutionary. If framed by the rhetoric of the modern world, the land ethic is a clear illustration of the naturalistic fallacy. If, however, we understand that language speaks, then Leopoldian ecology is a crucial element in the ongoing second scientific revolution spawned by Darwin and Clausius.

Leopold wrote near the end of *Sand County Almanac* that "It is only the scholar who appreciates that all history consists of successive excursions from a single starting-point, to which [humankind] returns again and again to organize yet another search for a durable scale of values. It is only the scholar who understands why the raw wilderness gives definition and meaning to the human enterprise."[61] What is remarkable about Leopold is that, without philosophical armament, he rose above the two culture split, which had separated human value from natural fact. Donald Worster observes that Leopold was not the philosopher that he might have been, a criticism that is accurate, especially since Leopold lacked any insight into time per se as the basis of an evolutionary paradigm. Yet he clearly realized both the individual and cultural implications of the Darwinian revolution, and also the egregious blunders that followed from the failure to heed the lessons of evolution, of nature in-its-order-of-operation. From the second world of culture Leopold was able to find his way back to the first world from which the human species had come.

## Snyder, Dwelling Poetically

Last but not least in our study of wilderness, civilization, and language we come to Gary Snyder, who has been aptly characterized as the poet laureate of that social and intellectual movement known as "deep ecology." However, such a characterization per se does not advance our inquiry, for it begs a number of questions. What is deep ecology? What is society? What is poetry? What is the role of the wilderness poet in society? These are extremely difficult questions which threaten to take us far afield. Furthermore, there are neither ready-made answers nor a framework within which to seek answers. Yet these questions also point to the paradigmatic challenge that Snyder's poetry poses for the modern mind, for the poet's language is ontogenetic, a medium within which meaning occurs, and a context within which knowledge can be understood. Yet the modern mind believes absolutely, as Heidegger argues, that poetry is "either a frivolous mooning and vaporing into the unknown, and a flight into dreamland, or is counted as a part of literature" and is therefore worthy of no one's attention save academics and the literary intelligentsia.[62] Modernists fail to realize that language speaks, and believe that they—the proud speakers and writers of language—are its masters and possessors, and that language is solely representational and expressive, an activity of an already thinking and speaking humankind. The analysis and interpretation of poetry is typically so infected. Such criticism remains, Heidegger tells us,

> confined by the notion of language that has prevailed for thousands of years. According to this idea language is the expression, produced by men [and women], of their feelings and the world view that guides them. Can the spell this idea has cast over language be broken? Why should it be broken? In its essence, language is neither expression nor an activity of [hu]man[kind]. Language speaks. We are now seeking the speaking of language in the poem.[63]

Upon realization that *language speaks* we also understand that the poet does not aspire to use descriptive language, as do all those who function within the confines of conventional speech and language. The thinking poet reaches toward a presence obscured by the obvious, toward what is absent or missing because of its concealment behind language, behind opinion, behind the governing system of ideology that rules the world: *the wilderness poet calls forth Being.*

What must be recognized absolutely, categorically, if Gary Snyder's poetry is to be understood as something other than a nostalgic longing for noble savages and forested mountainsides, clear waters and blue skies, is that language speaks. His poetry speaks, indeed resonates with the primal myths of the Paleolithic mind and archaic people, and through its saying reveals a world in which humankind might again be an integral part, a world in which mortals and immortals, the sacred and the profane, God and humans might again be whole. Snyder's "Turtle Island" is neither scientific description nor emotive expression, but rather the unconcealment of presences obscured behind the visible promontory of the world-in-force. His poetry is of course informed by the basic physical-chemical-biological-ecological facts of existence.[64] Yet the focus of the thinking poet is not upon the world as that is culturally given, but rather upon that dimension of Being concealed and lying behind that presence. The thinking poet necessarily abandons conventional categories and values to explore the hidden meanings of terms with revolutionary connotations and denotations. Through such speaking the old categories, such as the GNP, Man, Nation, Progress, Economic Growth, the Standard of Living, are bracketed, and Being revealed. As Joseph Campbell observes, such has always been the function of the great seers: to recognize "through the veil of nature, as viewed in the science of their times, the radiance, terrible yet gentle, of the dark, unspeakable light beyond, and through their words and images to reveal the sense of the vast silence that is the ground of us all and of all beings."[65]

What is *sui generis* about Snyder is not that he is a utopian, for visionaries abound, but rather that his poetry achieves a singular vision combining East (religion, philosophy, psychology) and West (ecology, anthropology) with ancient wisdom (especially Amerindian mythology) into what George Sessions and Bill Devall characterize as a *spiritual ecology*.[66] The term "spiritual ecology," however, begs for clarification, being so laden with possibilities of interpretation as to frustrate rather than facilitate our attempt to hear the poet's song clearly. One way of getting some perspective on the notion that Snyder is a spiritual ecologist is by attending to his "Eastern connection": that is, his abiding interest in Oriental philosophy, psychology, religion, and poetry. During his Japan years Snyder discovered the poetry of Miyazawa Kenji (1896-1933), and translated eighteen of his poems. Kenji, Snyder notes, was "born and lived his life among the farmers: school-teacher (Chemistry, Natural Sciences, Agriculture) and a Buddhist. His poems have many Buddhist allusions, as well as a scientific vocabulary."[67] Snyder might well have been describing himself in this latter sentence, for his poetry is difficult to understand without some familiarity with both ecology and Eastern philosophy and religion, as well as paleo-anthropology, mythology, and Amerindian lore.

Snyder's spiritual ecology was forged at least partially in the crucible of his Japanese experiences, and he remains a Zen Buddhist. The Eastern tradition, foreign to the linear, logical Western mind set, is crucial to Snyder's exploration and appreciation of nonlinguistic, nonegocentric, nonethnocentric modes of consciousness. In other words, the spiritual dimensions of Snyder's poetry well up and flow forth from premodern sensibilities, and Zen has been and remains a way of liberation for him. Accordingly, our interpretation of Snyder's poetry as a "spiritual ecology" entails other difficulties beyond defining that term itself.[68] If Snyder is influenced by an Eastern tradition, then the importance of silence looms prominently. If we are to hear the Earth Mother, welling up through the poet's song that sings of ancient animist sensibilities and myths, then we must first and above all else listen. As Lao Tsu reminds, the name that can be named is not the Tao, the Mother of the ten thousand things.[69]

The conventional connotation of the term "ecology" implies that to look at things ecologically is to see them as connected, as interrelating, as constituting a whole which is greater than the sum of its parts. That definition, however, suits either a functional (shallow) or a foundational (deep) ecologist, and so there must something more involved if Snyder is to be characterized as the poet laureate of deep ecology *and* a spiritual ecologist to boot. Anna Bramwell's notion of normative ecology is particularly suited to helping us understand Snyder as a spiritual ecologist. Although Bramwell does not include Snyder in her study, surely he would be considered by her as a poet-ecologist advocating a return to primitivism. However, there is nothing in Snyder's spiritual ecology that suggests a return to primitivism. He advances a premise that by exploring the "primitive" we can identify primary human potentials that are thwarted by modern society, and also asserts that by ordering our lives in accordance with or actualizing these potentials we will not only cease being burdens upon the earth's living systems, but will become more fully human. His poetic vision clearly and distinctly reveals the marks of a postmodern consciousness: he is a *thinking poet* whose poetry calls into being an ancient wisdom that resonates with contemporary ecology and whose saying transcends the dichotomies of Modernism.

Let us consider but a single example of dwelling poetically, and a poem particularly suited to the mountain setting of our inquiry into wilderness and civilization. Snyder writes in *The Old Ways* that "the original poetry is the sound of running water and the wind in the trees."[70] The poem "By Frazier Creek Falls" brings that insight alive.

Standing up on lifted, folded rock,
looking out and down—

The creek falls to a far valley,
hills beyond that
facing, half-forested, dry
—clear sky
strong wind in the stiff glittering needle clusters
of the pine—their brown
round trunk bodies
straight, still;
rustling trembling limbs and twigs

listen.

This living flowing land
is all there is, forever

We *are* it
it sings through us—

We could live on this Earth
without clothes or tools![71]

*Listen!* the poet tells us. This is the Eastern axis of Snyder's spiritual ecology: Listening quiets the mind, calms the senses, and opens us to intercourse with the earth. The mind forgets intellectual conventions: *Nature* as lifeless matter-in-mechanical-motion, *History* as the stage upon which human life is set, and the mores of culture. Go into the wilderness. Stand on the rock of granitic truth. Hear the *Ur* syllables, the seed syllables, of Mother Earth: *the wind! the moving water! the sighing boughs!* We are her children. She is our mother. *We are it, the flowing land!* This is the ecological axis of *Turtle Island.* "We could live on this Earth / without clothes or tools!" Hyperbole? or a poetic disclosing of a divine presence concealed from us by Modernism? a postmodern hierophany? a shaman's vision?

Snyder's poetic vision is rooted in earth consciousness, a rediscovery of the perennial philosophy or the wisdom of the ages, known to primal peoples across the face of earth during the Paleolithic era and even thereafter among archaic peoples that still practice hunting and gathering. While no one would accuse Snyder of being a Pangloss, neither is he a Cassandra. His vision is not one of a prophet from the wilderness who sees only destruction and pain, doom and apocalypse, but rather he envisions a world in which computer technicians might walk in the fall with migrating herds of elk. Snyder's poetry rings with the clarity of rams butting heads, and thus captures our attention, but flows with the sounds of mountain streams and thereby allows nature to speak. Snyder, like Thoreau, seeks the Red Face of Humankind, for a kind of Indian wisdom that eludes mere description and logic, and for words to disclose this presence. Like Muir, Snyder resonates with the plant people and wild California mountains through a primal bonding rather than scientific understanding: *Homo sapiens* is not the reason cosmos exists. Consequently, as with Leopold, Snyder's poetry cannot help but to promote perception of the land.

The prose section ("Plain Talk") of *Turtle Island* makes clear that Snyder is neither a nihilist nor pessimist, but rather a realist who looks to a future that begins from where we are now, yet has regrounded itself in a premodern past. Like Paul Shepard, Gary Snyder has firmly grasped the idea that culture is a mosaic, an evolutionary phenomenon. "There are many things in Western culture that are admirable. But a culture that alienates itself from the very ground of its own being—from the wilderness outside (that is to say, wild nature, the wild, self-contained, self-informing ecosystems) and from that other wilderness, the wilderness within—is doomed to a very destructive behavior, ultimately perhaps self-destructive behavior." Thus, the rediscovery of wildness is central to Snyder's spiritual ecology; *contact with the wild* leads to transformation through categorical change, as in "a new definition of humanism and a new definition of democracy that would include the nonhuman, that would have representation from those spheres." Snyder mentions the Amerindians as having categories large enough to admit *the other*. "In Pueblo societies a kind of ultimate democracy is practiced. Plants and animals are also people, and, through certain rituals and dances are given a place and a voice in the political discussions of the humans." Reaffirmation of wildness also, Snyder argues, widens our human horizon through increased insight and scientific (ecological) understanding. "You cannot communicate with the forces of nature in the laboratory." Once we bracket the conventional wisdom—Nature, History—we can then recognize that "there is more information of a higher order of sophistication and complexity stored in a few square yards of forest than there is in all the libraries of [hu]mankind. Obviously, that is a different order of information. It is the information of the universe we live in. It is the information that has been flowing for millions of years. In this total information context, [hu]man[kind] may not be necessarily the highest or most interesting product." Snyder is here encouraging us all to become seekers of Indian wisdom who seek knowledge not to dominate nature but to find our proper place in the web of life. Finally, through contact with the *Ursprung,* with the elemental and wild, modern man and woman might be whole again. "Our own heads: Is where it starts. Knowing that we are the first human beings in history to have so much of…culture and previous experience available to our study, and being free enough of the weight of traditional cultures to seek out a large identity; the first members of a civilized society since the Neolithic to wish to look clearly into the eyes of the wild and see our self-hood, our family, there."[72]

*Turtle Island's* prose also underscores Snyder's commitment to a *process* of socioeconomic transformation called "bioregionalism." A modernist would offer an explicit economic and political program for reform, and palliatives for the psychic travails of modern life. In contrast, Snyder envisions an old-new way of being where land use is sensitive to the whole fabric of life, and culture is rooted in the land rather than some *a priori* ideology the grounds for which have been long forgotten. Bioregionalism, as Snyder envisions it, would thus obviate so much of what has become an increasingly heavy and burdensome bureaucratic state. "Stewardship means, for most of us, find your place on the planet, dig in, and take responsibility from there—the tiresome but tangible work of school boards, county supervisors, local forests—local politics."[73] In an interesting 1979 interview ("Tracking Down the Natural Man") Snyder observes that "some of the solidity in *Turtle Island* is because of my sense of place, living here in Nevada County." So understood, *Turtle Island* redirects the reader's attention away from the merely contingent and evanescent facade of culture to an organic foundation beneath. Thinking ecologically is, he admits, difficult.

"We're so impressed by our civilization and what it's done, with our machines, that we have a difficult time recognizing that the biological world is infinitely more complex."[74] But *if modern people*, bound up in the second world, can rediscover that sense of place in the green world, that awareness of being grounded in nature that was characteristic of ancient people, then the second world might be transformed. Snyder observes in "The Bioregional Ethic" that "no political movement in the United States that came out of left field with so little beginning public support has had as much *effect on the whole American political and economic system* in such a short time as the environmental [ecology] movement."[75]

Ecological awareness is only one side of Snyder's poetics: the other complementary aspect is that of spiritual vision. Through rediscovery of a sense of place flow also spiritual benefits because "by being in place, we get the largest sense of community. We learn that community is of spiritual benefit and of health for everyone, that ongoing working relationships and shared concerns, music, poetry, and stories all evolve into the shared practice of a set of values, visions, and quests. That's what the spiritual path really is." To his credit, Snyder is not overly sanguine about the prospects for such change in consciousness, arguing that the evolution of consciousness—and he does not restrict consciousness to humankind—moves slowly.

> In our present over-speeded and somewhat abnormal situation, the long stability of traditional peasant cultures or primitive hunting and gathering cultures seems...dull. But from the spiritual standpoint, the evolution of consciousness goes at a different pace.... When we steer toward living harmoniously and righteously on the earth, we're also steering toward a condition of long-term stability in which the excitement, the glamour will not be in technology and changing fads. But it will be in a steady enactment and reenactment over and over again of basic psychological inner spiritual dramas, until we learn to find our way through to the next step.[76]

Modern people, however, are not entirely without resources. In the first place, "The objective eye of science, striving to see Nature plain, must finally look at 'subject' and 'object' and the very Eye that looks. We discover that all of us carry within us caves; with animals and gods on the walls; a place of ritual and magic."[77] And second, the poet's role is to aid and abet the process of spiritual growth, particularly through *healing songs.*

> ... I like to think that the concern with the planet, with the integrity of the biosphere, is a long and deeply-rooted concern of the poet for this reason: the role of the singer was to sing the voice of corn, the voice of the Pleiades, the voice of bison, the voice of antelope. To contact in a very special way an "other" that was not within the human sphere; something that could not be learned by continually consulting other human teachers, but could only be learned by venturing outside the borders and going into your own mind-wilderness, unconscious wilderness.[78]

Snyder's poetic vision is that of a spiritual ecologist. Caught up in the flow of our times, the modern mind thinks of wild nature as standing reserve, as totally and completely an alien "other," thus falsifying its own existence. Through an ethnopoetry, however, Snyder believes that we can mind the world like a mountain and see ourselves as living in a relationship of complementarity with rather than superiority to nature. "Thus [human] nature leads into nature—the wilderness—and the reciprocities and balances by which [humankind] lives on earth."[79]

## Conclusion

No absolute measure of it exists, but prior to the agricultural revolution humankind had done little to alter the natural world, while afterwards we became relentless and ceaseless agents of ecological change. "Fields," "weeds," and associated concepts were only the beginnings of a veritable tidal wave of anthropomorphic rationalizations used to justify, legitimate, and guide the transformation of wilderness into civilization. We cannot here explore the penetrating insights of scholars such as Paul Shepard and Marshall Sahlins, who have undressed the belief that history begins at Sumer. Erazim Kohák argues that "we could describe the development of Western thought as one of an ongoing, systematic depersonalization of both our conception of the world and of the world of our ordinary life itself."[80] We increasingly live in a world that has become robotic and crass, an *It world*, to use Martin Buber's appellation. And nature is nothing but an environment upon which is staged that drama which is Euro-culture.

Language is ontogenetic, and brings forth a human space in which *Homo sapiens* dwells. By standing within the hermeneutic circle we have seen that our wilderness thinkers have penetrated the vocabulary and rhetoric of Modernism (science, economics, religion, philosophy) and reanimated an ancient, premodern way of being in the world. So viewed, our wilderness philosophers and poets have managed to pass through the second world, which conceals from us the rooting of the human project in a natural soil, to the first world, that green world from which we have come. To the modernist, of course, such analysis seems mere romantic nonsense, readily dismissed as a kind of puerile longing for the halcyon days of our Paleolithic childhood, when in fact what is called for is increased management of an unruly natural world that is turning against us. To the postmodernist, such analysis confirms the validity of that thesis articulated in one form or another by all the essential wilderness writers: in wildness is the preservation of the world. For they look both behind and ahead, to a world that might be where humankind has rediscovered its fundamental co-relatedness with nature, and advances toward a future where harmony, integrity, and stability characterize the relations between culture and nature. Such would be, as Snyder so compellingly suggests, an old-new way of being. In other words, the postmodern mind grasps and accepts the reality of humankind's natural *artificiality*. That is the message of Darwin's century.

But what, asks the modernist, is this talk of the hermeneutical circle, nature fables, and dwelling poetically? Can such have any relevance to our attempts to save *the environment*? Surely we now know, after these thousands of years of culture, the value of nature and our relation to it, for it is the environment which sustains us, a limitless cornucopia of *resources*. Have we not caught that meaning precisely within our scientific, political, and economical web of language? Have we not through the power of science and technology, and the humanizing potential of liberal-democracy, rendered ourselves master and possessor of nature? But the question begged here—and understandably, for we are all enframed in time—is an awareness of the primordial question of language: there is no privileged access to a world beyond language, and thus reality can be *explicated* only through the medium of language. Can it be that the malaise, the pervasive illness of modern society, is grounded in our self-alienation from nature? from the Magna Mater? We now know that there are no re-sources apart from *the source*—that is evolutionary process, the cosmic wilderness, which grounds and sustains all existence. Yet our use of the word "resource" so affects our sensibilities that we no longer comprehend those things wild and free, existing naturally outside our culturally defined web of language and definition. We consequently conceive of *the source* of life only as a *resource*, as brute matter upon which is imposed human purpose and value.[81] And yet there are those to whom Hermes has spoken: these are the thinking poets and poetic thinkers. They have stood within the hermeneutic circle, and then passed through environment (the ecomachine) to the green world on its other side.

# Appendices

by Paul Shepard


## Appendix I
### On The Shift from Hunting-Gathering to Agriculture

Impressions after following the discussion for twenty-five years:

If agriculture spread mainly by aggrandizement against hunting-gathering peoples, the question remains about the "first" agriculture, which "must" have arisen from hunter-gatherers. After hundreds of millennia why should humans change to dependence on domestic plants and animals if it was bad for them?

First, the shift, once begun, was gradual. The time from the first evidence of domestic species to the first 'archaic high civilizations' was about 5,000 years. Almost everywhere the early 'farmers' continued to hunt and gather, as they still do today.

Second, how did it first begin? Students of the subject see a peculiar parabasis of geography, major climatic change, and the ecology of the peoples east of the Mediterranean. This web of relationships of environmental and cultural circumstances in advanced hunting cultures in the sub-tropics of the middle East and Mediterranean levant may hold the key.

Third, because of the characteristic fluidity of the band society, hunters typically shift location and move in small parties between larger groups. Bands do not remain constant in makeup. This enables hunters to utilize their environment so efficiently that the populations may briefly be denser than otherwise. This "fission and fusion" is a widespread trait.

Fourth, as the local climate changed, the resource base shifted away from big game, placing increased emphasis on harvesting wild seeds and hunting mid-sized mammals and small game. Bounded by sea, mountains and deserts, in a 'deteriorating' climate, these peoples responded by intensifying wild seed harvest and the monitoring (herding) of wild goats and sheep. This increased their sedentism.

Fifth, the same trait of easy mobility for individual members and families is the principal means of settling social tensions. Disputes typically increase when resources are strained and human numbers are dense. "As long as wild foods remained abundant the village succeeded in handling the numerous tensions of group life."[1]

Sixth, these people became the first farmers or horticulturalists, food collectors increasingly bound to their plants. This increased sedentism was linked to the genetic adaptations of the ancestors of wheat, barley, sheep, goats and dogs. The routines of husbandry were tied to the larger village life and political economy "to competition, warfare and the necessity of group defense."[2] For hunters the 'homeland' had not been a simple, defended territory. New distinctions were made between social and spatial boundaries. "Both kinds of boundaries are fluid for contemporary hunter-gatherers,"[3] but became rigidified in the first farmers.

Seventh, these circumstances also grounded the need for expansion by conquering nearby land, or displacing the hunters who lived there by organized aggression.

# Appendix II
### Exegesis of Claude Lévi-Strauss's Idea of Nature as Totemic Metaphor and Caste Metonym[1]

The separation of natural from cultural is an ancient part of human thought. It proceeds by comparing two systems of difference between which there is a formal analogy. The two systems form a symmetrical whole, reconciling diversity and difference with unity and coherence.

The balance is achieved by making unity where things seem naturally disparate and making diversity where the world seems naturally uniform. If the natural is accepted as a model of diversity, its cultural counterpart is a man-made interrelatedness (unity or homogeneity). Declared so to be related, all things 'believe' and 'practice' alike in myth—the bear and salmon can talk and marry. They are like castes—a cultural creation. Here nature is culturalized or socialized. In this totemic view, the different clans are bound together by the flow of genes and kinship through the exchange between groups (or marrying out) of women, who are the means of procreation and subsistence. Because they are perceived as naturally homogeneous (being members of the same species), they are a connection between different social groups, which are likewise naturally homogenous. This balances the cultural analogue of species interactions and is reinforced by meat-sharing.

Food chains are the paradigm of social connections. An otherwise overwhelming diversity of the world and potential danger of isolation and conflict between tribes is checked by the unifying, order-creating social model. Species in the ecosystem are the concrete reality, the basis of a metaphor for social groups.

In agriculture or caste societies occupational products replace natural species as the basis of social distinction. These people exchange goods, which are the concrete model of diversity. But their difference is due to the cultural division of castes or occupations. To balance this, women are held to be naturally diverse and cannot be exchanged except within their own social group or caste. Each caste is perceived as a species. Such groups take on the isolation of natural species. This extension of natural models only further impairs social and ecological connectedness, except for domestic plants and animals which are the cohorts of specific castes. An imbalance shifts toward difference and diversity as opposed to unity or relatedness. The social stresses to which this gives rise are partly compensated for by the rise of social amalgamates—chiefdoms, kingdoms and the state.

Where nature is the model, people see or declare themselves and species to be social groups, and the ecosystem to be a society. Where made objects are the model, groups of people perceive themselves as divided by a species-like difference. Totemic people "culturize a false nature truly," while agriculture people "naturalize a true culture falsely." In the latter case the extension of the group politically helps make up for the tendency for caste systems to block communication by non-intermarriage, a characteristic of territoriality and the confrontation of opposing political groups at their frontiers.

Wild species in the first group are a heuristic reference for making social structure and decisions. The second group incorporates some species as domesticates and metonymic members of society. The remainder (wild forms) have a reduced cognitive and metaphoric function. Most subsistence societies are intergrades of these two types. Although Lévi-Strauss does not pursue the totem-to-caste shift as a major transformation of the Neolithic, it seems apparent that, if true, it marks a departure from the paradigmatic role of "nature" in human thought. However, the underlying impulse, a genetically ingrained, evolutionary dimension of consciousness, will not go away. The reader is invited to speculate on the bizarre expressions of this instinct to which this post-mesolithic development has led.

# Notes

## Introduction, by Max Oelschlaeger

[1]A companion volume based on the discussions held during the conference week will be forthcoming under the title of *Voices for the Wilderness*.

[2]A different version of this essay was published in *Sierra* (September/October 1989). It will also appear in Snyder's forthcoming *The Practice of the Wild* (North Point Press). The other essays in this collection are previously unpublished.

## The Etiquette of Freedom, by Gary Snyder

[1]Tzvetan Todorov, *The Conquest of America* (New York: Harper and Row, 1985), p. 134.

## A Post-Historic Primitivism, by Paul Shepard

Acknowledgment: My thanks to Flo Krall for her careful reading, criticism and suggestions in the preparation of this paper.

[1]Herbert J. Muller, *The Uses of the Past*, Oxford University Press, New York, 1952, p. 38.

[2]Herbert Schneidau, *Sacred Discontent*, University of California, Los Angeles, 1976.

[3]Bogert O'Brien, "Inuit Ways and the Transformation of Canadian Theology," mss. 1979.

[4]Robert Hutchins, Preface to Mortimer J. Adler's Hundred Great Books Series, *The Great Ideas*, Encyclopedia Britannica, Chicago, 1952.

[5]Ivan Illich and Barry Sanders, *The Alphabetization of the Popular Mind*, North Point, San Francisco, 1988.

[6]Paul Shepard, *Nature and Madness*, Sierra Club Books, San Francisco, 1982.

[7]Edmund S. Carpenter, "If Wittgenstein Had Been an Eskimo," *Natural History* 89 (4), Feb., 1980.

[8]A. David Napier, *Masks, Transformation, and Paradox*, University of California, Los Angeles, 1986. And see Steven Lonsdale, *Animals and the Origin of Dance*, Thames and Hudson, New York, 1982.

[9]Edith Cobb, *The Ecology of Imagination in Childhood*, Columbia University Press, New York, 1978.

[10]J. H. Plumb, *Horizon* 41 (3), 1972.

[11]Scribners, New York, 1973.

[12]George Boas, *Essays on Primitivism and Related Ideas in the Middle Ages*, Johns Hopkins, Baltimore, 1948.

[13]Anthony Pagdon, *The Fall of Natural Man*, Cambridge, New York, 1982, p. 78.

[14]M. Navarro, *La Tribu Campa*, Lima, 1924, quoted in Gerald Weiss, "Campa Cosmology," *American Museum of Natural History Anthropological Papers*, 52, Part 5, New York, 1975.

[15]Dr. Will Durant, "A Last Testament to Youth," *The Columbia Dispatch Magazine*, Feb. 8, 1970.

[16]Ernest Thompson Seton, *Two Little Savages*, Doubleday, New York, 1903.

[17]Calvin Martan, *Keepers of the Game*, University of California, Los Angeles, 1978.

[18]Glyn Daniel, *The Idea of Prehistory*, Penguin, Baltimore, 1962, p. 57.

[19]Robert Ardrey, *The Hunting Hypothesis*, Athenaeum, New York, 1976.

[20]John Pfeiffer, *The Emergence of Man*, Harper and Row, New York, 1972.

[21]Nigel Calder, *Eden Was No Garden*, Holt, New York, 1967.

[22]Gordon Rattray Taylor, *Rethink, a Paraprimitive Solution*, Dutton, New York, 1973.

[23]José Ortega y Gasset, *Meditations on Hunting*, Scribners, New York, 1972.

[24]M. F. C. Bourdillon and Meyer Fortes, eds., *Sacrifice*, Academic Press, New York, 1980.

[25]Carleton Coon, *The Story of Man*, Knopf, New York, 1962, p. 187.

[26]W. E. H. Stanner, *White Man Got No Dreaming*, Australian National University Press, Canberra, 1979.

[27]Richard B. Lee and Irven Devore, eds., *Man the Hunter*, Aldine, Chicago, 1968.

[28]Sherwood L. Washburn, ed., *The Social Life of Early Man*, Aldine, Chicago, 1961.

[29]G. P. Murdock, *Ethnographic Atlas for New World Societies*, University of Pittsburgh Press, 1967.

[30]Marshall Sahlins, *Stone Age Economics*, Aldine, Chicago, 1972.

[31]Derek Freeman, letter, *Current Anthropology*, Oct., 1973, p. 379.

[32]Clifford Geertz, "The Impact of the Concept of Culture on the Concept of Man," in Stanley Diamond, *In Search of the Primitive: A Critique of Civilization*, Transaction Books, New Brunswick, 1974, p. 102.

[33]Melvin Konner, *The Tangled Wing: Biological Constraints on the Human Spirit*, Harper and Row, New York, 1983.

[34]Loren Eiseley, "Man of the Future," *The Immense Journey*, Random House, New York, 1957.

[35]Laurens van der Post, *Heart of the Hunter*, Harcourt Brace Jovanovich, New York, 1980.

[36]Roger M. Keesing, "Paradigms Lost: the New Ethnography and New Linguistics," *South West Journal of Anthropology* 28: 299-332, 1972.

[37]Ivan Illich, *Gender*, Pantheon, New York, 1982.

[38]Gina Bari Kolata, "!Kung Hunter-Gatherers: Feminism, Diet, and Birth Control," *Science* 185: 932-934, 1974.

[39]Claude Lévi-Strauss, *The Savage Mind*, University of Chicago Press, Chicago, 1966.

[40]C. H. D. Clarke, "Venator—the Hunter," mss., n.d.

[41]Irenaus Eibes-Eibesfeldt, *Love and Hate*, Holt, Rinehart and Winston, New York, 1971.

[42]Jane Howard, "All Happy Clans are Alike," *The Atlantic*, May, 1978.

[43]Gary Snyder, quoted in Peter B. Chowka, "The Original Mind of Gary Snyder," *East-West*, June, 1977.

[44]A. H. Ismail and L. B. Trachtman, "Jogging the Imagination," *Psychology Today*, March, 1973.

[45]P. S. Martin and H. E. Wright, Jr., eds., *Pleistocene Extinctions: the Search for a Cause*, Yale University Press, New Haven, 1967.

[46]Donald K. Grayson, "Pleistocene Avifaunas and the Overkill Hypothesis," *Science* 195: 691-693, 1977. Karl W. Butzer, *Environment and Archaeology*, Aldine, Chicago, 1971, p. 503ff. Michael A. Joachim, *Hunter-Gatherer Subsistence and Settlement: A Predictive Model*, Academic Press, New York, 1976. Marvin Harris, "Potlatch Politics and Kings' Castles," *Natural History*, May, 1974.

[47]Melvin J. Konner, "Maternal Care, Infant Behavior and Development Among the !Kung," in R. B Lee and I. Devore, *Kalahari Hunter Gatherers*, Harvard University Press, Cambridge, 1976.

[48]Joachim, op. cit., p. 22. Harry Jerison, *Evolution of the Brain and Intelligence*, Academic, New York, 1973.

[49]Richard Nelson, *Make Prayers to the Raven*, University of Chicago Press, Chicago, 1983, p. 225.

[50]Gary Urton, ed., *Animal Myths and Metaphors in South America*, University of Utah Press, Salt Lake City, 1985.

[51]Clarke, op. cit.

[52]Clarke, ibid.

[53]James Woodburn, "An Introduction to Hadza Ecology, in Lee and DeVore, *Man the Hunter*.

[54]Marek Zvelebil, "Postglacial Foraging in the Forests of Europe," *Scientific American*, May, 1986.

[55]C. Dean Freudenberger, "Agriculture in a Post-Modern World," mss. for conference, "Toward a Post-Modern World," Santa Barbara, California, January, 1987.

[56]Liberty Hyde Bailey, *The Holy Earth*, Scribners, New York, 1915, p. 83.

[57]Ibid., p. 151.

[58]Wes Jackson, *New Roots for Agriculture*, Friends of the Earth, San Francisco, 1980.

[59]Helen Spurway, "The Causes of Domestication," *Journal of Genetics* 53: 325, 1955.

[60]Jack M. Potter, *Peasant Society*, Little Brown, Boston, 1967.

[61]Jane Schneider, "Of Vigilance and Virgins: Honor, Shame, and Access to Resources in Mediterranean Societies," *Ethnology* 10: 1-24, 1971.

[62]Aldous Huxley, "Mother," *Tomorrow and Tomorrow and Tomorrow*, Harper and Row, New York, 1952.

[63]Shepard, *Nature and Madness*.

[64]Robert Redfield, *The Folk Cultures of Yucatan*, University of Chicago, Chicago, 1941.

[65]Joseph Campbell, *The Masks of God*, Viking Press, New York, 1959. vol. I, p. 180.

[66]Paul Shepard and Barry Sanders, *The Sacred Paw: The Bear in Nature, Myth and Literature*, Viking Press, New York, 1985.

[67]Tim Ingold, "Hunting, Sacrifice, and the Domestication of Animals," in *The Appropriation of Nature*, University of Iowa Press, Iowa City, 1987.

[68]Kolata, op. cit.

[69]Patricia Draper, "Social and Economic Constraints on Child Life among the !Kung," in Lee and DeVore, *Kalahari Hunter Gatherers*.

[70]C. H. Brown, "Mode of Subsistence and Folk Biological Taxonomy," *Current Anthropology* 26 (1): 43-53, 1985.

[71]"Campaign to Promote the Vegetarian Diet," (leaflet), Animal Aid Society, Tonbridge, England, n.d.

[72]Marvin Harris, *Sacred Cow, Abominable Pig*, Simon and Schuster, New York, 1985, p. 22.

[73]Robert Allen, "Food for Thought," *The Ecologist*, January, 1975.

[74]Stanley Boyd Eaton and Marjorie Shostak, "Fat Tooth Blues," *Natural History* 95 (6), July, 1986.

[75]Daniel, op. cit.

[76]Allen W. Johnson and Timothy Earle, *The Evolution of Human Societies: From Foraging Group to Agrarian State*, Stanford University Press, Stanford, 1987.

[77]John W. Berry and Robert C. Annis, "Ecology, Culture and Psychological Differentiation," *International Journal of Psychology* 9: 173-193, 1974.

[78]Robert Edgerton, *The Individual in Cultural Adaptation*, University of California, Los Angeles, 1971.

[79]N. K. Sandars, *Prehistoric Art in Europe*, Penguin, Baltimore, pp. 95-96.

[80]Marshall McLuhan, *Through the Vanishing Point*, Harper and Row, New York, 1968.

[81]David Lowenthall, "Is Wilderness Paradise Now?," *Columbia University Forum* 7 (2), 1964.

[82]Susan Sontag, *On Photography*, Dell, New York, 1973.

[83]Bertram Lewin, *The Image and the Past*, I.U.P., New York, 1968.

[84]Spurway, op. cit., 1968.

[85]Lévi-Strauss, op. cit., 1968.

[86]Fred Myer, *Pintupi Country, Pintupi Self*, Smithsonian Institution, Washington, D.C., 1986. Myer is following a path laid out by A. Irving Hallowell. For example see Hallowell, "Self, Society, and Culture in Phylogenetic Perspective," in Sol Tax, ed., *The Evolution of Man*, Aldine, Chicago, 1961.

[87]Gary Snyder, "On 'Song of the Taste,'" The Recovery of the Commons Project, Bundle #1, North San Juan, Ca., n.d.

[88]Gary Snyder, "Good, Wild, Sacred," *The Co-Evolution Quarterly*, Fall, 1983.

[89]Rene Dubos, "Environmental Determinants of Human Life," in David C. Glass, ed., *Environmental Influences*, Rockland University Press, 1968.

[90]Ibid., quoted from Lewis Mumford, *The Myth of the Machine*, Harcourt Brace, New York, 1966. Mumford probably got it from Loren Eiseley's "Man of the Future" in *The Immense Journey*.

[91]Walter J. Ong, "World as View and World as Event," *American Anthropologist* 71: 634-647. Dorothy Lee, "Codifications of Reality: Lineal and Non-Lineal," *Psychosomatic Medicine* 12 (2), 1969.

[92]Morris Berman, "The Roots of Reality," (a review of Humberto Maturana and Francisco Varelas, *The Tree of Knowledge*), *Journal of Humanistic Psychology* 29: 277-284, 1989.

[93]Emile Zuckenkandle, *Scientific American* 212: 63, 1960.

[94]Hyemeyohsts Storm, *Storm Arrows*, Harper and Row, New York, 1972.

[95]William Ayres Arrowsmith, "Hybris and Sophrosyne," *Dartmouth Alumni Magazine*, July, 1970.

[96]William Dupre, *Religion in Primitive Cultures*, Mouton, The Hague, 1975, p. 327.

[97]Kevin T. Jones, "Hunting and Scavenging by Early Hominids: A Study in Archaeological Method and Theory," Ph. D. Thesis, University of Utah, 1984.

[98]Janet Siskind, *To Hunt in the Morning*, Oxford University Press, New York, 1976, p. 109.

[99]Martin King Whyte, *The Status of Women in Preindustrial Societies*, Princeton University Press, Princeton, 1978.

[100]Arnold Modell, *Object Love and Reality*, I.U.P., New York, 1968. Chap. 5, "The Sense of Identity: The Acceptance of Separateness."

[101]Hara Estroff Marano, "Biology Is One Key to the Bonding of Mothers and Babies," *Smithsonian*, February, 1981.

[102]Melvin J. Konner, "Maternal Care, Infant Behavior and Development Among the !Kung," and Patricia Draper, "Social and Economic Constraints on Child Life Among the !Kung," in Lee and DeVore, *Kalahari Hunter Gatherers*.

[103]Sandars, op. cit., p. 26.

[104]Colin M. Turnbull, *The Human Cycle*, Simon and Schuster, New York, 1983.

## Ecocentrism, Wilderness, and Global Ecosystem Protection, by George Sessions

[1]John C. Hendee, George H. Stankey, and Robert C. Lucas, *Wilderness Management* (Washington: USDA Forest Service Misc. Publication No. 1365, 1978), pp. 140-41, quoted in Holmes Rolston III, *Philosophy Gone Wild: Essays in Environmental Ethics* (Buffalo: Prometheus Books, 1986), p. 119.

[2]The ideological split between Muir and Pinchot is discussed in Roderick Nash, *Wilderness and the American Mind* (New Haven: Yale University Press, 1982), 3rd ed., pp. 129-40; Stephen Fox, *John Muir and His Legacy: The American Conservation Movement* (Boston: Little, Brown and Co., 1981), pp. 109-30; Michael P. Cohen, *The Pathless Way: John Muir and the American Wilderness* (Madison: University of Wisconsin Press, 1984), pp. 160-61, 292ff.; for a more sympathetic interpretation of Pinchot's intentions, see John Rodman, "Four Forms of Ecological Consciousness Reconsidered," in Tom Attig and Donald Schemer, eds., *Ethics and the Environment* (Englewood Cliffs: Prentice-Hall, 1983), pp. 82-92.

[3]For a short introduction to Leopold's career and thought see Nash, *Wilderness*, pp. 182-99; see also J. Baird Callicott, *Companion to a Sand County Almanac* (Madison: University of Wisconsin, 1987); for a discussion of the rise of the

Age of Ecology and the meaning of "ecocentrism" see George Sessions, "The Deep Ecology Movement: A Review," *Environmental Review* 11, 2, (1987): 105-25; Sessions, "Ecocentrism and the Greens," *The Trumpeter* 5, 2, (1988): 65-69; Warwick Fox, "The Deep Ecology-Ecofeminism Debate and Its Parallels," *Environmental Ethics* 11, 2, (1989): 5-25.

[4]H. D. Thoreau, "Walking," in Charles Anderson, ed., *Thoreau's Vision: The Major Essays* (Englewood Cliffs: Prentice Hall, 1973); for a discussion of Muir's spiritual ecocentric orientation see Cohen, *Pathless Way*, chs. 1, 6, 7; Fox, *Muir and His Legacy*, pp. 43-53, 59, 79-81, 289-91, 350-55, 361; Frederick Turner, *Rediscovering America: John Muir in His Time and Ours* (San Francisco: Sierra Club Books, 1985).

[5]For a discussion of the dominant land-use and conservation attitudes and policies see Clayton R. Koppes, "Efficiency/Equity/Esthetics: Towards a Reinterpretation of American Conservation," *Environmental Review* 11, 2, (1987): 127-46; for an excellent recent discussion of the ecocentric revolution in Western religion and philosophy see Roderick Nash, *The Rights of Nature: A History of Environmental Ethics* (Madison: University of Wisconsin Press, 1989); also Thomas Berry, *The Dream of the Earth* (San Francisco: Sierra Club Books, 1988); Frederick Turner, *Beyond Geography: The Western Spirit Against the Wilderness* (New York: Viking Press, 1980).

[6]Thomas L. Fleischner, "Keeping It Wild: Toward a Deeper Wilderness Management," in Mitch Friedman, ed., *Forever Wild: Conserving the Greater North Cascade Ecosystem* (Bellingham, WA: Mountain Hemlock Press, 1988), pp. 73-79. See also the essays by Mitch Friedman, "How Much is Enough?: Lessons from Conservation Biology," and Edward Grumbine, "Ecosystem Management for Native Diversity," reprinted in *The Trumpeter* 5, 2, (1988): 47-52. *Forever Wild* is actually a theoretical discussion and a proposal for protecting the Greater North Cascades ecosystem. I have found these essays extremely helpful and have drawn heavily upon them in what follows.

[7]Edward Abbey, *Desert Solitaire* (New York: McGraw-Hill, 1968); see also Joseph Sax, *Mountains without Handrails* (Michigan: University of Michigan Press, 1980); Cohen, *Pathless Way*, pp. 305-310, 362-366.

[8]For a discussion of the development of the Sierra Club from a "hiking club" to an ecologically oriented organization, together with a discussion of Dave Brower and the wilderness conferences, see Michael P. Cohen, *The History of the Sierra Club: 1892-1970* (San Francisco: Sierra Club Books, 1988), especially pp. 187-332; for another history of the Club's transition and a discussion of Dave Brower as "Muir reincarnate," see Stephen Fox, *John Muir and His Legacy*, especially pp. 250-290; for a collection of the Wilderness conference papers, see William Schwartz, ed., *Voices for the Wilderness* (New York: Ballantine, 1969).

[9]Alston Chase, *Playing God in Yellowstone: The Destruction of America's First National Park* (Boston: The Atlantic Monthly Press, 1986), p. 33.

[10]For critiques of Chase's views, see Dave Foreman, Doug Peacock, and George Sessions, "Who's 'Playing God in Yellowstone?'" *Earth First!* 7, 11, (1986): 18-21; Holmes Rolston III, "Biology and Philosophy in Yellowstone," *Biology and Philosophy*, 1989, forthcoming. For criticism of the excessive

manipulation of wild animals by wildlife biologists and by Park officials for the amusement of tourists, see Morgan Sherwood, "The End of American Wilderness," *Environmental Review* 9, (1985): 197-209. But see also Arne Naess, "Should We Try to Relieve Clear Cases of Extreme Suffering in Nature?" (1988), unpublished ms.

[11]John Rodman, "Resource Conservation—Economics and After," (1986), unpublished ms.; quoted in Bill Devall and George Sessions, "The Development of Natural Resources and the Integrity of Nature: Contrasting Views of Management," *Environmental Ethics* 6, 2, (1984): 307.

[12]Chris Maser, "Sustainable Forestry," *The Trumpeter* 6, 2, (1989): 52-54; see also Chris Maser, *The Redesigned Forest* (San Pedro, CA: R&E Miles Publisher: 1988).

[13]Bob Nixon, "Focus on Forests and Forestry," *The Trumpeter* 6, 2, (1989): 38. This issue of *The Trumpeter* is devoted to forestry philosophy and practices; biologists, forestry planners, and others within the U.S. Forest Service are now beginning to speak out against the anti-ecological policies of their agency; see their newsletter, *Inner Voice: A Publication of the Association of Forest Service Employees for Environmental Ethics* 1, 1, Summer (1989). (P.O. Box 45, Vida, Oregon 97488).

[14]William Godfrey-Smith, "The Value of Wilderness," *Environmental Ethics* 1,4, (1979): 309-319.

[15]Warwick Fox, "Toward a Transpersonal Ecology: The Context, Influence, Meanings, and Distinctiveness of the Deep Ecology Approach to Ecophilosophy," Ph.D. dissertation submitted to Murdoch University, Western Australia, 1988: 261-263. Publication forthcoming (Boulder, CO: Shambala Press).

[16]Daniel Janzen, "The Future of Tropical Ecology," *Annual Review of Ecology and Systematics* 17, (1986): 305-6. Janzen's paper is cited in a paper that is very critical of deep ecology and wilderness preservation: Ramachandra Guha, "Radical American Environmentalism and Wilderness Preservation: A Third World Critique" *Environmental Ethics* 11, 1, (1989): 75-76. Guha tries to saddle the deep ecology position with the narrowly professional views Janzen expresses.

[17]Arne Naess, "Modesty and the Conquest of Mountains," in Michael Tobias, ed., *The Mountain Spirit* (Woodstock, New York: Overlook Press, 1979); see also Sax, *Mountains without Handrails*.

[18]The Zahniser quote appears in Fox, *John Muir*, p. 270.

[19]Bryan G. Norton, "Sand Dollar Psychology," *The Washington Post Magazine*, 1 June (1986): 13-14; quoted in Warwick Fox, "Toward a Transpersonal Ecology," p. 262.

[20]Paul Shepard, *Thinking Animals: Animals and the Development of Human Intelligence* (New York: Viking Press, 1978), pp. 246-252.

[21]Shepard's "natural ontogeny" argument appears in Paul Shepard, *The Tender Carnivore and the Sacred Game* (New York: Scribners, 1973) and *Thinking Animals*, but is developed most fully in Shepard, *Nature and Madness* (San Francisco: Sierra Club Books, 1982); for an earlier claim that humans are genetically programmed for wild environments, see Hugh H. Iltus, "Whose Fight is the Fight for Nature?" *Sierra Club Bulletin* 52, 9, (1967): 34-39; for

the introduction of the concept of "ecological self," see Arne Naess, "Self-Realization: An Ecological Approach to Being in the World," Roby Memorial Lecture, Murdoch University, Western Australia, March 1986; reprinted in *The Trumpeter* 4, 3, (1987): 35-42. Much of the emphasis of Warwick Fox's recent work is to elaborate upon Naess' concepts of ecological self and self-realization (see Fox, "Toward a Transpersonal Ecology," Part IV).

[22] Wayland Drew, "Killing Wilderness," *Ontario Naturalist*, (1972); reprinted in *The Trumpeter* 3, 1, (1986): 19-23; see also George Sessions, "1984: A Postscript," in Bill Devall and George Sessions, *Deep Ecology: Living as if Nature Mattered* (Salt Lake City: Gibbs Smith, Inc., 1985), pp. 254-256.

[23] Aldous Huxley, *Brave New World Revisited* (New York: Harper and Row, 1958), pp. 3-17; Julian Huxley's speech is referred to in Raymond F. Dasmann, *Planet in Peril: Man and the Biosphere Today* (New York: World Publishing Co., 1972), pp. ix-xi; see also Sir Julian Huxley, "The Age of Overbreed," *Playboy Magazine* (1965), reprinted in Playboy editors, *Project Survival* (Chicago: Playboy Press Book, 1971), pp. 43-64; for an early discussion of the awareness of the human overpopulation problem, see Raymond F. Dasmann, *The Last Horizon* (New York: Macmillan, 1963), pp. 210-227; see also Aldous Huxley, "The Politics of Ecology," *Center Magazine* (Santa Barbara, CA: Center for the Study of Democratic Institutions, 1963).

[24] See Holmes Rolston III, "Values in Nature"; "Values Gone Wild"; "Valuing Wildlands"; "Duties to Endangered Species," in Rolston, *Philosophy Gone Wild*: pp. 74-90; 118-142; 180-205; 206-220; for further analyses of the arguments for wild species protection, see the New Zealand ecophilosopher Alastair S. Gunn's "Preserving Rare Species," in Tom Regan, ed., *Earthbound: New Introductory Essays in Environmental Ethics* (New York: Random House, 1984), pp. 289-335; William Godfrey-Smith, "The Rights of Non-Humans and Intrinsic Value," in Don Mannison, Michael McRobbie, and Richard Routley, eds., *Environmental Philosophy* (Canberra: Australian National University, 1980), pp. 30-47; Norman Myers, *The Sinking Ark: A New Look at the Problem of Disappearing Species* (New York: Pergamon Press, 1979); Bryan G. Norton, ed., *The Preservation of Species* (Princeton, NJ: Princeton University Press, 1986); J. Baird Callicott, "On the Intrinsic Value of Nonhuman Species," in Callicott, *In Defense of the Land Ethic: Essays in Environmental Philosophy* (New York: State University of New York Press, 1989), pp. 129-155.

[25] For a historical account of the impact of the "life support system/ humans as endangered species arguments" on society, together with the ensuing "man-centered environmentalism," see Fox, *John Muir*, pp. 291-313; for the "rivet popping" analogy and the life support system arguments, see Paul and Anne Ehrlich, *Extinction: The Causes and Consequences of the Disappearance of Species* (New York: Random House, 1981), pp. xi-xiv, 77-100; see also Paul Ehrlich, "The Loss of Diversity: Causes and Consequences," E. O. Wilson, ed., *Biodiversity* (Washington, D.C.: National Academy Press, 1988), pp. 21-27.

[26] James Lovelock, *Gaia: A New Look at Life on Earth* (New York: Oxford University Press, 1979), pp. 41, 118f; for a critique of Lovelock, see Anthony Weston, "Forms of Gaian Ethics," *Environmental Ethics* 9, 3, (1987): 217-

230; for critiques of anthropocentric Teilhardian interpretations of Gaia, see George Sessions, "Review of C. Bonifazi's *The Soul of the World*," *Environmental Ethics* 3, 3, (1981): 275-281; George Sessions, "Deep Ecology, New Age, and Gaian Consciousness," *Earth First!* 7, 8, (1987): 27-30.

[27] See Tom Regan, "The Nature and Possibility of an Environmental Ethic," *Environmental Ethics* 3, (1981): 19-34; Paul W. Taylor, *Respect for Nature: A Theory of Environmental Ethics* (Princeton NJ: Princeton University Press, 1986), pp. 71-80. Taylor provides a carefully argued basis for a metaphysical position he calls "respect for nature" which, he claims, leads to regarding "the wild plants and animals of the Earth's natural ecosystems as possessing inherent worth." Taylor's system focuses only on the ethical value of individuals which does not allow him to say that species and ecosystems, as such, have inherent worth.

[28] Arne Naess, "The Shallow and the Deep, Long-Range Ecology Movements: A Summary," *Inquiry* 16, (1973): 95-100.

[29] John Rodman, "The Liberation of Nature?," *Inquiry* 20, (1977): 109; for a further critique of "ethical extensionism," see John Rodman, "Four Forms of Ecological Consciousness Reconsidered"; for a critique of Tom Regan's "animal rights" position, see J. Baird Callicott, "Review of Tom Regan, *The Case for Animal Rights*," in Callicott, *Defense of the Land Ethic*, pp. 39-47.

[30] Charles S. Elton, *The Ecology of Invasions by Animals and Plants* (London: Methuen, 1958); David Ehrenfeld, *The Arrogance of Humanism* (New York: Oxford University Press, 1978), pp. 207-208; Paul and Anne Ehrlich, *Extinction*, pp. 48-52.

[31] Callicott, "On the Intrinsic Value of Other Species," pp. 136-139, 153-155; see also Warwick Fox, "Toward a Transpersonal Ecology," pp. 283-284; Robin Attfield, *The Ethics of Environmental Concern* (New York: Columbia University Press, 1983); Thomas Merton, "The Wild Places," *Center Magazine* (Santa Barbara, CA: Center for the Study of Democratic Institutions, July, 1968); reprinted in Robert Disch, ed., *The Ecological Conscience: Values for Survival* (Englewood Cliffs, NJ: Prentice-Hall, 1970), pp. 37-43.

[32] For Cobb's critique of modern Western philosophy as anthropocentric, see John B. Cobb, Jr., "The Population Explosion and the Rights of the Subhuman World," in R. T. Roelofs and J. N. Crowley, et. al., eds., *Environment and Society* (Englewood Cliffs, NJ: Prentice-Hall, 1974); for an elaboration of Cobb's analysis, see George Sessions, "Anthropocentrism and the Environmental Crisis," *Humboldt Journal of Social Relations* 2, 1, (1974): 71-81; see also John Cobb, "Process Theology and Environmental Issues," *Journal of Religion* 4, (1980): 440-458; Charles Birch and John B. Cobb, Jr., *The Liberation of Life: From the Cell to the Community* (Cambridge: Cambridge University Press, 1983).

[33] Thomas Berry, *Teilhard in the Ecological Age* (Chambersberg, PA: Anima Books, 1982); Thomas Berry, "The Viable Human," *Revision* 9, 2, 1987: 75-81; Thomas Berry, *The Dream of the Earth*; Thomas Berry and Brian Swimme, *The Universe Story: An Autobiography from Planet Earth* (Clinton, WA: New Story Productions, 1989); for a discussion of Eiseley's evolutionary/ecological religious views, see Frederick Elder, *Crisis in Eden: A Religious Study of Man and Environment* (Tenn: Abingdon Press, 1970); Roderick S.

French, "Is Ecological Humanism a Contradiction in Terms?: The Philosophical Foundations of the Humanities Under Attack," in Robert C. Schultz and J. Donald Hughes, eds., *Ecological Consciousness: Essays from the Earthday X Colloquium University of Denver*, April 21-24, 1980 (Washington D.C.: University Press of America, 1981), pp. 43-66; Devall and Sessions, *Deep Ecology*, pp. 164-166.

[34] Arne Naess, "The Green Society and Deep Ecology," (The 1987 Schumacher Lecture), unpublished ms., p. 18. See also Naess, "Identification as a Source of Deep Ecological Attitudes," in Michael Tobias, ed., *Deep Ecology* (San Diego, CA: Avant Books, 1985), pp. 256-270; Arne Naess, *Ecology, Community, and Lifestyle: Outline of an Ecosophy* (Cambridge, Cambridge University Press, 1989).

[35] Naess, "Green Society and Deep Ecology," p. 19.

[36] For the selfish and destructive manipulation of wildlife by biologists, see Sherwood Anderson, "The End of American Wilderness"; Dian Fossey relates incidents of the indifference displayed by graduate students toward the welfare of the gorillas they were studying. See Farley Mowat, *Woman in the Mists* (New York: Warner Books, 1987).

[37] Godfrey-Smith, "The Value of Wilderness," p. 319.

[38] For historical discussions of the development of environmental thought, see Roderick Nash, *Wilderness*; Nash, *Rights of Nature*; George Sessions, "Ecocentrism and the Greens"; Anne and Paul Ehrlich, *Earth* (New York: Franklin Watts, 1987), ch. 7.

[39] Friedman, "How Much is Enough?," p. 34. I have drawn the following material from the excellent summaries by Friedman and Edward Grumbine, "Ecosystem Management for Native Diversity."

[40] Michael Soulé, "What is Conservation Biology?" *Bioscience* 35, (1985): 727-734; quoted in Friedman, ibid.

[41] O.H. Frankel and Michael Soulé, *Conservation and Evolution* (Cambridge: Cambridge University Press, 1981); Friedman, *Forever Wild*, pp. 1-2.

[42] Michael Soulé and D. Simberloff, "What Do Genetics and Ecology Tell Us About the Design of Nature Reserves?" *Biological Conservation* 35, (1986): 19-40.

[43] Friedman, "How Much is Enough?," p. 39; Norman Myers, "The Extinction Spasm Impending: Synergisms at Work," *Conservation Biology* 1, (1987): 14-21; A. P. Dobson, C. H. McLellan, and D. S. Wilcove, "Habitat Fragmentation in the Temperate Zone," in Michael Soulé, ed., *Conservation Biology: The Science of Scarcity and Diversity* (Sunderland, MA: Sinauer Press, 1986), pp. 237-256.

[44] Friedman, ibid.; A. Runte, *National Parks: The American Experience* (Lincoln: University of Nebraska Press, 1987); W. D. Newmark, "Legal and Biotic Boundaries of Western North American National Parks: A Problem of Congruence," *Biological Conservation* 33, (1985): 197-208; W. D. Newmark, "A Land-Bridge Island Perspective on Mammalian Extinctions in Western North American Parks," *Nature* 325, (1987): 430-432; R. M. May and D. S. Wilcove, "National Park Boundaries and Ecological Realities," *Nature* 324, (1986): 206-207; J. Terborgh and B. Winter, "Some Causes of Extinction," in B. A. Wilcox and Michael Soulé, *Conservation Biology: An Evolutionary-*

*Ecological Perspective* (Sunderland, MA: Sinauer Associates, 1980), pp. 19-133; R. F. Noss, "Wilderness Recovery and Ecological Restoration," *Earth First!* 5, 8, (1985): 18-19; R. F. Noss, "Recipe for Wilderness Recovery," *Earth First!* 6, (1986): 22, 25.

[45] Friedman, "How Much is Enough?," p. 37; C. Holtby, B. A. Wilcox, and Michael Soulé, "Benign Neglect: A Model of Faunal Collapse in the Game Reserves of East Africa," *Biological Conservation* 15, (1979): 259-270.

[46] Grumbine, "Ecosystem Management for Native Diversity," p. 46; Newmark, "Legal and Biotic Boundaries."

[47] Friedman, "How Much is Enough?," p. 43; Frankel and Soulé, *Conservation and Evolution*, 1981.

[48] Paul Shepard, "Ecology and Man: A Viewpoint," in Paul Shepard and Daniel McKinley eds., *The Subversive Science: Essays Toward an Ecology of Man* (New York: Houghton Mifflin, 1969); for essays by the early visionaries who warned about overpopulation and environmental deterioration, see Paul Ehrlich and John Holdren, *The Cassandra Conference: Resources and the Human Predicament* (College Station, TX: Texas A & M University Press, 1987); see also Leo Marx, "American Institutions and Ecological Ideals," *Science* 170 (1970): 945-952.

[49] Raymond B. Cowles, "Population Pressure and Natural Resources," in David Brower, ed., *The Meaning of Wilderness to Science: Proceedings, Sixth Biennial Wilderness Conference* (San Francisco: Sierra Club Books, 1960), pp. 79-94; for Brower's comments, see John McPhee, *Encounters with the Archdruid* (New York: Farrar, Straus and Giroux, 1971), pp. 74, 84-85, 226; see also Frank Graham, Jr., "Dave Brower: Last of the Optimists?" *Audubon*, Sept., (1982): 64-73; Gary Snyder, "Four Changes," in Garret DeBell, ed., *The Environmental Handbook* (New York: Ballantine, 1970), pp. 323-333 (revised in Gary Snyder, *Turtle Island* [New York: New Directions, 1974], pp. 91-102).

[50] Jay Forrester, *World Dynamics* (Cambridge, MA: Wright-Allen Press, 1971); Donella and Dennis Meadows, et. al., *The Limits to Growth: A Report for the Club of Rome's Project on the Predicament of Mankind* (New York: Universe Books, 1972); see also G. Tyler Miller, Jr., *Replenish the Earth: A Primer in Human Ecology* (Belmont, CA: Wadsworth Publishing Co., 1972).

[51] David Brower, "Toward an Earth International Park," *Sierra Club Bulletin* 52, 9, (1967): 20.

[52] Eugene P. Odum, *Fundamentals of Ecology* (Philadelphia: W. B. Saunders, 1971), p. 269. John Phillips, a philosopher and ecologist at St. Cloud State University in Minnesota, apparently developed his ideas in 1974 and presented them in "On Environmental Ethics," read at American Philosophical Association meeting, San Francisco, 1978, unpublished ms.

[53] Shepard, *Tender Carnivore*, pp. 260-273.

[54] Gary Snyder, "The Etiquette of Freedom," in Snyder, *The Practice of the Wild*, forthcoming.

[55] For the bioregional position, see Peter Berg, ed., *Reinhabiting a Separate Country: A Bioregional Anthology of Northern California* (San Francisco: Planet Drum Foundation, 1978); Gary Snyder, "Re-inhabitation," in Gary Snyder, *The Old Ways* (San Francisco: City Lights Books, 1977), pp. 57-66;

Thomas Berry, "Bioregions: The Context for Reinhabiting the Earth," in Berry, *Dream of the Earth*, pp. 163-170; Kirkpatrick Sale, *Dwellers in the Land: The Bioregional Vision* (San Francisco: Sierra Club Books, 1985).

[56]Nash, *Wilderness*, pp. 380-384.

[57]Nash, *Rights of Nature*, pp. 270-271 (footnote 28).

[58]For a proposal for restructured ecological cities, see Peter Berg, Beryl Magilavy, Seth Zuckerman, *A Green City Program* (San Francisco: Planet Drum Books, 1989); see also the excellent discussions and suggestions for an ecological approach to cities in Lewis Mumford, *The City in History: Its Origins, Its Transformations, and Its Prospects* (New York: Harcourt, Brace, and World, 1961); Theodore Roszak, *Where the Wasteland Ends: Politics and Transcendence in Postindustrial Society* (Garden City, New York: Doubleday, 1973); Theodore Roszak, *Person/Planet: The Creative Disintegration of Industrial Society* (Garden City, New York: Doubleday, 1978).

[59]Nash, *Rights of Nature*, pp. 168-169.

[60]Taylor, *Respect for Nature*, pp. 288, 310.

[61]Taylor, *Respect for Nature*, pp. 269-277; for the Naess/Sessions deep ecology platform, see Devall and Sessions, *Deep Ecology*, pp. 69-73; Arne Naess, "The Deep Ecological Movement: Some Philosophical Aspects," *Philosophical Inquiry* 8, (1986): 10-31; see also Arne Naess, "Deep Ecology and Life Style," in Neil Evernden, ed., *The Paradox of Environmentalism* (Ontario: York University, 1984), pp. 57-60.

[62]Taylor, *Respect for Nature*, pp. 53-58.

[63]For a powerful critique of the stewardship model as applied to agriculture, see Sara Ebenreck, "A Partnership Farmland Ethic," *Environmental Ethics* 5, 1, (1983): 33-45; for a discussion of "mixed communities" issues, see Arne Naess, "Self Realization in Mixed Communities of Humans, Bears, Sheep and Wolves," *Inquiry* 22, (1979): 231-242; Arne Naess and Ivar Mysterud, "Philosophy of Wolf Policies I: General Principles and Preliminary Exploration of Selected Norms," *Conservation Biology* 1, 1, (1987): 22-34.

[64]International Union for Conservation of Nature and Natural Resources (IUCN) with the United Nations Environmental Program (UNEP) and the World Wildlife Fund (WWF). *World Conservation Strategy: Living Resource Conservation for Sustainable Development* (Gland, Switzerland, 1980); Stan Croner, *An Introduction to the World Conservation Strategy* (San Francisco: Friends of the Earth Books, 1983); Brundtland Report issued by the World Commission on Environment and Development: Gro Harlem Brundtland, chairperson, *Our Common Future* (Oxford, Oxford University Press, 1987); Arne Naess, "Sustainable Development and the Deep Long-Range Ecology Movement," *The Trumpeter* 5, 4, (1988): 138-142 [revised version to appear in Ronald Engel, ed., *Ethics, Culture, and Sustainable Development*, forthcoming]; Arne Naess, "Ecosophy, Population, and Free Nature," *The Trumpeter* 5, 3, (1988): 113-119.

[65]Grumbine, "Ecosystem Management for Native Diversity," p. 48, 52-53.

[66]Grumbine, ibid.: 53; R. F. Noss, "Recipe for Wilderness Recovery"; J.J. Berger, *Restoring the Earth: How Americans Are Working to Renew Damaged Environments* (New York: Alfred Knopf, 1985).

[67]Naess, "Ecosophy, Population, and Free Nature," p. 118.

[68]Holmes Rolston III, "In Defense of Ecosystems," *Garden* (New York Botanical Garden) 12, 4, (1988): 2-5, 32.

[69]Paul Ehrlich, "Comments," *Defenders of Wildlife*, Nov/Dec., (1985); Anne and Paul Ehrlich, *Earth* (New York: Franklin Watts, 1987), p. 242.

[70]Thomas Fleischner, "Keeping It Wild," p. 79.

[71]For proposals to eliminate domestic grazing on public lands, see Denzel and Nancy Ferguson, *Sacred Cows at the Public Trough* (Bend, OR: Maverick Publications, 1983); Edward Abbey, "Free Speech: The Cowboy and His Cow," in Abbey, *One Life at a Time, Please* (New York: Henry Holt and Co., 1988), pp. 9-19.

[72]See *U. S. News and World Report*, "The Battle for the Wilderness," p. 107, 1, July 3, (1989): 16-21, 24-25; for the Earth First! wilderness proposals, see Dave Foreman and Howie Wolke, *The Big Outside* (Tucson, AZ: Ned Ludd Books, 1988).

[73]Grumbine, "Ecosystem Management for Native Diversity," p. 51.

[74]For excellent summaries of the current human overpopulation situation, see Anne and Paul Ehrlich, *Earth, Environment: An Introduction to Environmental Science* (Belmont, CA: Wadsworth Publishing Co., 1985) 4th ed., Part 3; Paul Ehrlich, Anne Ehrlich, and John Holdren, *Ecoscience: Population, Resources, Environment* (San Francisco: Freeman, 1977); see also Naess, "Ecosophy, Population, and Free Nature"; Naess, "Free Society and Deep Ecology."

[75]Raymond F. Dasmann, "National Parks, Nature, Conservation, and 'Future Primitive,' " *The Ecologist* 6 (1976); see also Raymond Dasmann, *Environmental Conservation* 4th ed., (New York: Wiley and Sons, 1979), chs. 16, 17.

[76]Arne Naess, "Japan's Second and Last Great Mistake," unpublished ms., 1989, pp. 1-10.

[77]Raymond F. Dasmann, *The Destruction of California* (New York: Macmillan, 1965).

[78]Geeta Dardick, "An Interview with Gary Snyder: When Life Starts Getting Interesting," *Sierra*, Sept/Oct. (1985): 68-73.

[79]Gerald Barney, *The Global 2000 Report to the U. S. President* (New York: Penguin, 1981); for criticism of Reagan's environmental policies, see Friends of the Earth, *Ronald Reagan and the American Environment: An Indictment* (San Francisco: Friends of the Earth Books, 1982); Jonathan Lash, Katherine Gillman and David Sheridan, *A Season of Spoils: The Story of the Reagan Administration's Attack on the Environment* (New York: Pantheon Books, 1984); Anne and Paul Ehrlich, *Earth*, ch. 8; for a critique of the Republican Party's abysmally poor track record on the environment, see Wallace Stegner, "The Legacy of Aldo Leopold," in J. Baird Callicott, ed., *Companion to a Sand County Almanac*, pp. 233-245.

[80]Fox, *John Muir*, pp. 306, 292-315, 322-329.

[81]Paul Ehrlich, *The Population Bomb* (New York: Ballantine Books, 1968), pp. 60-61, 152-157.

[82]Lester R. Brown, et. al. *State of the World, 1988: A Worldwatch Institute Report on Progress Toward a Sustainable Society* (New York, Norton, 1988); for a description of the "four levels" of environmental awareness: from

the "pollution level" to the "sustainable earth/deep ecology level," see G. Tyler Miller, *Living in the Environment*, p. 455.

[83]Karl Polanyi, *The Great Transformation* (Boston, Beacon Press, 1944), p. 72.

[84]Berry, "The Viable Human," pp. 77-78.

[85]Snyder, "Four Changes," p. 328.

[86]Anne and Paul Ehrlich, *Earth*, pp. 219-220.

[87]See W. R. Prescott, "The Rights of Earth: An Interview with Dr. Noel J. Brown," *In Context* no. 22, Summer (1989): 29-34.

[88]Naess, "Green Society and Deep Ecology," p. 16; World Charter for Nature. United Nations General Assembly (New York: United Nations, A/RES/37/7, Nov. 9, 1982); for further discussions of the World Charter for Nature, see Devall and Sessions, "The Development of Natural Resources," p. 321; Harold W. Wood, Jr., "The United Nations World Charter for Nature," *Ecology Law Quarterly* 12 (1985): 977-996.

[89]Arne Naess has made some provisional suggestions for approaches to Third World ecological problems in "Comments on the Article 'Radical American Environmentalism and Wilderness Preservation: A Third World Critique' by Ramachandra Guha," (1989), unpublished ms.

[90]Taylor, *Respect for Nature*, p. 312; for another discussion of the importance of "inner change" preceding social/political reform, see Jeremy Rifkin, *Time Wars: The Primary Conflict in Human History* (New York: Henry Holt and Co., 1987), p. 206.

## The Utility of Preservation and the Preservation of Utility, by Curt Meine

[1]Aldo Leopold, review of A.E. Parkins and J.R. Whitaker, *Our Natural Resources and Their Conservation, Bird-Lore* 39, 1 (January-February 1937): 74.

[2]Quoted in *Living Wilderness* 1, 1 (September 1935): 6.

[3]Susan Schrepfer, "Wildlife Ecologist," review of C. Meine, *Aldo Leopold: His Life and Work, Science* 241, 4870 (2 September 1988): 1237.

[4]Aldo Leopold, "Wherefore Wildlife Ecology," unpublished fragment, undated, University of Wisconsin Archives, Leopold Papers (LP) 10-6, 16.

[5]Susan Flader initiated the critical studies of Leopold with her seminal 1974 book *Thinking Like a Mountain: Aldo Leopold and the Evolution of an Ecological Attitude Toward Deer, Wolves, and Forests* (Columbia: University of Missouri Press). The 1987 centenary of Leopold's birth stimulated the publication of many articles and several books regarding Leopold and his work. Among the latter, see J. Baird Callicott, ed., *Companion to A Sand County Almanac* (Madison: University of Wisconsin Press, 1987); Curt Meine, *Aldo Leopold: His Life and Work* (Madison: University of Wisconsin Press, 1988); and Thomas Tanner, ed., *Aldo Leopold: The Man and His Legacy* (Ankeny, Iowa: Soil Conservation Society of America, 1987).

[6]Aldo Leopold, "To the Forest Officers of the Carson," 15 July 1913, *Carson Pine Cone*, July 1913, LP 10-11, 1.

[7]Leopold, "Address before the Albuquerque Rotary Club on Presentation of the Gold Medal of the Permanent Wild Life Protection Fund," c. July 1917, LP 10-8, 8.

[8] Leopold, "Social Consequences of Conservation" material, unpublished, c. 1930, LP 10-6, 16.

[9] Aldo Leopold, "Game and Wild Life Conservation," *Condor* 34, 2 (March-April 1932): 104.

[10] Ibid., 103-4.

[11] Aldo Leopold, "Game Cropping in Southern Wisconsin," *Our Native Landscape* (December 1933); no page given.

[12] Aldo Leopold, *Game Management* (New York: Scribner's Sons, 1933), 3.

[13] Ibid., 21.

[14] Ibid., 403.

[15] Ibid., 404.

[16] Ibid., 421-422.

[17] Aldo Leopold, "A Biotic View of Land," *Journal of Forestry* 37, 9 (September 1939): 727.

[18] Aldo Leopold, "The State of the Profession," *Journal of Wildlife Management* 4, 3 (July 1940), 344.

[19] Aldo Leopold, "The Outlook for Farm Wildlife," *Transactions of the Tenth North American Wildlife Conference* (Washington, D.C.: American Wildlife Institute, 1945), 168.

[20] Aldo Leopold, *A Sand County Almanac with Essays on Conservation from Round River* (New York: Ballantine Books, 1966), 20.

[21] Aldo Leopold, "The Wilderness and Its Place in Forest Recreation Policy," *Journal of Forestry* 19, 7 (November 1921): 718-19.

[22] Ibid., 719.

[23] Ibid., 721.

[24] See Donald Baldwin, *The Quiet Revolution: The Grass Roots of Today's Wilderness Preservation Movement* (Boulder: Pruett Publishing Company, 1972), 11-30; Roderick Nash, *Wilderness and the American Mind*, 3rd ed., (New Haven: Yale University Press, 1982), 185-86; Susan Flader, review of *The Quiet Revolution*, *Journal of Forest History* 18, 1-2 (April 1974): 36; Lawrence Rakestraw, "News, Comments, and Letters," *Journal of Forest History* 19, 1 (January 1975): 41; Roderick Nash, "Arthur Carhart: Wildland Advocate," *Living Wilderness* 44, 151 (December 1980): 32-34.

[25] Aldo Leopold, "The River of the Mother of God," unpublished ms., c. 1924, LP 10-6, 16.

[26] Aldo Leopold, "Conserving the Covered Wagon," *Sunset Magazine* 54, 3 (March 1925): 56.

[27] Ibid., 21.

[28] Aldo Leopold, "The Pig in the Parlor," *USFS Bulletin* 9, 23 (8 June 1925): 1-2.

[29] Aldo Leopold, "Wilderness as a Form of Land Use," *Journal of Land and Public Utility Economics* 1, 4 (October 1925): 398, 400.

[30] Ibid., 404.

[31] Ibid., 400.

[32] Aldo Leopold, "Why the Wilderness Society?," *Living Wilderness* 1, 1 (September 1935): 6.

[33] Aldo Leopold, "Planning for Wildlife," unpublished ms., 26 September 1941, LP 10-6, 16.

[34] Aldo Leopold, "Wilderness as a Land Laboratory," *Living Wilderness* 6, (July 1941): 3.

[35] Aldo Leopold, *A Sand County Almanac*, 264-265.

[36] Aldo Leopold, "Some Fundamentals of Conservation in the Southwest," c. March 1923, LP 10-6, 16. Published in *Environmental Ethics* 1, 2 (Summer 1979): 131-141.

[37] Ibid.

[38] Ibid.

[39] Aldo Leopold, "The Conservation Ethic," *Journal of Forestry* 31, 6 (October 1933): 639.

[40] Ibid., 641-642.

[41] Aldo Leopold, "Land Pathology," unpublished ms., 15 April 1935, LP 10-6, 16.

[42] Ibid.

[43] Ibid.

[44] Ibid.

[45] Aldo Leopold, "Notes on new theory of conservation," undated, c. 1941, LP 10-6, 16.

[46] Aldo Leopold, "The Farmer as a Conservationist," *American Forests* 45, 6 (June 1939), 298.

[47] Aldo Leopold, *A Sand County Almanac*, 210.

[48] Aldo Leopold, "Engineering and Conservation," unpublished ms., 11 April 1938, LP 10-6, 16.

[49] See J. Baird Callicott, "Leopold's Land Aesthetic," *Journal of Soil and Water Conservation* 38, 4 (July-August 1983): 329-332; and "The Land Aesthetic," *Environmental Review* 7 (Winter 1983): 345-358.

[50] Aldo Leopold, "Economics, Philosophy, and Land," unpublished lecture notes, 23 November 1938, LP 10-6, 16.

[51] Aldo Leopold, *A Sand County Almanac*, 190.

[52] Aldo Leopold, untitled fragment, c. 1946, LP 10-6, 16.

[53] Aldo Leopold, "Foreword."

[54] Aldo Leopold, *A Sand County Almanac*, 240.

[55] David Ehrenfeld, "Life in the Next Millennium: Who Will Be Left in the Earth's Community?," *Orion Nature Quarterly* 8, 2 (Spring 1989): 7.

[56] Ibid., 13.

[57] Ramachandra Guha, "Radical American Environmentalism and Wilderness Preservation: A Third World Critique," *Environmental Ethics* 11, 1 (Spring 1989): 79.

## Perceiving the Good, by Erazim Kohák

[1] Max Scheler, *Der Formalismus in der Ethik und die meteriale Wertethik*, (Berne, A. Francke AG Verlag; 1966). *Formalism in Ethics and Non-Formal Ethics of Value*, trans. Manfred S. Frings and Roger L. Funk, (Evanston, Northwestern University Press; 1973). Scheler articulates this theme most explicitly in the Preface to the Second Edition (p. xxiii) and repeats it throughout, as pp. 12-23, 48-80, *et passim*.

[2] Hans-Georg Gadamer, *Wahrheit und Methode*, (Tübingen, J. C. B. Mohr; 1960). *Truth and Method*, (New York, Crossroads Publishing Company; 1986), pp. 39-72.

[3]Edmund Husserl, *Allgemeine Einführung in die reine Phänomenologie*, (The Hague, Martinus Nijhoff; 1950), p. 11 *et passim. Ideas: General Introduction to Pure Phenomenology*, trans. W. R. Boyce Gibson, (New York, Collier; 1962), pp. 45-46. Dorion Cairns, *Guide for Translating Husserl*, (The Hague, Martinus Nijhoff; 1973), suggests *presentive intuition* as preferable translation. That is the term used in F. Kersten's translation, *Ideas Pertaining to a Pure Phenomenology*, (The Hague, Martinus Nijhoff; 1983), pp. 5-6.

[4]William James, *The Principles of Psychology*, (New York, Henry Holt and Co., 1890), vol. I, esp. pp. 271 and 459-64.

[5]Edmund Husserl, *Die Krisis der europäischen Wissenschaften*, (The Hague: Martinus Nijhoff; 1962), §§ 28-34, pp. 105-35 *et passim. The Crisis of European Sciences*, trans. David Carr, (Evanston, Northwestern University Press; 1970), pp. 103-35.

[6]Jan Patočka, "Edmund Husserl's Philosophy of the Crisis of Science and His Conception of a Phenomenology of the 'Life-World,' " trans. Erazim Kohák, *Husserl Studies*, vol. 2, no. 2, (1985), pp. 139-55. See esp. p. 153.

[7]While this theme runs throughout Merleau-Ponty's work, one of the clearest statements is the lecture, "The Primacy of Perception and its Philosophical Consequences," trans. James M. Edie, in James M. Edie, ed., *The Primacy of Perception*, (Evanston, Northwestern University Press; 1964), pp. 12-42.

[8]Martin Heidegger, *Sein und Zeit*, (Tübingen, Neomarius Verlag; 1958), pp. 52-75. *Being and Time*, trans. John Macquarrie and Edward Robinson, (New York, Harper and Brothers; 1962), pp. 78-106.

[9]Edmund Husserl, *Phänomenologische Untersuchungen zur Konstitution*, (The Hague, Martinus Nijoff; 1952), pp 143-71.

[10]Paul Ricoeur, *Le volontaire et l'involontaire*, (Paris: Aubier; 1963), p. 420. *Freedom and Nature: The Voluntary and the Involuntary*, trans. Erazim Kohák, (Evanston, Northwestern University Press; 1966), pp. 447-48.

[11]A bibliography of works devoted to the topic is massive. As one excellent example among many, we might cite Paul Taylor's rigorously argued volume, *Respect for Nature*, (Princeton, Princeton University Press; 1986), which introduces the concept of inherent worth and generates a set of rules for conduct that would be difficult to fault.

[12]This argument again appears in an entire range of writers, though the term is taken from Morris Berman, *The Reenchantment of the World*, (Ithaca: Cornell University Press; 1981), where it is elaborated on pp. 69-132.

[13]Cited in Erazim Kohák, *The Embers and the Stars*, (Chicago: University of Chicago Press; 1984), p. x.

[14]Of the many examples available I would select Richard Rorty's praise of "edifying" philosophy in his *Philosophy and the Mirror of Nature*, (Princeton, Princeton University Press; 1979), pp. 357-94. Rorty is clearly guided not by a trivial nihilism but by a noble attempt to provide a matrix of relative value for a world in which a quest for absolute worth appears to him vain for reasons cited.

[15]This is the "objectivism" with which Jan Patočka charges Edmund Husserl's friend Tomaś Masaryk whose diagnosis of the crisis of European humanity parallels Husserl's but who can recommend a turn to "objectivity," as against Husserl's turn to the subject, because he perceives that objectivity as value

laden and meaningfully ordered. So Jan Patočka, "Spiritual Crisis of Europe:
European Humanity in Husserl and Masaryk," trans. Erazim Kohák, in *On
Masaryk*, J. Novák, ed., (Amsterdam: Rodopi Press; 1988), pp. 97-110.
[16]Husserl, *Krisis der europäischen Wissenschaften*, p. 265; *Crisis of European
Sciences*, p. 262.
[17]Edmund Husserl, *Zur Phänomenologie der Intersubjektivität: Dritter Teil*,
(The Hague, Martinus Nijoff; 1973), esp. pp. 371-78 and 606-40. Note
reference to "dormant monads," p. 609.
[18]Ibid., pp. 378-406; see also Edmund Husserl, *Ideen II: Phänomenologische
Studien zur Konstitution*, Part III, pp. 211-46.
[19]Of texts presently available in English, the "Warsaw Lecture," cited above,
note 6, states the case most clearly. Extensive additional text will become
available shortly in Erazim Kohák, *Jan Patočka*, (Chicago: University of
Chicago Press; publication scheduled for summer, 1989.)
[20]Edmund Husserl, *Intersubjektivität III* (note 17 above), p. 377.
[21]Neil Evernden, *The Natural Alien*, (Toronto, University of Toronto Press;
1985), pp. 36ff.
[22]Richard Taylor, *With Heart and Mind*, (New York, St. Martin's Press; 1973).
I have made my contribution to the genre in *The Embers and the Stars*, note
13 above.
[23]Cf. "...dichterisch wohnet der Mensch..." in *Vorträge und Aufsätze*,
(Pfullingen, Neske; 1954), pp. 181-98; "...Poetically Man Dwells..." in
Albert Hoffstadter, ed. and trans., *Martin Heidegger: Poetry, Language,
Thought*, (New York, Harper and Row; 1971), pp. 211-29.
[24]*Was heißt Denken*, (Tübingen, Max Niemeyer Verlag; 1954), p. 50; *What Is
Called Thinking*, trans. F. D. Wieck and J. Glenn Grey, (New York, Harper
and Row; 1968), pp. 16-17.
[25]Cf. Emmanuel Levinas, "As If Consenting to Horror," *Critical Inquiry*, vol.
15, No. 2, (Winter 1989), pp. 485-88, esp. p. 487. See also contributions on
the topic by Hans-Georg Gadamer, Jürgen Habermas, Jacques Derrida,
Maurice Blanchot, Philippe Lacoue-Labarthe in ibid., pp. 407-84.
[26]"Brief über den Humanismus," *Wegmarken*, pp. 311-60; "Letter on Humanism," trans. Frank A. Capuzzi, in *Martin Heidegger: Basic Writings*, D. F.
Krell, ed., (New York, Harper and Row; 1977), pp. 189-242.
[27]Hans Jonas, *Das Prinzip Verantwortung*, (Frankfurt, Insel Verlag; 1979); *The
Imperative of Responsibility*, trans. Hans Jonas with David Herr, (Chicago,
The University of Chicago Press; 1984).
[28]While a number of such works appeared, the classic may well by Karl Jaspers,
*Die Atombombe und die Zukunft des Menschen*, (Munich: R. Piper and Co;
1958); *The Future of Mankind*, trans. E. B. Ashton, (Chicago, The University
of Chicago Press; 1961).

# A Brittle Thesis: A Ghost Dance: A Flower Opening,
by Michael P. Cohen

[1]"Question Authority," a paper given at "Solving Environmental Problems: The
Past as Prologue to the Present," a conference sponsored by the American
Society for Environmental History, April 27-30, 1989.

[2]Joseph W. Meeker, *The Comedy of Survival: Studies in Literary Ecology* (New York: Charles Scribner's Sons, 1974). We can invent ourselves according to two plot models, Meeker suggests: "The tragic man takes his conflict seriously, and feels compelled to affirm his mastery and his greatness in the face of his own destruction." (p. 22) The comic man "is durable even though he may be weak, stupid, and undignified." (p. 23) The lessons of comedy are humility and endurance. (p. 39) Comic man knows nobody cares about him. As picaresque hero, he is an outlaw, he is an opportunist, and he is "pushy." He perceives the world as dangerous. I believe that most of the heroes of the post World War II wilderness movement invented themselves on the tragic model, and that the movement itself has, as a result, invented itself through the use of the tragic plot.

[3]Clifford Geertz, *Works and Lives: The Anthropologist as Author* (Stanford, California: Stanford University Press, 1988). Hayden White, *Metahistory: The Historical Imagination in Nineteenth Century Europe* (Baltimore: Johns Hopkins University Press, 1973); *Topics of Discourse: Essays in Cultural Criticism* (Baltimore: Johns Hopkins Press, 1978).

[4]From the concluding paragraph of "The Pond in Winter": "I lay down the book and go to my well for water, and lo! there I meet the servant of the Brahmin, priest of Brahma and Vishnu and Indra, who still sits in his temple on the Ganges reading the Vedas, or dwells at the root of a tree with his crust and water jug."

[5]Raymond Dasmann, "The Prairies that Vanished," *The Destruction of California*, pp. 58-76.

[6]Michael P. Cohen, *The Pathless Way: John Muir and the American Wilderness* (Madison: University of Wisconsin Press, 1984), p. 232.

[7]Alfred W. Crosby, *Ecological Imperialism: the Biological Expansion of Europe, 900-1900* (New York: Cambridge University Press, 1986), pp. 152-154.

[8]"If we confine the concept of weeds to species adapted to human disturbance, then man is by definition the first and primary weed under whose influence all other weeds have evolved." Jack R. Harlan, *Crops and Man* (1975), quoted by Crosby, p. 269.

[9]See Patricia Nelson Limerick, *The Legacy of Conquest: The Unbroken Past of the American West* (New York: W.W. Norton & Co., 1987), pp. 37-41, for this idea, as applied to Western American History.

[10]Crosby, pp. 165, 170, 280.

[11]Crosby, p. 170.

[12]Dolores LaChapelle, *Sacred Land, Sacred Sex, Rapture of the Deep: Concerning Deep Ecology—and Celebrating Life* (Silverton: Finn Hill Arts, 1988), pp. 96-100 and *passim*. No doubt I distort Dolores's views, and am sorry if I do. She is looking to the East and Taoism for possibilities, and I, *at this point in my thinking*, am considering limitations. I have always wondered about Finn Hill Arts. In 1975, spending two months in the Hindu Kush of Afghanistan, and reading *Finegans Wake*, I completed the last climb of our expedition, and we called it Wake End Fin, little knowing that the wonderful valley of the Bashgall River, a tributary of the Indus, where we lived among the people and enjoyed their hospitality and their mountains, would be napalmed and bombed into desolation by the Soviets.

[13]An Earth First! bumpersticker.

[14]Alfred Crosby sees the Ghost Dance as an ecological fantasy where the Amerindian, "facing absolute defeat in their long struggle with the whites, fashioned a new religion that foretold an immediate change as great as that of the preceding three hundred years: A whole new world, with the Amerindian dead come alive again and the buffalo and the wapiti and all the game in their former profusion, would come up from the west and slide over the surface of the present world...." (p. 271) Tragic plot transformed into comic plot?

[15]Crosby, p. 280. His source is Bjorn Kurten, *The Age of Mammals* (London: Weidenfeld & Nicolson, 1971). One point of deep ecology is to call attention to this fact.

[16]Crosby, p. 196. We have been brutal and callous, but our germs killed indigenous populations before we could.

[17]Crosby, p. 280.

[18]William Cronon, *Changes in the Land: Indians, Colonists, and the Ecology of New England* (New York: Hill and Wang, 1983).

[19]*Changes in the Land*, p. 4.

[20]*Changes in the Land*, p. 15.

[21]*Changes in the Land*, p. 15.

[22]*Changes in the Land*, p. 11.

[23]David Brower, "A New Decade and a Last Chance: How Bold Shall We Be?" *Sierra Club Bulletin* v. 45 no. 1 (January, 1960), pp. 3, 4.

[24]*This Is the American Earth*, p. 62.

[25]*From Berlin to Jerusalem* (New York: Schocken, 1980), pp. 172-173.

[26]Edith Jane Hadley, *John Muir's Views of Nature and their Consequences*, (diss., University of Wisconsin, 1956), p. 453.

[27]*My First Summer in the Sierra* (Boston: Houghton Mifflin, 1911), p. 336.

[28]"A Near View of the High Sierra," in *The Mountains of California* (Boston: Houghton Mifflin, 1894).

[29]*First Summer*, pp. 326-341; *Pathless Way*, pp. 358-361.

[30]Leigh Ortenburger, *A Climber's Guide to the Teton Range* (San Francisco: Sierra Club Books, 1956), p. 50.

[31]*Wilderness and the American Mind*, p. 201. Nash is referring, of course, to Bob Marshall.

[32]*Pathless Way*, p. 96.

[33]This is an interesting aspect of the mountain. One climbs not simply the mountain, but its history. And one knows the history from stories one has listened to, late at night, around campfires, and from reading Leigh Ortenburger's guide, which is itself a set of geographically organized stories, plots, narratives of the climb one has not made, or one has made and remembers.

[34]He spoke, of course, as he wrote. So it was that I listened to Leigh, and was guided, while he allowed me to lead all of the classic pitches on the route. And I had to smile every so often because I did not know whether I was reading or listening while climbing the route.

[35]*Sacred Land*, p. 56.

[36]This is Muir's language in "A Near View of the High Sierra."

[37]*A Climbers Guide*, p. 123.

[38]Thomas H. Jukes to Bill Devall, September 17, 1969. See my *History of the Sierra Club*, p. 101. The occasion of this letter by Tom Jukes is complex. Norman Clyde was a guide for the Sierra Club, and for me. Jukes admired him, and wrote his obituary for the American Alpine Club. Jukes was himself a climber. (The Sierra Club published Leigh Ortenburger's guide to the Tetons.) As historian of the Club, I was guided by the stories of Club members of the past. Now I had been guided by Leigh, historian of the Tetons. As historian of wilderness, Nash was guided by Kimbrough. (Did they argue about history on the ascent?) Ortenburger and Nash discussed and debated the history of the region all along their walk down Garnet Canyon, after we had completed our climb.

[39]*Legacy of Conquest*, p. 25.

[40]Crosby, *Ecological Imperialism*, p. 308.

[41]Francis P. Farquhar, *History of the Sierra Nevada* (Berkeley: University of California Press, 1965), pp. 156-158.

## The Disembodied Parasite and Other Tragedies; or: Western Philosophy and How to Get Out of It, by Pete Gunter

[1]René Descartes. *Discourse on Method*. Trans. L. J. Lafleur. Indianapolis: Bobbs-Merrill, 1960, p. 46.

[2]Descartes, p. 45.

[3]Walter Prescott Webb. *The Great Frontier*. London: Secker and Warburg, 1953, 434. Cf. especially Ch. I, pp. 1-28.

[4]Webb, p. 37.

[5]Alfred North Whitehead. *Science and the Modern World*. New York: Macmillan, 1953, p. 8.

[6]Ilya Prigogine and Isabel Stengers. *Order Out of Chaos: Man's New Dialogue with Nature*. New York: Bantam Books, 1988, p. 349.

[7]Henri Bergson. "Dreams" in *Mind-Energy*. Trans. H. Wildon Carr. New York: Henry Holt and Company, 1920, pp. 104-133.

## Not Laws of Nature but Li (Pattern of Nature), by Dolores LaChapelle

[1]Luther Standing Bear, *Land of the Spotted Eagle* (Boston, 1933).

[2]Robert Temple, *The Genius of China: 3000 Years of Science, Discovery and Invention* (New York: Simon and Schuster, 1986), p. 9.

[3]E. R. Hughes, *The Great Learning and the Mean-in-Action* (New York: E. P. Dutton, 1934), pp. 4, 16, 17 and 18.

[4]E. R. Hughes. "Epistemological Methods in Chinese Philosophy," in C. A. Moore, ed., *The Chinese Mind* (Honolulu: East-West Center Press, 1967).

[5]Joseph Needham, *The Grand Titration: Science and Society in East and West* (London: Allen & Unwin, 1979), p. 299.

[6]Needham, pp. 301-305.

[7]Robert Payne, *The White Pony* (New York: New American Library), n.d., p. 24.

[8]Joseph Needham, *Science and Civilization in China*, v. 2 (Cambridge: Cambridge University Press), p. 581.

[9] Needham, *The Grand Titration*, p. 322.

[10] Ibid. p. 312. Further explanation of the word, *li*, is provided by Needham on p. 331 where he states that the three terms are homophones and defines them as follows: "*li*(a) good customs, ceremonial observances, ethical behaviour; *li*(b) organic pattern at all levels in the cosmos; *li*(c) calendrical science based on observational astronomy."

[11] Wei-ming Tu, "*Li* as a Process of Humanization," *Philosophy East and West* 22, 2 (April 1972), pp. 187-200.

[12] Robert M. Gimello, "The Civil Status of *Li* in Classical Confucianism," *Philosophy East and West* 14, 1, pp. 202-211.

[13] Needham, *The Grand Titration*, pp. 321-322.

[14] Toshihiko Izutsu, "The Temporal and A-temporal Dimensions of Reality in Confucian Metaphysics," *Eranos Jahrbach* 43 (1974), pp. 411-447.

[15] Dolores LaChapelle, *Sacred Land Sacred Sex: Concerning Deep Ecology and Celebrating Life* (Silverton: Finn Hill Arts, 1988), pp. 88-101.

[16] Needham, *Science and Civilization*, v. 2, p. 453.

[17] Wm. T. de Bary, *The Unfolding of Neo-Confucianism* (New York: Columbia University Press, 1975), p. 6.

[18] Needham, *Science and Civilization*, v. 2, p. 458.

[19] Stephen Toulmin, "Afterword: The Charm of the Scout," in C. Wilder and J. Weakland, eds., *Rigor and Imagination: Essays from the Legacy of Gregory Bateson* ( New York: Praeger Publishers, 1982), p. 365.

[20] Needham, *Science and Civilization*, v. 2, pp. 460-466.

[21] John M. Koller, *Oriental Philosophy* (New York: Scribners, 1985), p. 310.

[22] Needham, *Science and Civilization*, v. 2, p. 474.

[23] R. G. H. Siu, *Chi: A Neo-Taoist Approach to Life* (Cambridge: MIT press, 1974), p. 257.

[24] Needham, *Science and Civilization*, v. 2, p. 475.

[25] Ibid., p. 567.

[26] Ibid., p. 484.

[27] Izutsu, p. 440.

[28] Needham, p. 567.

[29] Ibid., p. 582.

[30] Ibid., p. 561.

[31] Gimello, "Civil Status of *Li*."

[32] Wei-ming Tu, "*Li* as a Process of Humanization."

[33] Silas Goldean (pseudonym), "The Principle of Extended Identity," *Pantheist Practice*, n.d. Arne Naess, in his diagram of deep ecology, puts Self-realization as his top or ultimate norm, from which all the norms below derive. This "Self" he identifies with "not only the ecosphere but even with the entire universe." Arne Naess, "Ecosophy T" in Bill Devall and George Sessions, *Deep Ecology* (Layton: Gibbs Smith, 1985), pp. 225-228.

[34] Gimello, "Civil Status of *Li*."

[35] Vance Martin and Mary Inglis, eds., *Wilderness: The Way Ahead* (Middleton, WI: Lorian Press, 1984).

[36] Konrad Lorenz, "The Fashionable Fallacy of Dispensing with Description" and "The Enmity Between Generations and Its Probable Ethological Causes,"

in Richard Evans, *Konrad Lorenz: The Man and His Ideas* (New York: Harcourt Brace and Jovanovich, 1975), pp. 152-179 and 218-269.

[37]C. G. Jung, *The Undiscovered Self* (Atlantic Monthly Press, 1957).

[38]Hughes, "Epistemological Methods in Chinese Philosophy."

[39]Gregory Bateson, *Mind and Nature: A Necessary Unity* (New York: E. P. Dutton, 1979), p. 13.

[40]James Gleick, *Chaos: Making a New Science* (New York: Penguin, 1988), p. 6.

[41]Ibid., p. 5.

[42]Ibid., pp. 299 and 308.

[43]Bill Mollison, *Permaculture*, p. 83.

[44]Ibid., p. 89.

[45]Gleick, p. 104.

[46]Izutsu, "Temporal and A-temporal Dimensions."

[47]Ibid.

[48]Gregory Bateson, "The Logical Categories of Learning and Communication," in G. Bateson, *Steps to an Ecology of Mind* (New York: Ballantine Books, 1972), pp. 279-308.

[49]Izutsu, "Temporal and A-temporal Dimensions."

[50]Dolores LaChapelle, *Earth Wisdom* (Silverton: Finn Hill Arts, 1984), pp. 19-23.

[51]L. Wieger, *Chinese Characters* (New York: Dover, 1965), p. 239.

[52]The Japanese Shugendo sect hold ceremonies on Mt. Haguro in which two monks spend time on the mountain to discover the "root traces." Togawa explains: "Heaven and earth and the natural seasons 'move' according to a prescribed order, and according to this 'movement,' new phenomena or 'root-traces' are born . . . out of this primordial potentiality." See H. Byron Earhart, "Four Ritual Periods of Haguro Shugendo," *History of Religions* (v. 5), pp. 93-113.

[53]Andrew L. March, "An Appreciation of Chinese Geomancy," *Journal of Asia Studies* 27, 2 (February 1969), pp. 253-267.

[54]Roy Rappaport, *Pigs for the Ancestors* (New Haven: Yale University Press, 1968).

[55]G. Reichel-Dolmatoff, "Cosmology as Ecological Analysis: A View from the Rain Forest," *Man: Journal of the Royal Anthropological Institute* (September 1978).

[56]Rappaport, *Pigs for the Ancestors*

[57]Roy Rappaport, *Ecology, Meaning and Religion* (Richmond: North Atlantic Books, 1979), pp. 100-101.

[58]Ibid., pp. 238.

[59]Bertha P. Dutton, *American Indians of the Southwest* (Albuquerque: University of New Mexico, 1983).

[60]Mollison, p. 97.

[61]Kuang-ming Wu, *Chuang Tzu: World Philosopher at Play* (New York: Scholar's Press, 1982), p. 115.

## The Blessing of Otherness and the Human Condition, by Michael Zimmerman

[1]Cf. my essays, "Toward a Heideggerian Ethos for Radical Environmentalism," *Environmental Ethics*, 5 (Summer, 1983), 99-131; "Implications of Heidegger's Thought for Deep Ecology," *The Modern Schoolman*, LXIV (November, 1986), 19-43. Concerning deep ecology, cf. Bill Devall and George Sessions, *Deep Ecology: Living as if Nature Mattered* (Salt Lake City: Peregrine Smith Books, 1985). This book contains a very helpful bibliography of works in deep ecology.

[2]Cf., for example, Martin Heidegger, *The Question Concerning Technology and Other Essays*, trans. William Lovitt (New York: Harper & Row, 1977); *Discourse on Thinking* (translation of *Gelassenheit*), trans. John M. Anderson and E. Hans Freund (New York: Harper & Row, 1966).

[3]Cf. Martin Heidegger, "Letter on Humanism," in *Basic Writings*, trans. David Ferrell Krell (New York: Harper & Row, 1977).

[4]Cf. Karl Löwith, *Der Weltbegriff der neuzeitlichen Philosophie* (Heidelberg: Carl Winter, 1960); "Zu Heideggers Seinsfrage: Die Nature des Menschen und die Welt der Natur," *Aufsätze und Vorträge, 1930-1970* (Stuttgart: Verlag W. Kohlhammer, 1971).

[5]Cf. Hans Jonas, *The Gnostic Religion: The Message of the Alien God and the Beginnings of Christianity* (Boston: Beacon Press, 1963).

[6]On the issue of Heidegger's relation to National Socialism, cf. Michael E. Zimmerman, "Philosophy and Politics: The Case of Heidegger," *Philosophy Today*, XXXIII, No. 1 (Spring, 1989), pp. 3-20; and Zimmerman, "The Thorn in Heidegger's Side: The Question of National Socialism," *The Philosophical Forum*, XX, No. 4 (Summer, 1989), 326-365. For a larger treatment of the political dimension of Heidegger's thought, cf. Zimmerman, *Heidegger's Confrontation with Modernity: Technology, Politics, and Art* (Bloomington: Indiana University Press, 1990).

[7]On the topic of the "green" dimension of National Socialism, cf. Anna Bramwell, *Ecology in the 20th Century: A History* (New Haven: Yale University Press, 1989).

[8]Martin Heidegger, *Hölderlins Hymne "Germanien" und "Der Rhein,"* ed. Suzanne Ziegler, Vol. 39, Gesamtausgabe (Frankfurt am Main: Vittorio Klostermann, 1980), p. 195ff.

[9]On this topic, cf. Michel Haar, *Le chant de la terre* (Paris: Editions de l'Herne, 1987).

[10]Martin Heidegger, *Die Grundbegriffe der Metaphysik. Welt—Endlichkeit—Einsamkeit*, ed. Friedrich-Wilhelm von Herrmann, Vol. 29/30, Gesamtausgabe (Frankfurt am Main: Vittorio Klostermann, 1983), p. 282ff.

[11]Susan Griffin, *Woman and Nature: The Roaring Inside Her* (New York: Harper & Row, 1978).

[12]Ibid., pp. 95-96.

[13]Ibid., p. 104.

[14]Ibid., p. 150.

[15]Ibid.

[16]Ibid., p. 190.

[17] Ibid., p. 175.

[18] Ibid., p. 227.

[19] Cf., for example, Ariel Kay Salleh, "Deeper than Deep Ecology: The Eco-Feminist Connection," *Environmental Ethics*, 6 (Winter, 1984), 339-345; Jim Cheney, "Eco-Feminism and Deep Ecology," *Environmental Ethics*, 9 (Summer, 1987), 115-145.

[20] For essays critical of essentialist eco-feminism, cf. Karen J. Warren, "Feminism and Ecology: Making Connections," *Environmental Ethics*, 9 (Spring, 1987), 3-20; Michael E. Zimmerman, "Feminism, Deep Ecology, and Environmental Ethics," *Environmental Ethics*, 9 (Spring, 1987), 21-44; Warwick Fox, "The Deep Ecology-Ecofeminism Debate and Its Parallels," *Environmental Ethics*, 11 (Spring, 1989), 5-25. Cf. also Dolores LaChapelle, *Sacred Land, Sacred Sex, Rapture of the Deep: Concerning Deep Ecology—and Celebrating Life* (Silverton: Finn Hill Arts, 1988).

[21] Recently, the nature-photographer and author Steven J. Meyers has written two excellent essays on humanity's place in nature. Cf. *On Seeing Nature* (Golden, Colorado: Fulcrum, 1987) and *Lime Creek Odyssey* (Boulder, Colorado: Fulcrum, 1989).

[22] On the phenomenon of recollectivization in National Socialism, cf. Erich Neumann, *The Origins and History of Consciousness*, trans. R. F. C. Hull (Princeton: Princeton/Bollingen, 1970), and Erich Fromm, *Escape from Freedom* (New York: Avon Books, 1965).

[23] On the problems involved with the tendency of the ego to dissociate itself from the body, cf. Ken Wilber, *Up From Eden: A Transpersonal View of Human Evolution* (Boulder: Shambhala, 1983).

[24] LaChapelle, in *Sacred Land*, makes an important attempt to reconcile the perceived need to regain a relationship with the wild, on the one hand, with the need to move beyond the level of ego-consciousness, without sacrificing what is positive about such ego-consciousness. LaChapelle, in other words, seems to be searching for her own way to describe what Jung spoke of as authentic "individuation," that integrated mode of selfhood in which the collective, natural, and egoistic dimensions of a person are harmonized.

[25] Hans Peter Duerr, *Dreamtime: Concerning the Boundary between Wilderness and Civilization*, trans. Felicitas Goodman (New York: Basil Blackwell, 1985), p. 125. This remarkable book has 133 pages of text and 324 pages of footnotes and bibliography! My thanks to Dolores LaChapelle for having recommended this book.

[26] Duerr, *Dreamtime*, p. 64.

[27] For more on Heidegger's concept of authentic temporality, cf. my book, *Eclipse of the Self: The Development of Heidegger's Concept of Authenticity*, 2nd edition (Athens: Ohio University Press, 1986).

[28] Ibid., p. 121. For an outstanding comparison of Heidegger and Eckhart, cf. John D. Caputo, *The Mystical Element of Heidegger's Thought* (Athens: Ohio University Press, 1978).

[29] Ibid., p. 102.

[30] I am indebted to Steven J. Meyers for reminding me to emphasize the continuity between humans and the rest of the natural world.

[31] Duerr discusses Castaneda's books at some length. In my view, one of the

best of those books is *Journey to Ixtlan* (New York: Simon and Schuster, 1972).

[32]Ibid., p. 133.

[33]Dolores LaChapelle, *Earth Wisdom* (Silverton, Colorado: Finn Hill Arts, 1978) and *Sacred Land, Sacred Sex, Rapture of the Deep.*

[34]Cf. Paul Shepard, *Man in the Landscape* (New York: Ballantine, 1967); *The Tender Carnivore and the Sacred Game* (New York: Scribners, 1973); *Nature and Madness* (San Francisco: Sierra Club Books, 1982).

## Wilderness, Language, and Civilization, by Max Oelschlaeger

[1]Marjorie Grene, "The Paradoxes of Historicity," in Brice R. Wachterhauser, ed. *Hermeneutics and Modern Philosophy* (Albany: State University of New York Press, 1986), 182.

[2]See Anna Bramwell, *Ecology in the 20th Century: A History* (London: Yale University Press, 1989), especially 248.

[3]Paul Ricoeur, *Main Trends in Philosophy* (New York: Holmes and Meier Publishers, Inc., 1979), 269.

[4]Among other recent works see Loyal Rue, *Amythia: Crisis in the Natural History of Western Culture* (Tuscaloosa: University of Alabama Press, 1989), especially 30-59; Frederick Turner, *Natural Classicism: Essays on Literature and Science* (New York: Paragon House Publishers, 1985), especially 3-58; and Dolores LaChapelle, *Sacred Land Sacred Sex—Rapture of the Deep: Concerning Deep Ecology—And Celebrating Life* (Silverton: Finn Hill Arts, 1988), especially 22-27. Also see David Bohm, *Wholeness and the Implicate Order* (New York: Routledge and Kegan Paul Inc., 1980), *passim*, for a penetrating study of the interrelations between language, consciousness, and reality.

[5]See above, Paul Shepard, "A Post-Historic Primitivism," 79.

[6]Pete A. Y. Gunter, "The Big Thicket: A Case Study in Attitudes toward Environment," in William T. Blackstone, ed., *Philosophy and Environmental Crisis* (Athens: University of Georgia Press, 1974), 131-34.

[7]Neil Evernden, *The Natural Alien: Humankind and Environment* (Toronto: University of Toronto Press, 1985), 124.

[8]Martin Heidegger, *Poetry, Language, Thought,* trans. Albert Hofstadter, (New York: Harper and Row, Publishers, 1971), 6. See Michael Zimmerman, "Toward a Heideggerean *Ethos* for Radical Environmentalism," *Environmental Ethics* 5 (Summer 1983), for a revealing account of the relation of Heideggerean thought to deep ecology.

[9]See Stanley Diamond, *In Search of the Primitive: A Critique of Civilization,* 3rd printing, (New Brunswick: Transaction Books, 1987).

[10]Dolores LaChapelle, "Ritual Is Essential," in Bill Devall and George Sessions, *Deep Ecology* (Salt Lake City: Peregrine Smith Books, 1985), 250. Also see Dolores LaChapelle, *Earth Wisdom* (Silverton: Finn Hill Arts, 1978), 137-50; and LaChapelle, *Sacred Land,* 116-302.

[11]Some of the ideas herein are covered in greater detail in my forthcoming *The Idea of Wilderness* (New Haven: Yale University Press, 1991).

[12]See John Ray, *The Wisdom of God Manifested in the Works of Creation*, 12th ed., (London: John Rivington, John Ward, Joseph Richardson, 1759); and William Paley, *Natural Theology: Or, Evidences of the Existence and Attributes of the Deity, Collected from the Appearances of Nature* (London: R. Fauldner, 1802). Also see Clarence J. Glacken, *Traces on the Rhodian Shore: Nature and Culture in Western Thought from Ancient Times to the End of the Eighteenth Century* (Berkeley: University of California Press, 1967), for a penetrating discussion of physico-theology specifically, and more generally for an analysis of the emergence of Western culture's characteristic attitudes toward nature.

[13]Ralph Waldo Emerson, *Selected Writings of Ralph Waldo Emerson*, ed. William H. Gilman, (New York: New American Library, 1965), 217; 223.

[14]I attempt to avoid sexist language in my own writing, and I believe that such an effort is intellectually and ethically obligatory, although we cannot here argue those cases.

[15]One will recall both Kant's critique of the arguments for God's existence, and his dictum that concepts without percepts are empty, percepts without concepts blind, implying that natural phenomena cannot in principle be reduced solely to mind.

[16]See Donald Worster, *Nature's Economy: The Roots of Ecology* (San Francisco: Sierra Club Books, 1977), *passim* for a thorough analysis of these issues. Also see Ernst Mayr, *Toward a New Philosophy of Biology: Observations of an Evolutionist* (Cambridge: Harvard University Press, 1988), 161-264 for discussion of the implications of the evolutionary paradigm for the argument from design.

[17]Henry David Thoreau, *A Week on the Concord and Merrimack Rivers*, in Brooks Atkinson, ed., *Walden and Other Writings of Henry David Thoreau* (New York: The Modern Library, 1937), 301.

[18]*A Week*, 333.

[19]Turner, *Natural Classicism*, 178.

[20]*A Week*, 344; 356.

[21]Henry David Thoreau, *The Journal of Henry David Thoreau*, vol. 5, ed. Bradford Torrey and Francis H. Allen, (Gibbs M. Smith, Inc., 1984), 135. Originally published 1906.

[22]All *Walden* citations are from Henry David Thoreau, *Walden and Other Writings*, ed. Joseph Wood Krutch, (New York: Bantam Books, 1982), 332.

[23]Henry David Thoreau, *The Maine Woods*, foreword Richard F. Fleck, (New York: Harper and Row, Publishers, 1987), 95. Fleck and many others agree that the raw and elemental nature of Thoreau's trek through the Maine woods to Mt. Ktaadn galvanized his thinking into an evolutionary mode. (xxv)

[24]*Journal* 3, 164.

[25]*Walden*, 332-33.

[26]*Walden*, 331-32.

[27]*Journal* 4, 174.

[28]Lynn White, Jr., "The Historical Roots of Our Ecologic Crisis," in Ian G. Barbour, ed., *Western Man and Environmental Ethics* (Reading: Addison-Wesley Publishing Company, 1973), 24-25. "Historical Roots" originally appeared in *Science* 155 (March 1967).

[29]Norman Gottwald, *The Hebrew Bible: A Socio-Literary Introduction* (Philadelphia: Fortress Press, 1985), 607. Also see Hans-Georg Gadamer, *Truth and Method* (New York: Crossroad Publishing Company, 1988), 460-98.

[30]With reinforcement from the deep ecology movement, Muir's contemporary biographers have led the way in reappraising his work. See Stephen Fox, *John Muir and His Legacy: The American Conservation Movement* (Boston: Little, Brown and Company, 1981), reissued as *The American Conservation Movement: John Muir and His Legacy* (Madison: University of Wisconsin Press, 1985); Michael P. Cohen, *The Pathless Way: John Muir and American Wilderness* (Madison: University of Wisconsin Press, 1984); and Frederick Turner, *Rediscovering America: John Muir in His Time and Our Own* (New York: Viking Penguin Inc., 1985), also available in a Sierra Club Books edition, 1985.

[31]See e.g., Roderick Nash, *Wilderness and the American Mind*, 3rd ed., (New Haven, Yale University Press, 1982), 125.

[32]See above, Erazim Kohák, "Perceiving the Good," for a penetrating discussion of the reality of lived experience.

[33]And he saw too the psychological impoverishment of the death orthodoxy, the belief that humankind was endowed with some supernatural and perdurable soul. Rather, every day was a resurrection day. He celebrated the renewal of the everlasting process of nature in human consciousness.

[34]For discussion of the term "History" see above, Shepard, "Post-Historic Primitivism," 40-47. See Mircea Eliade, *The Sacred and the Profane: The Nature of Religion*, trans. Willard R. Trask, (New York: Harcourt Brace Jovanovich, Publishers, 1959), *passim*, for related discussion.

[35]Charles Hartshorne, *The Divine Relativity: A Social Conception of God* (New Haven: Yale University Press, 1948), 90.

[36]John Muir, *John of the Mountains: The Unpublished Journals of John Muir*, ed. Linnie Marsh Wolfe, (Madison: University of Wisconsin Press, 1979), 137-38; 153-54; 137-38.

[37]Rue, *Amythia*, 176.

[38]J. E. Barnhart, *Religion and the Challenge of Philosophy* (Totowa: Littlefield, Adams and Co., 1975), 152-153.

[39]John Muir, *Travels in Alaska* (Boston: Houghton Mifflin Company, 1979), 67-68. This passage can be read as consistent with a panentheistic perspective. However, on balance Muir is best read as a pantheist, given his denial of the "death orthodoxy" (i.e., the immortality of the soul in Heaven), his rejection of the catastrophist geology, and his bracketing of the idea that God created the human species to rule over the earth and its creatures.

[40]*Unpublished Journals*, 138.

[41]Samuel Alexander's *Space, Time, and Deity* (2 vols), (Gloucester: Peter Smith, 1979), originally published in 1920, remains largely unrecognized among contemporary scholars, who prefer the more comfortable evolutionary theism of Whitehead and Hartshorne. For a short statement of the arguments for evolutionary theism see David Ray Griffin, *God and Religion in the Postmodern World: Essays in Postmodern Theology* (Albany: State University of New York Press, 1989), 69-82. The fundamental problem with

all arguments that attempt to reconcile theism and evolution remains that stated by Ilya Prigogine: once we accept the reality of time, then there can be nothing outside that process. In other words, the reality of evolution or the irreversibility of time is inseparable from the meaning of existence. See Ilya Prigogine and Isabelle Stengers, *Order Out of Chaos: Man's New Dialogue with Nature* (New York: Bantam Books, Inc., 1984), *passim*. Also see Ilya Prigogine, *From Being to Becoming: Time and Complexity in the Physical Sciences* (New York: W. H. Freeman and Company, 1978); and Ilya Prigogine, "Man's New Dialogue with Nature," *Perkins Journal* 36 (Summer 1983).

[42] See above, Michael P. Cohen, "A Brittle Thesis: A Ghost Dance: A Flower Opening," 204.

[43] Alasdair MacIntyre, "Pantheism," in vol. 6 of *The Encyclopedia of Philosophy*, ed. Paul Edwards, (New York: Collier Macmillan Publishers, 1967), 35.

[44] See Evernden, *Natural Alien*, and Erazim Kohák, *The Embers and the Stars: A Philosophical Inquiry into the Moral Sense of Nature* (Chicago: University of Chicago Press, 1984), for critiques of those who desacralize the natural world by viewing it as an objective phenomenon only.

[45] Robinson Jeffers, *The Double Axe and Other Poems* (New York: Liveright, 1977), 105.

[46] J. Baird Callicott, *In Defense of the Land Ethic: Essays in Environmental Philosophy* (Albany: State University of New York Press, 1989), and Curt Meine, *Aldo Leopold: His Life and Work* (Madison: University of Wisconsin Press, 1988), are insightful studies that should help scholars reconcile these many interpretations.

[47] Alfred North Whitehead, *Science and the Modern World* (New York: Macmillan Publishing Co., Inc., 1967), 47.

[48] Joseph Campbell, *Mythologies of the Great Hunt*, vol. 1, part 2 of *The Way of the Animal Powers* (New York: Harper and Row, Publishers, 1988), xxiii.

[49] In this cultural awakening academic philosophy has played a minuscule role, being lost in the void of analysis and metaethics; as Richard Rorty argues in *Philosophy and the Mirror of Nature* (Princeton: Princeton University Press, 1979), the systematic rather than the edifying philosopher dominates the philosophical world. Only a handful of twentieth-century philosophers, like Heidegger, Whitehead, and Bergson, have done anything more than pay lip service to the land.

[50] Antihunters who assess Leopold's hunting as either inconsistent with or contradictory to his land ethic are mistaken (ignoring the inner incoherence of these positions in their own right) in that they ignore these most fundamental considerations of human nature. Antihunting critiques of Leopold also ignore the actual historical role that sport hunters have played in the American conservation movement. See John F. Reiger, *American Sportsmen and the Origins of Conservation* (New York: Winchester Press, 1975).

[51] Aldo Leopold, *Game Management* (Madison: University of Wisconsin Press, 1986), 403. *Game Management* was originally published in 1933.

[52] *Game Management*, 5. Of course, contemporary studies of archaic cultures provide ample empirical support for this observation, but these studies were

not available to Leopold. In part his observation was grounded in what might be termed an evolutionary deduction: unrestrained killing now leads to starvation later.

[53]*Game Management*, 391.

[54]See above, Shepard, "Post-Historic Primitivism," 79.

[55]Leopold's viewpoint is entirely consistent with the thesis suggested by Paul Shepard and others that the Neolithic revolution does not mark the rise of humankind but the beginning of the Fall. See Paul Shepard, *The Tender Carnivore and the Sacred Game* (New York: Charles Scribner's Sons, 1973).

[56]The sacred cows that roam India, denuding without restraint an already battered landscape, are perhaps the ultimate parody of an animal rights and antihunting philosophy; only an ecologically grounded viewpoint can even aspire to "the truth."

[57]Such a position itself is a corollary to a postmodernist paradigm and its fundamental presuppositions that (i) time is real and irreversible, (ii) process is evolutionary (i.e., emergent novelty is real and moves toward increasingly organized complexity), and (iii) inquiry is process revealing itself (i.e., rational process is an actual constituent of reality).

[58]Prigogine, *Order Out of Chaos*, 312. See also Rue, *Amythia*, 176-77.

[59]Aldo Leopold, *A Sand County Almanac, with Essays on Conservation from Round River* (New York: Random House, 1970), 189. *Sand County Almanac* was originally published in 1949.

[60]Wallace Stegner, "The Legacy of Aldo Leopold," in J. Baird Callicott, ed., *Companion to A Sand County Almanac, Interpretive and Critical Essays* (Madison: University of Wisconsin Press, 1987), 245.

[61]*Sand County Almanac*, 279.

[62]Heidegger, *Poetry, Language, Thought*, 213.

[63]*Poetry, Language, Thought*, 196-97.

[64]Most philosophers, ironically, have not taken the philosophical implications of science seriously, and instead have engaged in a rear guard rehashing of methodological issues long since irrelevant to the ongoing course of scientific inquiry.

[65]Campbell, *Mythologies of the Primitive Hunters*, 10.

[66]Devall and Sessions, *Deep Ecology*, 171. Sessions in particular has influenced me to see and read Snyder as the poet laureate of deep ecology. See George Sessions, "Gary Snyder: Post-Modern Man," forthcoming in a *festschrift* to be published on the occasion of Snyder's sixtieth birthday. Sessions interprets Snyder's postmodern vision as a fusion of theory and practice that brings together the old ways with Eastern philosophy and contemporary ecology.

[67]Gary Snyder, *The Back Country* (New York: New Directions, 1968), 82. Kenji's poems cut across time and place; for example his poem entitled "Some Views Concerning the Proposed Site of a National Park" undresses the hypocrisy that masquerades as enlightenment in so much of our own National Park "movement."

[68]Any critic necessarily interposes an interpretation between reader and poem, and furthermore inevitably adumbrates the meaning and significance of the poet's work. Further, Snyder has been a protean writer, and engaged during

the last twenty years in producing what may be his definitive poetic vision, a poem entitled "Mountains and Rivers without End." Surely this material will directly bear on any interpretation of his work.

[69]See the *Tao te Ching*, verse 1, Chapter 1.

[70]Gary Snyder, *The Old Ways: Six Essays* (San Francisco: City Lights Books, 1977), 35-36.

[71]Gary Snyder, *Turtle Island* (New York: New Directions, 1969), 41.

[72]*Turtle Island*, 106; 106; 104; 107; 108; 102.

[73]*Turtle Island*, 101.

[74]Gary Snyder, *The Real Work: Interviews and Talks 1964-1979*, ed. Wm. Scott McLean, (New York: New Directions Books, 1980), 85; 87.

[75]*Real Work*, 139. See Robert C. Paehlke, *Environmentalism and the Future of Progressive Politics* (New Haven: Yale University Press, 1989), for related reading, especially Chapter 6, Environmentalism as a System of Values. Paehlke's analysis of the basic values and concepts that lie at the heart of environmental politics (e.g., decentralization, appreciation of the web of life, sustainable economics) coincides remarkably with Snyder's, and both believe that environmentally committed constituencies are going to play an increasingly important role in the Western liberal-democracies.

[76]*Real Work*, 141; 86-87.

[77]Gary Snyder, *Earth House Hold: Technical Notes & Queries To Fellow Dharma Revolutionaries* (New York: New Directions Books,1957), 132.

[78]*The Old Ways*, 36-37.

[79]*Earth House Hold*, 127.

[80]Kohák, *Embers and the Stars*, 210

[81]See Holmes Rolston, III, *Philosophy Gone Wild: Essays in Environmental Ethics* (Buffalo: Prometheus Books, 1986), *passim*.

## Appendices, by Paul Shepard

### Appendix 1

[1]Allen W. Johnson and Timothy Earle, *The Evolution of Human Societies: From Foraging Group to Agrarian State*, Stanford University, 1987, p. 82.

[2]Ibid., p. 150.

[3]Richard B. Lee, "!Kung Spatial Organization: An Ecological and Historical Perspective," *Human Ecology* I (2), 125-147, 1972.

### Appendix 2

[1]Claude Lévi-Strauss, *The Savage Mind*, University of Chicago, 1966, Chap. IV.

# Notes on Contributors

**Michael Peter Cohen** is the author of *The Pathless Way: John Muir and the American Wilderness* (1984) and *The History of the Sierra Club: 1892-1970* (1988). He teaches English at Southern Utah State College (Cedar City, Utah).

**Pete A. Y. Gunter** is the author of *Bergson and the Evolution of Physics* (1969), *The Big Thicket: A Challenge for Conservation* (1972), and *River in Dry Grass* (1984). He teaches philosophy at the University of North Texas (Denton, Texas).

**Erazim Kohák** is the author of *Idea and Experience* (1978), *The Embers and the Stars: A Philosophical Inquiry into the Moral Sense of Nature* (1984), and *Jan Patočka* (forthcoming). He teaches philosophy at Boston University (Boston, Massachusetts).

**Dolores LaChapelle** is the author of *Earth Festivals* (1976), *Earth Wisdom* (1978), and *Sacred Land, Sacred Sex* (1988). She directs the Way of the Mountain experiential learning center in Silverton, Colorado.

**Curt Meine** is the author of *Aldo Leopold: His Life and Work* (1988). He teaches in the Land Resources Program at the University of Wisconsin (Madison, Wisconsin).

**Max Oelschlaeger** is the author of *The Environmental Imperative* (1977) and *The Idea of Wilderness* (forthcoming, Yale University Press, 1991). He teaches philosophy at the University of North Texas (Denton, Texas).

**George Sessions** is the author (with Bill Devall) of *Deep Ecology: Living as if Nature Mattered* (1985). He teaches philosophy at Sierra College (Rocklin, California).

**Paul Shepard** is the author of *Man in the Landscape* (1967), *The Tender Carnivore and the Sacred Game* (1973), *Thinking Animals* (1978), *Nature and Madness* (1982), and (with Barry Sanders) *The Sacred Paw* (1985) among other books. He teaches human ecology and natural philosophy at Pitzer College and the Claremont Graduate School (Claremont, California).

**Gary Snyder** is the author of *The Back Country* (1968), *Earth House Hold* (1969), *Turtle Island* (1974), *The Old Ways* (1977), and *Axe Handles* (1983) among other books. Since 1985 he has been a member of the English Department faculty at U. C. Davis (Davis, California).

**Michael Zimmerman** is the author of *Eclipse of the Self: The Development of Heidegger's Concept of Authenticity* (1978) and *Technology, Politics, and Art: Heidegger's Confrontation with Modernity* (1990). He teaches philosophy, psychology, and psychiatry at Tulane University (New Orleans, Louisiana).